ENCOUNTERS WITH CHAOS

Denny Gulick

University of Maryland
College Park

McGraw-Hill, Inc.

New York   St. Louis   San Francisco   Auckland   Bogotá   Caracas
Lisbon   London   Madrid   Mexico   Milan   Montreal   New Delhi
Paris   San Juan   Singapore   Sydney   Tokyo   Toronto

Credits for Figures

p. 27: Courtesy, A. Garfinkel and R. Harper
    53: Courtesy, University of Maryland Chaos Group
    55: Courtesy, University of Maryland Chaos Group
    60: Courtesy, Carson D. Jeffries
    92: Courtesy, T. N. Palmer
    93: Courtesy, Carl D. Murray
  173: Courtesy, University of Maryland Chaos Group
  174: Courtesy, University of Maryland Chaos Group
  199: Courtesy, University of Maryland Chaos Group
  202: Courtesy, University of Maryland Chaos Group
  204: Courtesy, A. Garfinkel and R. Harper
  214: Courtesy, University of Maryland Chaos Group
  218: Courtesy, University of Maryland Chaos Group
  264: Courtesy, University of Maryland Chaos Group
  283: Courtesy, Springer-Verlag

This book was typeset in Times Roman by the author
using MathWriter™ 2.0 on a Macintosh® IIsi.
The editors were Richard Wallis and Margery Luhrs;
the production supervisor was Anthony DiBartolomeo.
R. R. Donnelley & Sons Company was printer and binder.

ENCOUNTERS WITH CHAOS

1 2 3 4 5 6 7 8 9 0 DOC DOC 9 0 9 8 7 6 5 4 3 2

ISBN 0-07-025203-3

Library of Congress Cataloging-in-Publication Data

Gulick, Denny.
   Encounters with chaos / Denny Gulick.
       p.   cm.
   Includes bibliographical references and index.
   ISBN 0-07-025203-3
   1. Chaotic behavior in systems.   2. Fractals.   I. Title.   91-42191
   Q172.5.C45G85   1992
   003' .7—dc20

To my wife, Frances, and our children

David, Barbara, and Sharon

ABOUT THE AUTHOR

Denny Gulick is Professor of Mathematics at the University of Maryland, College Park, where he has taught since 1965. After receiving a B.A. degree from Oberlin College, he obtained his Ph.D. in 1963 from Yale University, where he specialized in functional analysis and worked with Charles E. Rickart. Besides his research in functional analysis, he has co-authored textbooks for calculus and precalculus. A few years ago he began to study chaotic dynamics, and offered the first undergraduate course on mathematical chaos at the College Park campus of the University of Maryland.

When not doing mathematics, Professor Gulick enjoys playing the cello in string quartets and playing tennis. He has also taken an active part as advisor for his children's swimming teams. He and his family have made several recent visits to Japan as guests of the Japanese, in an endeavor to promote international friendship and understanding among children through doll missions that were inaugurated by his grandfather sixty-five years ago.

CONTENTS

PREFACE

The concept of chaotic behavior now pervades virtually all the sciences. In simple terms, we can say that an object exhibits chaos if its motion is not random, but there are no predictable patterns to it. Chaotic dynamics is the study of such behavior, and despite the unpredictability of chaotic motion, mathematical analysis can often describe it.

In recent years the topic of chaotic dynamics has become increasingly popular. Applications of chaotic dynamics have extended to disciplines as diverse as electric circuits, weather prediction, orbits of satellites, chemical reactions, and the spread of disease. A fundamental reason for this popularity is the power of the computer, with its ability to produce complex calculations and to create fascinating graphics. The computer has allowed scientists and mathematicians to solve problems in chaotic dynamics that hitherto seemed untractable, and to analyze scientific data that in earlier times appeared to be either random or flawed.

Mathematics lies at the foundation of chaotic dynamics. The very concepts that describe chaotic behavior are mathematical in nature, whether they be analytic, geometric, algebraic, or probabilistic. Some of these concepts are elementary; others are sophisticated. There are many books that discuss chaos in an expository manner, as there are many treatises on chaos theory, written at the graduate level. In writing *Encounters with Chaos*, our goal has been to provide a readable introduction to chaotic dynamics at a modest level of sophistication -- specifically, for anyone who has a knowledge of calculus. The book includes the important mathematical concepts associated with chaotic dynamics, and supports the definitions and results with motivation, examples, and applications where feasible. An insert with 36 color plates illustrating important topics in the book appears just before Section 1.1.

Encounters with Chaos is divided into five chapters. Chapter 1 presents functions of one variable, and explores the fundamental notions of periodicity and bifurcation. Chapter 2 introduces the concepts of chaos and strong chaos for functions of one variable, along with the related topics of conjugacy and transitivity. These first two chapters lay the groundwork for Chapters 3–5, and offer examples of most of the mathematical behavior exhibited by chaotic systems in higher dimensions. They also rely on mathematics only through the calculus.

In Chapters 3–5 we study chaotic dynamics for functions of more than one variable. The analysis involves vectors and matrices at an elementary level. A brief review of matrices is included in Section 3.1 for those who desire it. Chapter 3 focuses on chaotic dynamics for functions of two variables and discusses the famous Hénon and Smale functions to illustrate the concepts. In Chapter 4 we turn to fractal geometry and its relation to chaotic dynamics. In addition to a presentation of the notions of fractal dimension and fractals, there is a discussion of Julia sets and the Mandelbrot set, which are among the most celebrated fractals. The chapter ends with an introduction to iterated function systems, made famous by Michael Barnsley in his attempt to enhance image compression. Finally, Chapter 5 analyzes solutions of systems of differential equations that relate to chaotic motion. Special attention is placed on the dynamics of the pendulum and the system of differential equations employed by Edward Lorenz to model weather prediction. This last chapter presumes a basic knowledge of systems of differential equations that is normally obtained in an elementary differential equations course.

The book contains more than enough material for a one-semester course. In our one-semester course at the University of Maryland, we have covered virtually all of Chapters 1–3 and selected topics from Chapters 4 and 5. To permit coverage as flexible as possible, we have made Chapters 4 and 5 relatively independent from one another. In order to save some time, one can omit proofs to certain theorems (like those in Sections 1.7 and 1.8), which are included for completeness.

At the end of each section is a supply of exercises designed to reinforce topics and give added insight to the concepts presented in the section. In addition to the regular exercises there are, where feasible, exercises utilizing computer programs which we have included in the appendix. In order to make the programs as accessible as possible, we have written them in TRUE BASIC. Answers to selected odd-numbered exercises are placed at the end of the book, and full solutions to all exercises are included in an instructor's manual prepared by Professor Jim Hoste.

I acknowledge with great appreciation the many people who have assisted me in the preparation of this book. I would like to express my deepest gratitude to colleagues Celso Grebogi and James Yorke, who helped introduce me to chaotic dynamics, guided me when I was lost, and made the vast graphics resources of the University of Maryland Chaos Group available to me. To the reviewers of the manuscript I give my hearty thanks: Daniel Drucker, Jim Hoste, Guan-Hsong Hsu, Edward Packel, Richard Parris, Philip Straffin, and Steven Strogatz. Special thanks go to Bau-Sen Du and Helena Nusse, who read earlier versions as well as my nearly final version and made literally hundreds of pertinent suggestions; Helena Nusse also provided several figures. I appreciate suggestions made by colleagues Joseph Auslander, Kenneth Berg, David Dyer, Michael Fitzpatrick, Alan Garfinkel, Arturo Lopes, Chris Rorres, and Dan Rudolph. Once again I have the pleasure of using wonderful graphics produced by Art Matrix in Ithaca, New York. Next, I would like to thank those who helped me technically. Special thanks go to Jim Hoste, who prepared the instructor's manual for the book, and unearthed errata and obscurities as he proceeded. Loren Argabright introduced me to MathWriter II, the excellent word-processing software for the Macintosh II with which I have prepared the book. I thank both Barbara Gulick, who helped in effecting corrections on the Macintosh, and Edgar Rummel, who found inconsistencies in the final preparation of the manuscript. I also appreciate the efforts of the staff at McGraw-Hill, including Margery Luhrs and Mel Haber. I give many thanks to my long-time friend Richard Wallis, who as mathematics editor gave me abundant assistance in the preparation of this book.

Finally my heartfelt thanks go to my wife and children for their help in seeing me through the ordeal of writing and typesetting a book virtually all by myself. The completion of this project would not have been possible without their understanding.

Denny Gulick
College Park, Maryland

INTRODUCTION

Encounters with Chaos is an introduction to the study of the new field called chaotic dynamics. In chaotic dynamics one analyzes objects subject to an unpredictable, but not random, behavior. We say that such behavior is chaotic. Recently we have learned that objects around us, such as planets in our solar system, molecules in the atmosphere, water particles in a stream, and electrical impulses in the heart and brain, can display chaotic behavior. In order to gain a little perspective on the new area of chaotic dynamics, let us turn the clock back a century.

Until a hundred years ago it was thought that the motion of planets could be completely understood provided that the equations modeling their motion were accurately prescribed. Indeed, during the seventeenth century Johannes Kepler used the calculations of Tycho Brahe to convince himself that the planets in our solar system move in elliptical orbits around the sun. From Kepler's fundamental principles of motion, Isaac Newton proved that any planet under the influence of only its sun moves in an elliptical orbit. Thus Newton resolved the "two-body problem," which refers to the analysis of the motion of one body under the influence of a single other body.

Rarely, however, is a planet subject to the force of a single other object. Indeed, the earth's motion is affected in a nontrivial way by the gravitational forces of the sun, moon, and even other planets in our solar system, most notably Jupiter. Likewise, the orbit of our own moon is influenced by the earth as well as the sun. Therefore Newton's results on the two-body problem gave at best only an approximate solution to the motion of a planet subject to more than one gravitational force.

During the 1890's the great French mathematician Henri Poincaré also studied mathematical aspects of planetary motion. He focused on the "three-body problem," in which the motion of one body is influenced by precisely two other bodies. During his work on this problem he developed a new kind of mathematics which is today called topology (or rubber-sheet geometry). After much effort, Poincaré proved that there is no simple solution to the three-body problem. In other words, Poincaré proved that it is in general impossible to give a simple prescription for the orbit of one body influenced by two other bodies.

In addition, Poincaré realized that if one takes two different readings of the position of a planet at a given moment, then no matter how close the readings are, after enough time the orbits of the planet corresponding to the two different readings will separate away from one another. Accurate long-range prediction of the orbit of the moon is thus impossible. This is one of the basic features of chaotic behavior. At the turn of the century, this idea was particularly offensive to scientists and mathematicians, not only because it contradicted their intuition of the regularity of nature and the universe, but also because they had no tools with which to analyze motion that is unpredictable.

With the advent of the computer, analysis of the motion of planets and other moving systems has taken on a new life. Computers and supercomputers can make trillions of calculations in the blink of an eye. As a result, scientists and mathematicians have been able to gain an understanding of physical and theoretical entities that until recently were beyond their reach; they have also been able to create fractal pictures and landscapes that adorn calendars and covers of magazines.

Now we turn to a few areas of science in which chaotic behavior has recently been observed. During the early 1960's the noted MIT meteorologist Edward Lorenz was trying to model weather patterns by means of mathematical equations. Of course in order to make reasonable predictions of weather one needs to use innumerable variables (such as temperature, humidity, and wind velocity, at thousands of locations both on earth and in the atmosphere), so one inevitably is faced with a great many differential equations. In order to make calculations possible, Lorenz refined a known system of three differential equations relating to convection, and turned to his Royal McBee LGP-30 electronic computer that was equipped with 6-place accuracy.

One time Lorenz substituted a 3-place approximation for the 6-place number he had earlier used as an initial condition. To his astonishment he found that the predicted weather pattern obtained from the 3-place initial condition was far different from the earlier predicted weather pattern. Thinking that it was a fluke, he repeated the process several times. Each time the McBee responded in a similar fashion. Additional analysis led Lorenz to conclude that a slight alteration in the initial conditions could result in enormous differences in the output. Thus weather patterns exhibit a particular kind of chaotic behavior that has been termed the "butterfly effect," because in theory the flapping of butterfly wings in Rio de Janiero now could bring on a tornado in Texas several weeks later. It follows that although one might be able to predict the weather with reasonable accuracy in the short term, long-range predictions of weather are futile. Nevertheless, short-range predictions are meaningful and become more accurate as new and more variables are inserted into the system of equations. In fact, the weather bureau uses up to a million variables in their systems of equations.

Recently biological scientists have utilized notions of chaotic dynamics in the study of such diverse areas as the growth of insect and fish populations, infectious disease, and the heartbeat. In studying childhood diseases like rubella and chicken-pox, they have tried to ascertain whether there are predictable patterns for the spread of the disease. Their findings seem to show that before a vaccine appears, the patterns of the spread of the diseases are neither random nor predictable. In other words, a certain kind of chaos appears in these patterns. Presently scientists and mathematicians are using notions of chaotic dynamics to study the transmission of the human immunodeficiency virus (HIV) that can lead to acquired immunodeficiency syndrome (AIDS). In addition, experts have recently determined that heartbeats and brain waves can exhibit either regular or chaotic behavior, depending on conditions imposed on them. These findings have opened up (theoretical) possibilities for early detection of heart failure or epileptic episodes. There are many more areas of research that are utilizing notions of chaotic dynamics. They include chemical reactions, earthquake prediction, sunspot activity, satellite orbits, efficiency of combustion chambers, dripping faucets, and stock market prices.

Why should one study chaos? There are several reasons. First, the concept of chaos now pervades virtually all sciences. Second, ideas related to chaos represent not only an intellectual achievement for scientists and mathematicians but also a way of thinking new to many people of all walks of life. Analyzing motion that is unpredictable, but not random, can shed new information about subjects such as the orbit of Pluto and the effect of ingesting cocaine.

A third reason for studying chaos comes from mathematics. Chaos entails very beautiful mathematical ideas that are accessible to those who have studied calculus. In addition, there are many mathematical problems in the study of chaos that are easy to state but which continue to baffle the experts. In contrast to most mathematics that undergraduate students encounter, which was fully developed scores of years ago, many of the results appearing in the study of chaos are modern, having appeared within only the past twenty years. Chaos is alive and is exciting, enhanced by the computation and graphics capabilities of the computer and by applications to the widely diverse areas mentioned above. It is one of the richest new branches of mathematics.

We hope that those who read this book will appreciate not only the beauty of chaotic dynamics but also the diversity of applications of the theory of chaos. The subject is growing, as a mathematical area and in disciplines outside of mathematics. A look at a few issues of *Nature* or *New Scientist* magazines should convince a reader of the widespread application of chaos. Finally, we hope that readers of this book will view the world a little differently, understanding that certain kinds of motion are inherently unpredictable yet not random. This understanding, we believe, will be the legacy of chaos.

COLOR PLATES

1: The Hénon attractor. Yellow represents the attractor, black the basin of attraction, red the local stable manifold of the nonzero fixed point, and blue the points escaping to infinity, colored according to the speed of divergence. (See page 173.)

2: The Lorenz attractor. (See page 282.)

3–10: Initial positions for a periodically forced damped pendulum, with four basins of attraction represented by red, yellow, green, and blue. Darker shades of color represent points that settle down more slowly to the appropriate periodic motion. Successive plates magnify centers of the preceding plates, with final magnification approximately 100,000. (See page 274.)

11–12: Initial positions for a periodically forced damped pendulum with two basins of attraction represented by red and blue. Lighter shades of color represent points that settle down more slowly. Plate 12 has more red, because of a slightly stronger periodic force. (See pages 273–274.)

13: The Ikeda attractor, yellow representing the attractor when the control parameter ρ is less than a certain value called the crisis value, and red the additional points in the attractor when ρ rises above the crisis value. (See page 210.)

14: The Tinkerbell attractor, represented by the curve. Black represents the basin of attraction. Other shades outside escape to infinity, colored according to the speed with which they recede. (See page 202.)

15–17: Julia sets for $f(z) = z^2 + c$. (See page 214.)
    Plate 15: $c = 0.32 + 0.043i$
    Plate 16: $c = -0.121 + 0.739i$ (This Julia set is called the "Douady rabbit.")
    Plate 17: $c = 0.36 + 0.1i$ (This Julia set is called a dragon.)

18: The Mandelbrot set. (See page 217.)

19–24: Successive blow-ups at the boundary of the Mandelbrot set, with final magnification approximately 130,000. (See page 217.)

25–30: Successive blow-ups at another location on the boundary of the Mandelbrot set, with final magnification approximately 130,000. (See page 217.)

31–36: Magnifications at six locations on the boundary of the Mandelbrot set. (See page 217.)

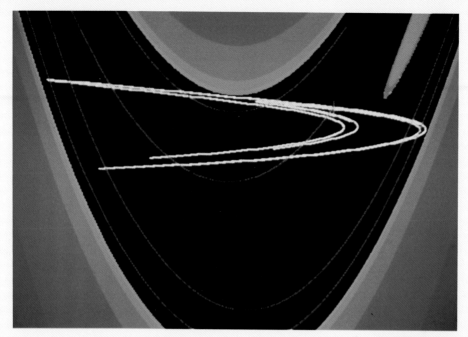

COLOR PLATE 3

COLOR PLATE 4

COLOR PLATE 5

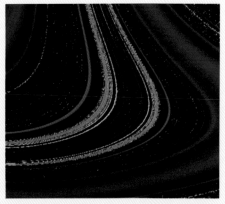

COLOR PLATE 6

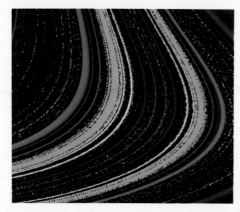

COLOR PLATE 7

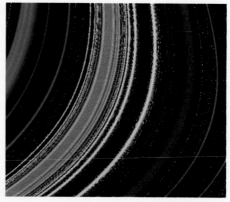

COLOR PLATE 8

COLOR PLATE 9

COLOR PLATE 10

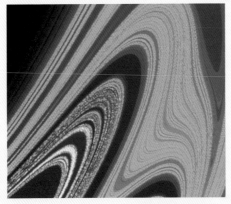

COLOR PLATE 11

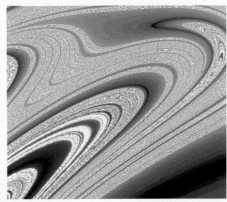

COLOR PLATE 12

COLOR PLATE 13

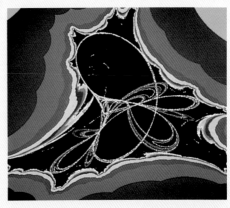

COLOR PLATE 14

COLOR PLATE 15

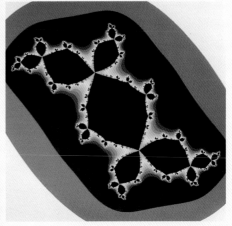

COLOR PLATE 16

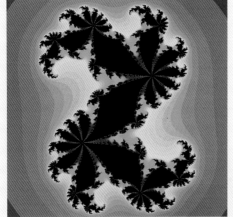

COLOR PLATE 17

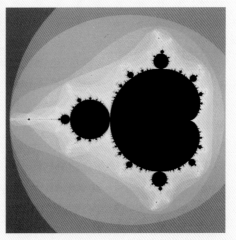

COLOR PLATE 18

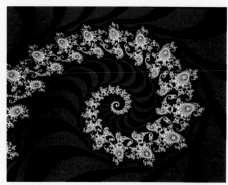

COLOR PLATE 19

COLOR PLATE 20

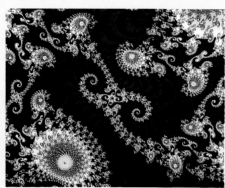

COLOR PLATE 21

COLOR PLATE 22

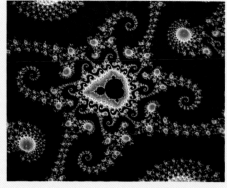

COLOR PLATE 23

COLOR PLATE 24

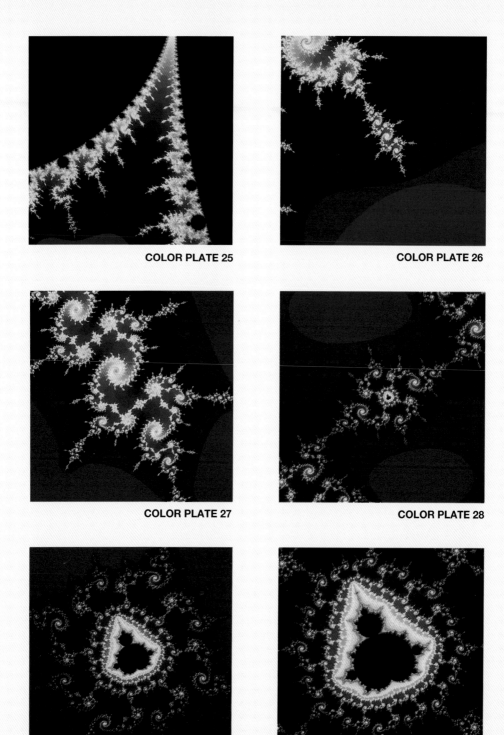

COLOR PLATE 25

COLOR PLATE 26

COLOR PLATE 27

COLOR PLATE 28

COLOR PLATE 29

COLOR PLATE 30

COLOR PLATE 31

COLOR PLATE 32

COLOR PLATE 33

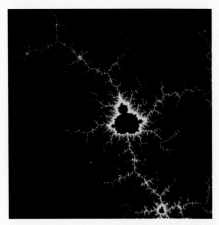

COLOR PLATE 34

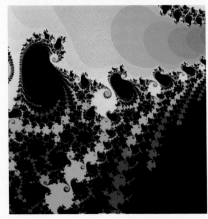

COLOR PLATE 35

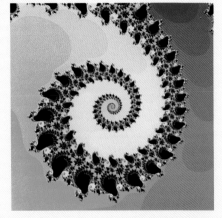

COLOR PLATE 36

CHAPTER
1

PERIODIC POINTS

Suppose that f is a function and x is in the domain of f. Then the iterates of x, which consist of x, $f(x)$, $f(f(x))$, $f(f(f(x)))$, ... , form the orbit of x. The kinds of orbits that are possible for various values of x and for different functions constitute the basis for the study of chaos.

Chapter 1 is devoted to the study of those x's for which some iterate is again x. Such a point is called a periodic point. For example, if $f(x) = x^2 - 1$, then we have $f(0) = -1$ and $f(-1) = 0$, so that $f(f(0)) = 0$ and thus 0 is a periodic point of f. Sections 1.1–1.3 lay the foundation for a detailed analysis of periodic points. Sections 1.4–1.5 focus on families of functions, and discuss how periodic points can vary from function to function within a family of functions. In those sections we examine in detail two illustrious families of functions: the tent family and the quadratic family. In Section 1.6 we discuss various kinds of bifurcations, where the quality or number of periodic points changes. Section 1.7 is devoted to consequences of a function having periodic orbits containing but three elements. Finally, in Section 1.8 we discuss the largest number of periodic orbits a given function can have, each of which attracts all nearby points.

Chapter 1 is a prelude to the study of chaotic functions, which we define and discuss in Chapter 2.

1.1 ITERATES OF FUNCTIONS

Suppose that we key .5 into a calculator, and then repeatedly depress the x^2-button. The calculator would display the numbers

$$0.5, \quad 0.25, \quad 0.0625, \quad 0.00390625, \quad 0.0000152858, \quad ... \tag{1}$$

one after another. What is the calculator giving us with this sequence of numbers? If we let $f(x) = x^2$ and let $x_0 = 0.5$, then the sequence in (1) consists of

$$x_0, \quad f(x_0), \quad f(f(x_0)), \quad f(f(f(x_0))), \ldots$$

These numbers are called iterates of x_0 for f.

DEFINITION 1.1. Let f be a function and let x_0 be in the domain of f. Then

$$f(x_0) = \text{the \textbf{first} iterate of } x_0 \text{ for } f$$

$$f(f(x_0)) = \text{the \textbf{second} iterate of } x_0 \text{ for } f$$

More generally, if n is any positive integer, and a_n is the nth iterate of x_0 for f, then $f(a_n)$ is the $(n+1)$st iterate of x_0 for f.

For convenience we will adopt the notation

$$f^{[2]}(x_0) \text{ for } f(f(x_0)), \quad f^{[3]}(x_0) \text{ for } f(f(f(x_0)))$$

and more generally,

$$f^{[n]}(x_0) \text{ for the } n\text{th iterate of } x_0 \text{ for } f$$

We call the sequence $\{f^{[n]}(x_0)\}_{n=0}^{\infty}$ of iterates of x_0 the **orbit** of x_0. Sometimes we will write x_n for $f^{[n]}(x_0)$. In that case, $\{x_n\}_{n=0}^{\infty}$ is the orbit of x_0. In the literature, the usual notation for the nth iterate of f is f^n, rather than $f^{[n]}$.

Caution: Throughout the book it will be understood that all iterates of each function under discussion will lie in the domain of the function. As a result, we normally will not explicitly mention this assumption.

Now we are ready to illustrate a few iterates for several functions:

$f(x)$	x	orbit of x
$f(x) = x^2$	1	1, 1, 1, 1, 1, ...
$f(x) = x^2 - 1$	-1	$-1, 0, -1, 0, -1, \ldots$
$f(x) = x^2 + 1$	-2	$-2, 5, 26, 677, 458330, \ldots$
$f(x) = x^2 + \dfrac{1}{4}$	0	$0, 0.25, 0.3125, 0.347656\cdots, 0.370864\cdots, \ldots$
$f(x) = 4x - 4x^2$	$\dfrac{1}{3}$	$0.333333\cdots, 0.888888\cdots, 0.395061\cdots, 0.955951\cdots, \ldots$

As you see, even with simple quadratic functions, the orbits seem to have very different behaviors. Indeed, the orbit of 1 for the function x^2 is constantly 1. Next, the orbit of -1 for the function $x^2 - 1$ oscillates between -1 and 0. By contrast, the orbit of -2 for $x^2 + 1$ is unbounded, and it is not clear how the iterates of 0 and 1/3 in the last two examples behave. In time we will prove that the iterates of 0 for $x^2 + 1/4$ converge to 1/2, and the iterates of 1/3 for $4x - 4x^2$ spread themselves over the interval (0, 1) in a seemingly unpredictable manner.

Among the many functions whose orbits one can describe are the sine and cosine functions. For the sine function (with radian measure), the orbit of, say, 2 begins

$$2, \ 0.9092\cdots, \ 0.7890\cdots, \ 0.7097\cdots, \ 0.6516\cdots, \ 0.6064\cdots,$$

Continuing, we find that the 50th iterate is 0.2350... , and the 100th iterate is 0.1692... , with successive iterates decreasing. This leads us to conjecture that $f^{[n]}(2)$ approaches 0 as n increases without bound. In Example 2 of Section 1.2 we will prove that this conjecture is true. Meanwhile you might calculate iterates of 2 for $\cos x$, and see if they seem to have a reasonable limit.

Iterates form the basis of the Newton-Raphson method for approximating a zero of a function. (You may have encountered the method in calculus.) We start with a function f, and assume that near a zero z of f the derivative of f is nonzero. Next, we select an initial value x_0 that ideally is reasonably close to z. We define x_1 by

$$x_1 = x_0 - \frac{f(x_0)}{f'(x_0)}$$

For any positive integer n, we use x_n to define x_{n+1} by the formula

$$x_{n+1} = x_n - \frac{f(x_n)}{f'(x_n)} \tag{2}$$

Formula (2) constitutes the **Newton-Raphson method** (also called the **Newton method**). For many of the functions encountered in calculus (such as polynomial functions), judicious selection of the initial value x_0 can readily give good estimates of a zero of the given function. Figure 1.1 shows one such example.

If we let

$$g(x) = x - \frac{f(x)}{f'(x)}$$

then $x_1 = g(x_0)$, $x_2 = g(x_1)$, and in general, $x_{n+1} = g(x_n)$. Thus the sequence $\{x_n\}_{n=0}^{\infty}$ generated by the Newton-Raphson method consists of the iterates of x_0 for g.

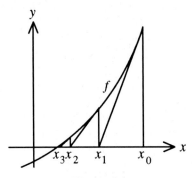

Figure 1.1

EXAMPLE 1. Let $f(x) = x^2 - 7$. Use the Newton-Raphson method to approximate a zero of f until successive approximations are within 10^{-4} of each other.

Solution. The zero z of f satisfies $z^2 - 7 = f(z) = 0$, so that $z = \sqrt{7}$. To approximate z by the Newton-Raphson method, let $x_0 = 3$, and let g be defined by

$$g(x) = x - \frac{f(x)}{f'(x)} = x - \frac{x^2 - 7}{2x} = \frac{1}{2}\left(x + \frac{7}{x}\right)$$

The initial few iterates of 3 for g are

$$x_0 = 3 \qquad x_1 = 2.66666\cdots \qquad x_2 = 2.64583\cdots \qquad x_3 = 2.64575\cdots$$

Since $|x_2 - x_3| < 10^{-4}$, the approximation we desire is $2.64575\cdots$, which you can check is an estimate of $\sqrt{7}$ accurate to 5 places. ❏

In Section 1.2 we will learn why the iterates of *any* positive number approach $\sqrt{7}$ for the function g. (See Exercise 26 in Section 1.2.)

Finally we mention that an unfortunate choice of x_0 can lead to the failure of the Newton-Raphson method to approximate a zero of a given function. See Exercise 11 for an example. Thus one needs to be careful when using the method.

Graphical Analysis of Iterates

If we are able to render a reasonably precise graph of a given function, then we may be able to analyze graphically the orbits of various members of the domain.

First we draw the graph of f, along with the line $y = x$. To exhibit the orbit of x_0, first locate x_0 on the x axis. Notice that $(x_0, f(x_0))$ lies not only on the graph of f but also on the vertical line through $(x_0, 0)$ (Figure 1.2(a)). The horizontal line through $(x_0, f(x_0))$ crosses the line $y = x$ at the point $(f(x_0), f(x_0)) =$

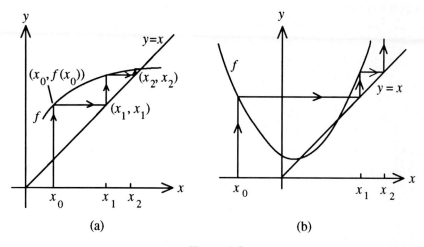

Figure 1.2

(x_1, x_1). By applying the same process with x_1 replacing x_0, we obtain the point (x_2, x_2). Continuing in the same manner, we can determine the location of (x_3, x_3), (x_4, x_4), and thus in theory we can produce the orbit of x_0. This process is called a **graphical analysis** of the orbit of x_0, and is carried out in Figure 1.2.

Next, let $f(x) = \sin x$, the graph of which appears in Figure 1.3. With a little experimentation you should be able to convince yourself that the arrows in the graphical analysis end up in the narrow passages between the graphs of f and $y = x$, and converge to 0.

Caution: Graphical analysis can indicate how orbits behave, and can guide one to a proof of a given result. However, except in special cases, we will rely on rigorous mathematical proofs and not solely on graphical analysis.

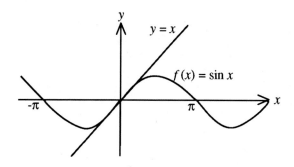

Figure 1.3

EXERCISES 1.1

1.   Let $g(x) = \cos x$. Use the computer program ITERATE in the appendix to try to determine $\lim_{n \to \infty} g^{[n]}(x)$ for

   a. $x = 2$ b. $x = -1$ c. $x = \pi/4$

2.   Let $f(x) = 3x - 3x^2$. Calculate the first dozen iterates of $1/2$, and see if you can detect $\lim_{n \to \infty} f^{[n]}(1/2)$.

In Exercises 3–5, use graphical analysis to determine whether the iterates of the given point x seem to converge or not. If the iterates seem to converge, guess the limit.

3. $f(x) = \cos x$; $x = 3$; $x = \pi/2$ 4. $f(x) = x - x^3/3$; $x = -1$

5. $f(x) = \begin{cases} 2x, & \text{for } 0 \le x \le 1/2 \\ 2x - 1, & \text{for } 1/2 < x \le 1 \end{cases}$; $x = 1/3$

6.   Find a function f such that $f(0)$ is the maximum value and $f^{[2]}(0)$ is the minimum value of f.

7.   Consider the function L defined by $L(x) = ax + b$, where $a \ne 1$. Determine the values of a for which $\lim_{n \to \infty} L^{[n]}(x)$ exists (as a number) for each real number of x, and find the value of that number in terms of a and b.

In Exercises 8–10, use the Newton-Raphson method, with the given value of x_0, to approximate a zero of the function f to within 10^{-4}.

8. $f(x) = \cos x - x$; $x_0 = 1$ 9. $f(x) = x^3 - 2x - 5$; $x_0 = 2$

10. $f(x) = x^4 + \sin x$; $x_0 = -1$

11.   Let $f(x) = 1 + x(x - 1)^6$. See what happens when you try the Newton-Raphson method with each of the following initial values.

   a. 0   b. 1   c. 1.05   d. 1.1

1.2 FIXED POINTS

A point whose iterates are the same point is called a fixed point. Fixed points are very important in the study of the dynamics of functions.

DEFINITION 1.2. Let p be in the domain of f. Then p is a **fixed point** of f if $f(p) = p$.

Graphically, a point p in the domain of f is a fixed point of f if and only if the graph of f touches (or crosses) the line $y = x$ at (p, p) (Figure 1.4).

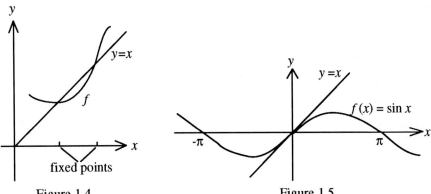

Figure 1.4   Figure 1.5

From Figure 1.5 we might conjecture that the origin is the only point at which the graph of $\sin x$ and the line $y = x$ touch each other. We will prove that this is true. In the solution we will use the Mean Value Theorem, which says that if f is continuous on $[a, b]$ and differentiable on (a, b), then there is a c in (a, b) such that

$$f'(c) = \frac{f(b) - f(a)}{b - a}, \text{ or equivalently, } f(b) - f(a) = f'(c)(b - a)$$

EXAMPLE 1. Let $f(x) = \sin x$. Show that 0 is the unique fixed point of f.

Solution. To begin, we observe that $f(x) \neq x$ if $|x| > 1$, since $|\sin x| \leq 1$ for all x. Next, if $0 < x \leq 1$, then the Mean Value Theorem implies the existence of z between 0 and x such that

$$\sin x = \sin x - \sin 0 = f'(z)(x - 0) = x \cos z$$

Since $0 < \cos z < 1$ for such z, it follows that $0 < \sin x = x \cos z < x$. The fact that $f(-x) = -f(x)$ implies that $x \neq f(x)$ for all $x < 0$. Finally, $\sin 0 = 0$, so we conclude that 0 is the unique fixed point of f. ❏

The next theorem will be very important to us. For convenience we will write $f^{[n]}(x) \to p$ for "$f^{[n]}(x)$ approaches p" (as n increases without bound).

THEOREM 1.3. Suppose that f is continuous at p, and let x be in the domain of f. If $f^{[n]}(x) \to p$ as n increases without bound, then p is a fixed point of f.

Proof. By hypothesis, $f^{[n]}(x) \to p$, so that $f^{[n+1]}(x) \to p$. Since $f^{[n]}(x) \to p$, the continuity of f at p yields $f(f^{[n]}(x)) \to f(p)$. However $f^{[n+1]}(x) = f(f^{[n]}(x))$, so that by substituting $f^{[n+1]}(x)$ for $f(f^{[n]}(x))$, we find that $f^{[n+1]}(x) \to f(p)$. The uniqueness of the limit of a given sequence implies that $f(p) = p$. Consequently p is a fixed point of f. ■

From calculus we know that a bounded sequence $\{x_n\}_{n=0}^{\infty}$ that is increasing converges to the least number z such that $x_n \leq z$ for all n. A similar statement holds for a bounded decreasing sequence, and hence for any monotone (that is, increasing or decreasing) sequence.

COROLLARY 1.4. Suppose that f is a continuous function defined on a closed interval. Assume that $\{f^{[n]}(x)\}_{n=0}^{\infty}$ is a bounded, monotone sequence. Then there is a fixed point p such that $f^{[n]}(x) \to p$ as n increases without bound.

Proof. By the comment above, bounded monotone sequences always converge. Thus this result is an immediate consequence of Theorem 1.3. ■

Now we will use Corollary 1.4 to prove that the iterates of any real x for the sine function converge to 0 — a result deduced by graphical analysis in Section 1.1.

EXAMPLE 2. Let $f(x) = \sin x$. Show that the iterates of any x converge to 0.

Solution. Let x be an arbitrary number. To show that the sequence $\{f^{[n]}(x)\}_{n=0}^{\infty}$ is bounded, we observe that $-1 \leq \sin x \leq 1$ for each number x. Thus the sequence lies in $[-1, 1]$, and hence is bounded. Next we observe that

$$f(-x) = \sin(-x) = -\sin x = -f(x)$$

Therefore if we can show that the sequence converges to 0 for each x in $[0, 1]$, then the same happens for each x in $[-1, 0]$, and hence for all x. Thus we only need to show that the sequence converges for each x in $[0, 1]$.

Since $f(0) = 0$, let $0 < x \leq 1$. We will show next that $\{f^{[n]}(x)\}_{n=0}^{\infty}$ is a decreasing sequence. As in the solution of Example 1, the Mean Value Theorem yields a z between 0 and x such that

$$\sin x < x \cos z < x$$

Therefore $0 < \sin x < x < 1$ for $0 < x \leq 1$, which means that

$$f(x) < x \text{ for } 0 < x \leq 1$$

It follows that for any $n \geq 0$,

$$f^{[n+1]}(x) = f(f^{[n]}(x)) = \sin f^{[n]}(x) < f^{[n]}(x)$$

We conclude that $\{f^{[n]}(x)\}_{n=0}^{\infty}$ is a decreasing sequence when $0 < x \leq 1$. Since the sequence is also bounded, Corollary 1.4 implies that the sequence must converge to a fixed point, which by Example 1 is 0. Consequently $\{f^{[n]}(x)\}_{n=0}^{\infty}$ converges to 0 for all x.   ❏

Theorem 1.3 provides information concerning the Newton-Raphson method described in Section 1.1. Recall that the method involves calculating a sequence $\{x_n\}_{n=0}^{\infty}$ created by letting x_0 be an initial value, and defining

$$x_{n+1} = x_n - \frac{f(x_n)}{f'(x_n)}$$

Here we assume that $f'(x_n) \neq 0$ for all n. In Section 1.1 we indicated that if $\{x_n\}_{n=0}^{\infty}$ converges, then its limit is a zero of f. Now we can support this assertion. Let

$$g(x) = x - \frac{f(x)}{f'(x)}$$

Then $x_1 = g(x_0)$, $x_2 = g(x_1) = g^{[2]}(x_0)$, and in general, $x_n = g^{[n]}(x_0)$. Thus $\{x_n\}_{n=0}^{\infty}$ is the sequence of iterates of x_0 for g. Theorem 1.3 tells us that if the sequence converges, then it converges to a fixed point z of g. In that case,

$$z = g(z) = z - \frac{f(z)}{f'(z)}$$

so that

$$\frac{f(z)}{f'(z)} = 0, \text{ or equivalently, } f(z) = 0$$

This means that z is a zero of f, as we wished to prove.

Attracting and Repelling Fixed Points

By applying graphical analysis we can see diverse behavior for the iterates of various points. Indeed, in Figure 1.6(a) the iterates of x approach the fixed point

p, whereas in Figure 1.6(b) the iterates tend toward ∞. The iterates of x in Figure 1.6(c) have each of these characteristics, depending on the x. We are led to the following definition.

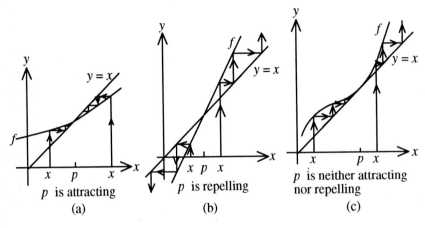

Figure 1.6

DEFINITION 1.5. Let p be a fixed point of f.

a.   The point p is an **attracting fixed point** of f provided that there is an interval $(p - \varepsilon, p + \varepsilon)$ containing p such that if x is in the domain of f and in $(p - \varepsilon, p + \varepsilon)$, then $f^{[n]}(x) \to p$ as n increases without bound. (Such a point is also called **asymptotically stable** in the literature.)

b.   The point p is a **repelling fixed point** of f provided that there is an interval $(p - \varepsilon, p + \varepsilon)$ containing p such that if x is in the domain of f and in $(p - \varepsilon, p + \varepsilon)$ but $x \neq p$, then $|f(x) - p| > |x - p|$.

It follows from the definitions above that the fixed point in Figure 1.6(a) is attracting, and that the one in Figure 1.6(b) is repelling. That not every fixed point is attracting or repelling is demonstrated in Figure 1.6(c), where points to the left of p are attracted to p and points to the right of p are repelled from p. Other kinds of fixed points that are neither attracting nor repelling can occur.

Caution: There is no standard definition in the literature for attracting and repelling fixed points. We have chosen definitions that seem reasonable for our purposes.

For most functions, it is not so easy to prove directly from Definition 1.5 that a given fixed point p is attracting (or repelling, or neither). However, if f is differentiable at p, then a useful criterion exists, which we will state and prove in Theorem 1.6. In the proof we will need to apply the Axiom of Mathematical Induction, which is frequently called the Law of Induction.

LAW OF INDUCTION: Assume that for each integer greater than or equal to an initial integer n_0, a statement, formula or equation, $S(n)$, is given. Suppose that

    i. $S(n_0)$ is true.

    ii.   For any integer $n \geq n_0$, if $S(n)$ is true, then $S(n + 1)$ is true.

Then $S(n)$ is true for all integers $n \geq n_0$.

    Step (ii) in the axiom is frequently called the **inductive step**. Now we are ready to state and prove Theorem 1.6.

THEOREM 1.6.   Suppose that f is differentiable at a fixed point p.

a.   If $|f'(p)| < 1$, then p is attracting.

b.   If $|f'(p)| > 1$, then p is repelling.

c.   If $|f'(p)| = 1$, then p can be attracting, repelling, or neither.

Proof. To begin our proof of (a), we notice that since $|f'(p)| < 1$, the definition of derivative implies that there is a positive constant $A < 1$ and an open interval $J = (p - \varepsilon, p + \varepsilon)$ such that if x is in J and $x \neq p$, then

$$\left| \frac{f(x) - f(p)}{x - p} \right| \leq A$$

Therefore $|f(x) - f(p)| \leq A\, |x - p|$, for all x in J. For each such x, this means that

$$|f(x) - p| = |f(x) - f(p)| \leq A|x - p| \tag{1}$$

so that $f(x)$ is in J because $0 < A < 1$. Thus $f(x)$ is at least as close to p as x is. Let x be fixed in J. If $f^{[n]}(x) = p$ for some n, then $f^{[n]}(x) \to p$ as n increases without bound, so we will assume henceforth that $f^{[n]}(x) \neq p$ for all n. Next we will use the Law of Induction to prove that

$$|f^{[n]}(x) - p| \leq A^n\, |x - p| \text{ for all } n \geq 1 \tag{2}$$

By (1), the inequality holds for $n = 1$. Next, we assume that (2) holds for a given $n > 1$. Then $f^{[n]}(x)$ is in J since $0 < A^n < A < 1$. Therefore by (1) with $f^{[n]}(x)$ substituted for x, and then by (2), we find that

$$|f^{[n+1]}(x) - p| = |f(f^{[n]}(x)) - p| \leq A\, |f^{[n]}(x) - p| \leq A(A^n|x - p|)$$

so that $|f^{[n+1]}(x) - p| \leq A^{n+1}|x - p|$. By the Law of Induction we deduce that (2) holds for all integers $n \geq 1$. Since $A^n \to 0$ as n increases without bound, it follows that $f^{[n]}(x) \to p$ for every x in J. Thus (a) is proved. The proof of (b) is analogous. Part (c) is addressed in Exercise 10. ∎

We can put Theorem 1.6 to immediate use.

EXAMPLE 3. Let $\mu > 0$ be a constant, and let

$$f(x) = \mu x(1 - x) = \mu x - \mu x^2, \text{ for } 0 \le x \le 1$$

a. Find the values of μ for which 0 is an attracting fixed point.
b. Find the values of μ for which there is a nonzero fixed point.
c. Find the values of μ for which the nonzero fixed point is attracting.

Solution. Notice that x is a fixed point of f if $x = \mu x - \mu x^2$. Thus either $x = 0$ or else $1 = \mu - \mu x$, which implies that $x = 1 - 1/\mu$. If $0 < \mu \le 1$, then we have $1 - 1/\mu \le 0$, so there is only one fixed point in the interval $[0, 1]$, namely 0. By contrast, when $\mu > 1$, there are two distinct fixed points in $[0, 1]$: 0 and $1 - 1/\mu$. Next we will determine which fixed points are attracting and which are repelling. Since $f'(x) = \mu - 2\mu x$, it follows that

$$f'(0) = \mu \quad \text{and} \quad f'(1 - 1/\mu) = \mu - 2\mu(1 - 1/\mu) = 2 - \mu$$

Theorem 1.6 tells us that 0 is attracting if $0 < \mu < 1$ and is repelling if $1 < \mu$. It also tells us that $1 - 1/\mu$ is attracting if $1 < \mu < 3$, and is repelling if $\mu > 3$. Finally, it is possible to show that 0 is attracting if $\mu = 1$ (Exercise 15), and that $1 - 1/\mu$ is attracting if $\mu = 3$ (Exercise 16). ❏

Basins of Attraction

If a fixed point p of f is attracting, then all points near to p are "attracted" toward p, in the sense that their iterates converge to p.   The collection of *all* points whose iterates converge to p is called the basin of attraction of p.

DEFINITION 1.7. Suppose that p is a fixed point of f. Then the **basin of attraction** of p consists of all x such that $f^{[n]}(x) \to p$ as n increases without bound, and is denoted by B_p.

EXAMPLE 4. Let $f(x) = x^2$. Find the basin of attraction B_0 of the fixed point 0.

Solution. If $|x| < 1$, then $f^{[n]}(x) = x^{(2^n)} \to 0$ as n increases without bound, so that x is in B_0. By contrast, if $|x| \ge 1$, then $|f^{[n]}(x)| \ge 1$, so that x is not in B_0. Thus B_0 consists of all x such that $|x| < 1$, that is, $B_0 = (-1, 1)$. (We could also draw the same conclusion by using graphical analysis.) ❏

We remark that if p is a repelling fixed point, then its basin of attraction can consist of the single point p, as happens for 0 if $f(x) = 2x$. At the other end of the spectrum, the basin of attraction of the fixed point 0 of $\sin x$ consists of all real numbers, which in effect is what we showed in Example 2.

Eventually Fixed Points

Finally we introduce the notion of eventually fixed point, which will be of use in later examples.

DEFINITION 1.8. Let x be in the domain of f. Then x is an **eventually fixed point** of f if there is a positive integer n such that $f^{[n]}(x)$ is a fixed point of f.

A fixed point is trivially an eventually fixed point. However, if $f(x) = \sin x$, then $f(\pi) = 0$ and $f(0) = 0$, so that π is an eventually fixed point that is not a fixed point. In order not to create confusion, when we refer to x as an eventually fixed point, we will generally assume that x is *not* a fixed point.

EXAMPLE 5. Let T be defined by

$$T(x) = \begin{cases} 2x & \text{for } 0 \le x \le 1/2 \\ 2 - 2x & \text{for } 1/2 < x \le 1 \end{cases}$$

Show that $1/8$ is an eventually fixed point.

Solution. A routine check shows that

$$T(\frac{1}{8}) = \frac{1}{4}, \quad T(\frac{1}{4}) = \frac{1}{2}, \quad T(\frac{1}{2}) = 1, \quad T(1) = 0, \quad T(0) = 0$$

Therefore $1/8$ is an eventually fixed point. ❏

The function T is called a **tent function**, because of the shape of its graph (see Figure 1.7). Example 5 says that T has an eventually fixed point. One can show by an analogous argument that if $x = k/2^n$, where k and n are positive integers and $0 < k/2^n \le 1$, then x is an eventually fixed point of T (see Exercise 27). We will use this result when we study T in more detail in Section 1.4.

We mention that if x is an eventually fixed point of f (so that some iterate of x is a fixed point p of f), then x is automatically in the basin of attraction of p. The converse is false, however, because the iterates of points can converge to p without eventually being p.

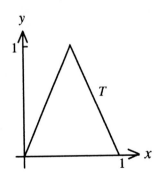

Figure 1.7

EXERCISES 1.2

In Exercises 1–6, find the fixed points, and determine whether each is attracting or repelling.

1. $f(x) = 4x - x^2$ 2. $f(x) = x^3 - x/3$

3. $f(x) = \sqrt{x}$ 4. $f(x) = e^{x-1}$

5. $f(x) = \arcsin x$ 6. $f(x) = 1/x$

7. Let $g(x) = x^2 + 1/4$. Show that if $|x| > 1/2$, then $|g^{[n]}(x)| \to \infty$ as n increases without bound.

8. Let $f(x) = \cos x$.
   a. Show that there is exactly one fixed point p, and that it is attracting.
   b. Find the basin of attraction of p.
   c. With the value of p obtained in part (a), use the computer program NUMBER OF ITERATES to determine the smallest n such that $|f^{[n]}(2) - p| < .001$.

9. a. Let $f(x) = \sin x$. Use the computer program NUMBER OF ITERATES to determine the smallest n such that $0 < f^{[n]}(2) < .1$.
   b. The 10,000th iterate of 2 for $\sin x$ is larger than .001. Why do you suppose that the convergence to the fixed point 0 is so slow?

10. a. Let $f(x) = \arctan x$. Show that $f'(0) = 1$ and 0 is an attracting fixed point.
    b. Let $g(x) = x^2 + 1/4$. Show that $g'(1/2) = 1$ and $1/2$ is a fixed point that is neither attracting nor repelling.

c.   Let $h(x) = x^3 + x$. Show that $h'(0) = 1$ and 0 is a repelling fixed point.

11.   Let $f(x) = |x|/2$. Show that 0 is an attracting fixed point of f. Can you infer that 0 is attracting from Theorem 1.6? Explain why or why not.

12.   Let m be a real number, and consider the line given by $L(x) = mx + (1 - m)$.
      a.   Show that 1 is a fixed point of L.
      b.   Find all values of m for which 1 is attracting, and all values of m for which 1 is repelling.
      c.   Find a value of m such that 1 is neither attracting nor repelling.

13.   Let a and b be constants, and let $L(x) = ax + b$.
      a.   Let $|a| \neq 1$. Find the fixed point p of L, and determine the values of a for which p is attracting, and those for which p is repelling.
      b.   Let $|a| = 1$. Show that any fixed point(s) that L has are neither attracting nor repelling.

14.   Let $f(x) = e^x$. Show that f has no fixed points.

15.   Let $f(x) = x - x^2$ for $0 \leq x \leq 1$. Show that 0 is an attracting fixed point.

16.   Let $f(x) = 3x - 3x^2$. Show that $2/3$ is an attracting fixed point.

17.   Let $f(x) = 3x(1 - x)$. Using the computer program NUMBER OF ITERATES, determine the minimum number of iterates of $1/2$ that it takes to approximate the fixed point $2/3$ to within

      a.   .1   b.   .05   c.   01   d.   .005

18.   Let $f(x) = \tan x$. Show that f has an infinite number of fixed points, and classify the fixed points as attracting, repelling, or neither.

19.   Let $f(x) = (\tan x)/2$. Show that the iterates of x converge to 0 whenever x is in $(-\pi/3, \pi/3)$.

20.   Let $f(x) = x^2 + 1/4$. Use the Law of Induction to show that
      a. $f^{[n]}(x) \rightarrow 1/2$ as n increases without bound if $-1/2 \leq x \leq 1/2$.
      b. $f^{[n]}(x) \rightarrow \infty$ as n increases without bound if $|x| > 1/2$.

21.   Let $f(x) = 4x^2 + 1/16$. Use the Law of Induction to show that
      a. $f^{[n]}(x) \rightarrow 1/8$ as n increases without bound if $-1/8 \leq x \leq 1/8$.
      b. $f^{[n]}(x) \rightarrow \infty$ as n increases without bound if $|x| > 1/8$.

In Exercises 22–25, use either algebra or graphical analysis to find the largest open interval in the basin of attraction of the fixed point 0.

22. $f(x) = \sin^2 x$ 23. $f(x) = x^4$

24. $f(x) = x^5 + x^3$ 25. $f(x) = (.5)x(1 - x)$

26.   Let $f(x) = x^2 - 7$, and let $g(x) = x - f(x)/f'(x)$. Show that if x is any positive number, then the iterates of x approach the zero of f. (Thus the Newton-Raphson method is successful for f when *any* initial positive value of x is picked.)

27.   Let T be the tent function defined in Example 5. Show that if x is in the interval $(0, 1)$ and has the form $x = k/2^n$, where k and n are positive integers, then x is an eventually fixed point.

28.   Let $f(x) = x^2$. Show that if $0 < |x| < 1$, then x is in the basin of attraction of 0 but is not an eventually fixed point.

29.   Let $f(x) = x^2 + 1/8$.
       a.   Find the two fixed points of f, and show that one of them is attracting.
       b.   Find the basin of attraction of the attracting fixed point found in (a).

30.   Suppose that $f \geq 0$, and that p is a fixed point of f. Let g be defined by $g(x) = (f(x))^{1/2}$ for all x in the domain of f. Determine under what conditions p is a fixed point of g.

31.   Suppose that f has the graph pictured in Figure 1.8, with fixed point p.
       a.   Determine whether p is attracting, repelling, or neither.
       b.   Find a formula for a function that has the shape pictured in Figure 1.8.

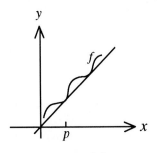

Figure 1.8

32.   Suppose that f is continuous and strictly increasing on $(-\infty, \infty)$, and 0 is the only fixed point of f. Assume also that the graph of f lies above the

line $y = x$ for $x < 0$, and lies below the line $y = x$ for $x > 0$. Show that 0 is an attracting fixed point.

33. Suppose that p is a fixed point of f, and that the graph of f is tangent to the line $y = x$ at the point $(p, f(p))$. Show that if $(p, f(p))$ is an inflection point, then p is either attracting or repelling.

34. We will say that the fixed point p is attracting-repelling if p is attracting for points to one side of p and is repelling for points to the other side (and close to p).

    a.   Let $f(x) = x(1 - x)$. Find an attracting-repelling fixed point for f.

    b.   Let $g(x) = x^2 + 1/4$. Find an attracting-repelling fixed point for g.

    c.   Find a continuous function h and a fixed point p such that p is neither attracting, nor repelling, nor attracting-repelling.

1.3 PERIODIC POINTS

Periodicity is a notion common in everyday language. For example, Halley's comet has a period of approximately 76 years. Similarly, the longer a pendulum is, the longer its period is. The notion of periodicity is central to the study of dynamics.

DEFINITION 1.9. Let x_0 be in the domain of f. Then x_0 **has period n** (or is a **period-n point**) if $f^{[n]}(x_0) = x_0$, and if in addition, $x_0, f(x_0), f^{[2]}(x_0),\dots,$ $f^{[n-1]}(x_0)$ are distinct. If x_0 has period n, then the orbit of x_0, which is

$$\{x_0,\ f(x_0),\ f^{[2]}(x_0),\ \dots,\ f^{[n-1]}(x_0)\}$$

is a **periodic orbit** and is called an **n-cycle**.

By Definition 1.9, fixed points are periodic points — with period 1. If a point has period 1, then we will refer to it as a fixed point (rather than a periodic point).

To illustrate a 2-cycle, let $h(x) = -x^3$. Then $\{-1, 1\}$ is a 2-cycle because $h(-1) = 1$ and $h(1) = -1$. Next, we will exhibit a 3-cycle for the tent function T.

EXAMPLE 1. The tent function T is given by

$$T(x) = \begin{cases} 2x & \text{for } 0 \le x \le 1/2 \\ 2 - 2x & \text{for } 1/2 < x \le 1 \end{cases}$$

Show that $\{2/7, 4/7, 6/7\}$ is a 3-cycle for T.

Solution.   A routine check yields

$$T\left(\frac{2}{7}\right) = \frac{4}{7}, \quad T\left(\frac{4}{7}\right) = 2 - 2\left(\frac{4}{7}\right) = \frac{6}{7}, \quad \text{and} \quad T\left(\frac{6}{7}\right) = 2 - 2\left(\frac{6}{7}\right) = \frac{2}{7}$$

confirming that $\{2/7, 4/7, 6/7\}$ is a 3-cycle for T.   ❏

Not only does T have a 3-cycle; it has n-cycles for every positive integer n. This is one of the reasons why the tent function is featured in the study of dynamics. We will take a closer look at the tent function in Section 1.4.

Graphically, an n-cycle of a function is represented by a closed loop. Figure 1.9(a) shows the 2-cycle $\{-1, 1\}$ for the function $-x^3$, and Figure 1.9(b) shows the 3-cycle for the tent function in Example 1 above.

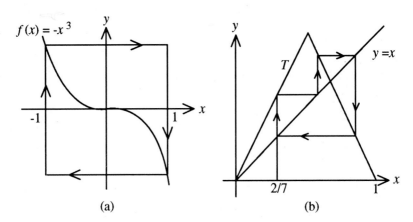

(a)   (b)

Figure 1.9

It is important to notice that if $f(x) = z$ and $f(z) = x$, then

$$f^{[2]}(x) = f(f(x)) = f(z) = x$$

so that x is a fixed point of $f^{[2]}$. By the same token, z is a fixed point of $f^{[2]}$. Thus, if $\{x, z\}$ is a 2-cycle for f, then x and z are both fixed points of $f^{[2]}$. Conversely, if x is a fixed point of $f^{[2]}$ that is not a fixed point of f, then there is a point z different from x such that $\{x, z\}$ is a 2-cycle of f, so that x is a period-2 point of f. Therefore

$$\{x, z\} \text{ is a 2-cycle for } f \text{ if and only if } f(x) = z,$$
$$\text{where } x \text{ and } z \text{ are distinct fixed points of } f^{[2]} \qquad (1)$$

For example, assume that $f(x) = x^2 - 1$, so that $f^{[2]}(x) = (x^2 - 1)^2 - 1 = x^4 - 2x^2$.

Obviously 0 is a fixed point of $f^{[2]}$ that is *not* a fixed point of f. Thus there must be a z such that $\{0, z\}$ is a 2-cycle for f. Since $f(0) = -1$ we deduce that $\{0, -1\}$ is a 2-cycle for f. More generally, $\{x_0, x_1, x_2, \ldots, x_{n-1}\}$ is an n-cycle of f if and only if x_k is a fixed point of $f^{[n]}$, for $k = 0, 1, 2, \ldots, n-1$.

Attracting Periodic Points

Suppose that x is a period-n point of f. Then x is a fixed point for $f^{[n]}$. Therefore we have a natural way of defining attracting and repelling periodic points.

DEFINITION 1.10. Let x be a period-n point for a function f. Then x is an **attracting** period-n point if x is an attracting fixed point of $f^{[n]}$; also x is a **repelling** period-n point if x is a repelling fixed point of $f^{[n]}$.

Suppose that f is continuous at a period-n point x. If x is attracting (repelling), then each point in $\{x, f(x), f^{[2]}(x), \ldots, f^{[n-1]}(x)\}$ is an attracting (repelling) period-n point, so we say that the n-cycle $\{x, f(x), f^{[2]}(x), \ldots, f^{[n-1]}(x)\}$ is **attracting (repelling)**.

In particular, if $n = 2$ then the period-2 point x is attracting if and only if there is an interval $(x - \varepsilon, x + \varepsilon)$, such that whenever y is in $(x - \varepsilon, x + \varepsilon)$,

$$f^{[2n]}(y) \to x \quad \text{and} \quad f^{[2n+1]}(y) \to f(x)$$

as n increases without bound. Figure 1.10 shows a number x that is attracted to a 2-cycle of f.

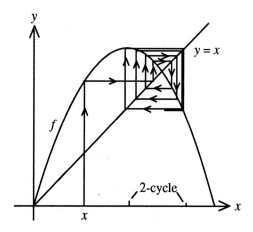

Figure 1.10

EXAMPLE 2. Let $f(x) = -x^{1/3}$. Show that 1 is an attracting period-2 point of f.

Solution. First notice that $f(1) = -1$ and $f(-1) = 1$. Therefore the point 1 has period 2. Next, observe that

$$f^{[2]}(x) = f(f(x)) = -(-x^{1/3})^{1/3} = x^{1/9}, \text{ so that } (f^{[2]})'(1) = \frac{1}{9}$$

Theorem 1.6 then implies that 1 is an attracting fixed point of $f^{[2]}$, so that 1 is an attracting period-2 point by Definition 1.10. ❏

One could also prove that 1 is an attracting point of period 2 by showing that there is an interval J around 1 such that whenever x is in J, $|f^{[2]}(x) - 1| < |x - 1|$.

In Section 1.2 we gave a criterion for attracting and repelling fixed points that involves the derivative. Similarly, there is a criterion for attracting and repelling cycles that involves the derivative. Before we state it in Theorem 1.12, we have a preliminary result.

THEOREM 1.11. Let $\{x, z\}$ be a 2-cycle of f. If $f^{[2]}$ is differentiable at x and at z, then

$$(f^{[2]})'(x) = f'(x)f'(z) = (f^{[2]})'(z) \tag{2}$$

Proof. Using the Chain Rule and the fact that $f(x) = z$, we find that

$$(f^{[2]})'(x) = (f \circ f)'(x) = [f'(f(x))][f'(x)] = f'(x)f'(z)$$

By symmetry we have $(f^{[2]})'(z) = f'(x)f'(z)$. ■

THEOREM 1.12. Let $\{x, z\}$ be a 2-cycle for f.
a.   If $|f'(x)f'(z)| < 1$, then the 2-cycle is attracting.
b.   If $|f'(x)f'(z)| > 1$, then the 2-cycle is repelling.

Proof. The result follows directly from Theorems 1.6 and 1.11, and the definition of an attracting (repelling) 2-cycle. ■

If $|f'(x)f'(z)| = 1$, then we cannot conclude anything about whether the cycle $\{x, z\}$ is attracting, repelling or neither. For example, let $f(x) = 1/x$. Then $f^{[2]}(x) = x$, so that $\{x, 1/x\}$ is a 2-cycle for each $x \neq 0$. Evidently the 2-cycle is neither attracting nor repelling, although $|f'(x)f'(z)| = |(f^{[2]})'(x)| = 1$ for all $x \neq 0$.

When $|f'(x)f'(z)| \neq 1$, the criterion can be effective in telling us if $\{x, z\}$ is attracting or repelling.

EXAMPLE 3. Let $f(x) = x^2 - 3x + 2$. Show that $\{0, 2\}$ is a repelling 2-cycle.

Solution. Since $f(0) = 2$ and $f(2) = 0$, it follows that $\{0, 2\}$ is a 2-cycle. The fact that $f'(x) = 2x - 3$ implies that $f'(0) = -3$ and $f'(2) = 1$, so that

$$f'(0)f'(2) = (-3)(1) = -3$$

Therefore Theorem 1.12 implies that $\{0, 2\}$ is a repelling 2-cycle of f. ❏

Figure 1.11 displays the graph of f, with 2-cycle $\{0, 2\}$. By analyzing the iterates of x, which is close to 0, we can see why the 2-cycle is repelling.

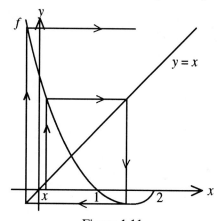

Figure 1.11

If $\{x, f(x), \ldots, f^{[n-1]}(x)\}$ is an n-cycle, then by the Chain Rule,

$$(f^{[n]})'(x) = [f'(f^{[n-1]}(x))] \, [f'(f^{[n-2]}(x))] \cdots [f'(f(x))] \, [f'(x)] \qquad (3)$$

If the absolute value of the right-hand side of (3) is < 1 (> 1), then the n-cycle is attracting (repelling). We remark that if x is a fixed point, then (3) becomes

$$(f^{[n]})'(x) = [f'(x)]^n \qquad (4)$$

An **eventually periodic point** is a point some iterate of which is periodic. For example, let $f(x) = x^2 - 1$. It follows that 1 is an eventually periodic point, since $f(1) = 0$, $f(0) = -1$, and $f(-1) = 0$. Henceforth we will generally assume (as we do with eventually fixed points), that when we refer to a point as eventually periodic, the point is not periodic.

Defining the basin of attraction of an attracting cycle is more complicated.

However, informally the **basin of attraction** of an attracting cycle is the collection of points whose iterates are eventually arbitrarily close to the points in the cycle. Again using $f(x) = x^2 - 1$, we can show that the basin of attraction for the 2-cycle $\{-1, 0\}$ consists of all numbers in the interval $((1 - \sqrt{5})/2, (1 + \sqrt{5})/2)$ except those whose iterates are eventually the fixed point $(1 - \sqrt{5})/2$. (See Exercise 9.)

To give a further illustration of the notions appearing in this section, we introduce the function B, which is a close relative of the tent function:

$$B(x) = \begin{cases} 2x & \text{for } 0 \leq x \leq 1/2 \\ 2x - 1 & \text{for } 1/2 < x \leq 1 \end{cases}$$

(Figure 1.12). Then B is called the **baker's function**, in reference to the kneading of bread dough. (The function is related to the sawtooth function that is prominent in engineering.) Exercises 10–19 are devoted to several properties, among which are the following:

1. A number x in $[0, 1]$ is eventually periodic if and only if x is rational.
2. A number x in $[0, 1]$ is called a **dyadic rational** if it has the form $k/2^m$ for some nonnegative integers k and m. A number x in $[0, 1]$ is eventually fixed if and only if x is a dyadic rational.
3. If p is an odd, positive integer, then k/p is periodic, for $k = 1, 2, \ldots, p - 1$.

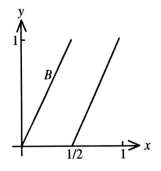

Figure 1.12

Although the baker's function is not continuous at $1/2$, it has many interesting properties, and has been the subject of much attention in the study of dynamics.

Time Series and Periodic Points

Many of a person's or animal's biological functions can be analyzed by means of a **time series**, which is a graph that registers a particular variable such as voltage or pressure as a function of time. Voltage is plotted against time in an electrocardiogram

(EKG) and in an electroencephalogram (EEG).

Figure 1.13 illustrates the EKG of a cat, before and after ingestion of a dose of cocaine. The EKG appearing in Figure 1.13(a) is normal, the pattern consisting of a tall and a shorter spike that repeats regularly as time passes. The heartbeat is said to be periodic. By contrast, the EKG appearing in Figure 1.13(b) is abnormal and irregular, and could represent the onslaught of a life-threatening cardiac fibrillation.

In a similar vein, Figure 1.14(a) displays the EEG of a cat before ingestion of a dose of cocaine, whereas Figure 1.14(b) shows the EEG after ingestion. However, in contrast to the EKG that is normal and periodic before cocaine and irregular after-

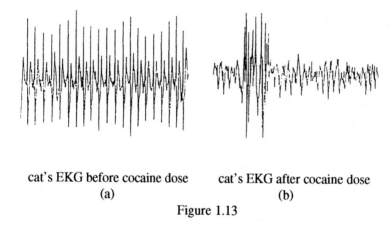

cat's EKG before cocaine dose   cat's EKG after cocaine dose
            (a)   (b)

Figure 1.13

wards, the normal EEG that is unpredictable and irregular before cocaine became much more regular (though not really periodic) after cocaine. The latter EEG is characteristic of a life-threatening brain dysfunction or an epileptic seizure.

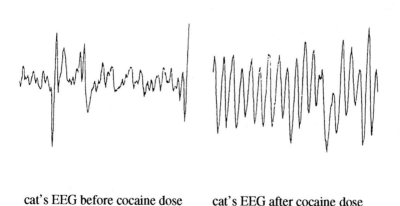

cat's EEG before cocaine dose   cat's EEG after cocaine dose
            (a)   (b)

Figure 1.14

EXERCISES 1.3

1. Let $f(x) = -\dfrac{1}{2}x^2 - x + \dfrac{1}{2}$. Show that 1 is an attracting period-2 point.

2. Show that $\{2/9, 4/9, 8/9\}$ is a repelling 3-cycle for the tent function T.

3. Let $f(x) = 1/x$. Show that if $x \neq -1$, 0, or 1, then x is a period-2 point.

4. Let $f(x) = 1/(1 - x)$. Show that if $x \neq 0$ or 1, then x is a period-3 point.

5. Let $f(x) = 3.2x - 3.2x^2$. Use a calculator or the program ITERATE to find a 2-cycle, and show that it is attracting.

6. Let $f(x) = 3.84x - 3.84x^2$. Use a calculator or the program ITERATE to find a period-3 point, and determine whether it is attracting or not.

7. Let T be the tent function. Find a point that is not periodic but is eventually periodic with period
    a. 3   b. 4   c. 5

8. Let $f(x) = \cos x$. Determine whether f has any period-n points with $n > 1$.

9. Let $f(x) = x^2 - 1$. Show that the basin of attraction of the 2–cycle $\{-1, 0\}$ consists of all numbers in the interval $((1 - \sqrt{5})/2, (1 + \sqrt{5})/2)$ except for those whose iterates are eventually the fixed point $(1 - \sqrt{5})/2$.

Exercises 10–19 involve the baker's function B.

10. Sketch the graphs of $B^{[2]}$ and $B^{[3]}$.

11. Find the fixed points and the period-2 points of B.

12. Determine whether the following points are fixed, eventually fixed, periodic or eventually periodic, and indicate their periods if they are periodic.
    a.   3/7   b.   3/16   c.   1/10   d.   1/11

13. For each positive integer n, determine the number of fixed points of $B^{[n]}$.

14. Show that x in $[0, 1]$ is an eventually fixed point of B if and only if x is a dyadic rational.

15. Show that x in $[0, 1]$ is an eventually periodic point of B if and only if x is a rational.

16. Show that if p is an odd positive integer, then k/p is periodic, for $k = 1, 2,$ ..., $p - 1$. Is the converse true? Explain your answer.

17. Show that if $1/2 < x \leq 1$, then $B(x) \leq x^2$.

18. Let m be an arbitrary positive integer, and assume that x is not a dyadic rational. Show that there is an integer $n \geq m$ such that
    a. $B^{[n]}(x) < 1/2$ b. $B^{[n]}(x) > 1/2$

19. Let c be a constant, and let $f(x) = x^3 - 3x + c$. Determine the values of c for which $\{0, c\}$ is a 2-cycle. Is such a 2-cycle attracting for any such value of c? Explain your answer.

20. Let $f(x) = ax^3 - bx + 1$, where a and b are constants. Determine the values of a and b for which $\{0, 1\}$ is an attracting 2-cycle.

21. Let m be a positive integer. Prove that if $(f^{[m]})'(x) = 0$, then there is an iterate x_k of x such that $f'(x_k) = 0$.

22. Let $|f'(x)| < 1$ for all x. Show that f cannot have any period-2 points.

23. a. Let f be increasing. Show that there are no period-n points for $n > 1$.
    b. Let f be decreasing. Show that there are no period-n points for $n > 2$.
    c. Find an example of a decreasing function f that has a fixed point and a period-2 point.

24. Let f be a linear function. Show that there are no period-n points for $n > 2$.

1.4 FAMILIES OF FUNCTIONS

In Example 3 of Section 1.2 we discussed the function f defined by

$$f(x) = \mu x(1 - x), \text{ for } 0 \leq x \leq 1$$

where μ was a fixed positive constant. We found that f has one or two fixed points, depending on whether $0 < \mu \leq 1$ or $\mu > 1$. Thus the number of fixed points depends on the value of μ. In order to emphasize the fact that the function depends on μ, we will henceforth designate f by Q_μ, so that

$$Q_\mu(x) = \mu x(1 - x), \text{ for } 0 \leq x \leq 1$$

The family $\{Q_\mu\}$ for $\mu > 0$ is the **quadratic family**, so named because each of the functions in the family is a quadratic function.

A collection of functions such as $\{Q_\mu\}$ is called a **parametrized family of functions** (or a **one-parameter family**), and μ is the **parameter** for the family. Other parametrized families that we will encounter are

$$g_\mu(x) = x^2 + \mu, \text{ for all } x$$

$$T_\mu(x) = \begin{cases} 2\mu x \text{ for } 0 \le x \le 1/2 \\ 2\mu(1 - x) \text{ for } 1/2 < x \le 1 \end{cases} \quad \text{where } 0 < \mu \le 1$$

$$E_\mu(x) = \mu e^x, \text{ for all } x$$

$$S_\mu(x) = \mu \sin x, \text{ for } 0 \le x \le \pi$$

Notice that μ is constant and x is the variable for each function in the parametrized families listed above. For the family $\{T_\mu\}$, μ is restricted to the interval $(0, 1]$ in order that the range of the functions T_μ will be contained in the domain $[0, 1]$.

There are names for the families listed above. The family $\{T_\mu\}$ is the **tent family** (because the functions T_μ are from the same mold as the tent function T), $\{E_\mu\}$ is the **exponential family**, and $\{S_\mu\}$ is the **sine family**. We give the family $\{g_\mu\}$ no special name; later we will show that the family $\{g_\mu\}$ is a close relative of the quadratic family $\{Q_\mu\}$. When we wish to refer to a general parametrized family rather than a specific one, we will denote it by $\{f_\mu\}$.

The way in which the orbits in a parametrized family change as the parameter varies is called the **dynamics of the family**. In the present section we will study the dynamics of $\{g_\mu\}$ and $\{T_\mu\}$; we will devote the entire Section 1.5 to the dynamics of $\{Q_\mu\}$.

The Family $\{g_\mu\}$

The family $\{g_\mu\}$ consists of the functions defined by

$$g_\mu(x) = x^2 + \mu, \text{ for all } x$$

which are among the simplest nonlinear differentiable functions. The dynamics of the family $\{g_\mu\}$ vary according to the value of μ. If $0 \le \mu$, we can describe in detail the orbit $\{g_\mu^{[n]}(x)\}_{n=0}^{\infty}$ for any real number x, whereas if $\mu < 0$, the orbits can be very complicated. As a result, in this section we will limit the discussion to the members of $\{g_\mu\}$ for which $0 \le \mu$.

Among nonnegative parameters for $\{g_\mu\}$, two values are especially noteworthy: $\mu = 0$ and $\mu = 1/4$. For $\mu = 0$ we have g_0, which is the simplest function in the family and is defined by $g_0(x) = x^2$ (Figure 1.15(a)). A moment's reflection reveals that the fixed points of g_0 are 0 and 1. Next, we notice that $|g_0(x)| = x^2 < |x|$ if $|x| < 1$, and $|g_0(x)| = x^2 > |x|$ if $|x| > 1$. It follows that 0 is an attracting fixed point whose basin of attraction is $(-1, 1)$, and that 1 is a repelling fixed point. Since all iterates of x approach 0 if $|x| < 1$ and are unbounded if $|x| > 1$, there can be no periodic points besides 0 and 1.

Next we turn to $g_{1/4}$ (Figure 1.15(b)). Notice that

$$g_{1/4}(x) - x = x^2 + \frac{1}{4} - x = \left(x - \frac{1}{2}\right)^2 \begin{cases} = 0 \text{ if } x = 1/2 \\ > 0 \text{ if } x \neq 1/2 \end{cases}$$

Therefore $g_{1/4}$ has one and only one fixed point: 1/2. Moreover, the graph of $g_{1/4}$ lies above the line $y = x$ except at $x = 1/2$, and is tangent at the point $(1/2, 1/2)$. Using graphical analysis (or the Mean Value Theorem), one can show that the basin of attraction of the fixed point 1/2 is $[-1/2, 1/2]$, and that 1/2 repels points to the right. We call a fixed point that attracts on one side and repels on the other an **attracting-repelling fixed point**.

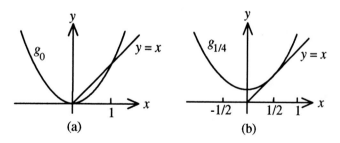

Figure 1.15

Because the graph of g_μ shifts upward as μ increases, and because the graph of $g_{1/4}$ is tangent to the graph of $y = x$, a glance at Figure 1.15(b) suggests that the graph of g_μ intersects the graph of $y = x$ if $0 < \mu \leq 1/4$ and does not intersect it if $\mu > 1/4$. Therefore we will divide the analysis of $\{g_\mu\}$ for the remaining positive values of μ into two groups: $0 < \mu < 1/4$ and $\mu > 1/4$.

Case 1. $0 < \mu < 1/4$

The number x is a fixed point of g_μ if and only if $x = g_\mu(x) = x^2 + \mu$, which is equivalent to $x^2 - x + \mu = 0$. Solving for x, we obtain

$$x = \frac{1}{2} - \frac{1}{2}\sqrt{1 - 4\mu} \quad \text{or} \quad x = \frac{1}{2} + \frac{1}{2}\sqrt{1 - 4\mu}$$

as the fixed points of g_μ. Next we will determine which (if any) of these points is attracting. Since $g'_\mu(x) = 2x$, it follows that

$$\left| g'_\mu\left(\frac{1}{2} \pm \frac{1}{2}\sqrt{1-4\mu}\,\right)\right| = \left|1 \pm \sqrt{1-4\mu}\,\right|$$

If we let

$$p_\mu = \frac{1}{2} - \frac{1}{2}\sqrt{1-4\mu} \quad \text{and} \quad q_\mu = \frac{1}{2} + \frac{1}{2}\sqrt{1-4\mu}$$

then $|g'_\mu(p_\mu)| = \left|1 - \sqrt{1-4\mu}\,\right| < 1$ for $0 < \mu < 1/4$. By Theorem 1.6, p_μ is an attracting fixed point. By contrast, $|g'_\mu(q_\mu)| = \left|1 + \sqrt{1-4\mu}\,\right| > 1$ for $0 < \mu < 1/4$, so that again by Theorem 1.6, q_μ is a repelling fixed point. It turns out that if $0 < \mu < 1/4$, then the basin of attraction of p_μ is the open interval $(-q_\mu, q_\mu)$, and the iterates of each number x such that $|x| > q_\mu$ march off toward ∞. Thus every number other than $\pm q_\mu$ has the property that its iterates approach p_μ or are unbounded. We mention that as the parameter μ approaches $1/4$, the fixed points p_μ and q_μ are drawn toward each other, and actually coalesce when $\mu = 1/4$ (compare Figure 1.16(a) with 1.16(b)).

Case 2. $\mu > 1/4$

As μ increases beyond $1/4$, the dynamics of g_μ change dramatically, because the entire graph of g_μ lies above the line $y = x$ (Figure 1.16(c)). Thus there is no fixed point. We can prove this formally by noticing that if $\mu > 1/4$, then

$$g_\mu(x) - x = x^2 - x + \mu > x^2 - x + \frac{1}{4} = \left(x - \frac{1}{2}\right)^2 \geq 0$$

so that $g_\mu(x) > x$ for all x. Moreover, since the iterates of each number x form an

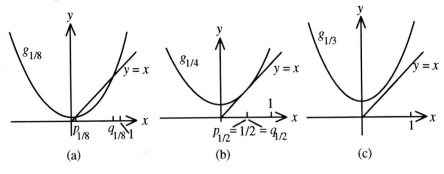

(a)   (b)   (c)

Figure 1.16

increasing sequence that diverges to ∞, it follows that g_μ has no periodic points.

Something very special has happened for $\mu = 1/4$, because as μ increases and passes through $1/4$, g_μ first has two fixed points and then none. We say that the family has a bifurcation at $1/4$. More generally we have the following definition.

DEFINITION 1.13. A parametrized family $\{f_\mu\}$ has a **bifurcation at** μ_0, or **bifurcates at** μ_0, if the number or nature (attracting vs. repelling) of periodic points of f_μ changes as μ passes through μ_0. In this case μ_0 is said to be a **bifurcation point** for the family.

The term "bifurcate" comes from Latin words meaning "two branches." From Definition 1.13 we infer that $\{g_\mu\}$ bifurcates at the number $1/4$. Bifurcation points signal changes in dynamics of a parametrized family. We will discuss bifurcation points for each of the parametrized families that we encounter, and will devote Section 1.6 to bifurcations.

The Tent Family $\{T_\mu\}$

Recall that the tent family consists of the functions T_μ defined by

$$T_\mu(x) = \begin{cases} 2\mu x & \text{for } 0 \leq x \leq 1/2 \\ 2\mu(1-x) & \text{for } 1/2 < x \leq 1 \end{cases}$$

Figures 1.17(a)–(c) display T_μ for $\mu = 2/7$, $1/2$ and $5/6$. As μ increases, the height of the graph of T_μ rises, because of the factor μ in the formula for T_μ. From this observation and the three graphs in Figure 1.17 we deduce that if $0 < \mu < 1/2$, then T_μ intersects the line $y = x$ once (at 0), whereas if $1/2 < \mu < 1$, then there are two points of intersection. We are led to analyze separately the members of $\{T_\mu\}$ for which $0 < \mu < 1/2$, $\mu = 1/2$, and $1/2 < \mu < 1$. Finally we will study T_1, which is the original tent function T and which has some very interesting features.

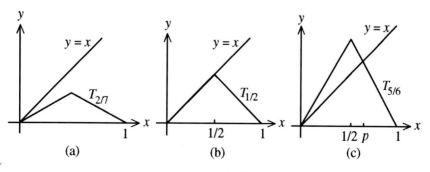

Figure 1.17

Case 1. $0 < \mu < 1/2$

The graph in Figure 1.17(a) shows that 0 is the only fixed point of T_μ. Since $0 < \mu < 1/2$, it follows from the definition of T_μ that if $0 < x \le 1/2$, then

$$0 \le T_\mu(x) = 2\mu x < x$$

and if $1/2 < x \le 1$, then

$$0 \le T_\mu(x) = 2\mu(1 - x) < 1 - x < \frac{1}{2} < x$$

Consequently for any x in $[0, 1]$, the sequence $\{T_\mu^{[n]}(x)\}_{n=0}^{\infty}$ is bounded and decreasing. By Corollary 1.4, the sequence converges to the fixed point 0. Therefore 0 is an attracting fixed point whose basin of attraction is $[0, 1]$.

Case 2. $\mu = 1/2$

First we notice that if $0 \le x \le 1/2$, then $T_{1/2}(x) = 2(1/2)x = x$, so that x is a fixed point of $T_{1/2}$ (Figure 1.17(b)). Next, we calculate that if $1/2 < x \le 1$, then

$$0 \le T_{1/2}(x) = 2(1/2)(1 - x) = 1 - x \le 1/2$$

so that $T_{1/2}(x)$ is a fixed point of $T_{1/2}$. Consequently every point in $[0, 1]$ either is a fixed point of $T_{1/2}$ or has a fixed point for its first iterate.

Case 3. $1/2 < \mu < 1$

In addition to the fixed point 0, there is a second fixed point p that lies in $(1/2, 1]$, as you can see in Figure 1.17(c). To evaluate p we solve the equation

$$p = T_\mu(p) = 2\mu(1 - p)$$

which yields

$$p = \frac{2\mu}{1 + 2\mu}$$

As μ increases from $1/2$ toward 1, p increases from $1/2$ toward $2/3$. Because $|T_\mu'(x)| = 2\mu > 1$ on $[0, 1]$ except at $1/2$, both 0 and p are repelling fixed points. The period-2 points of T_μ are the fixed points of $T_\mu^{[2]}$, which is given by

$$T_\mu^{[2]}(x) = \begin{cases} 4\mu^2 x & \text{for } 0 \le x \le 1/4\mu \\ 2\mu(1 - 2\mu x) & \text{for } 1/4\mu < x \le 1/2 \\ 2\mu(1 - 2\mu + 2\mu x) & \text{for } 1/2 < x \le 1 - 1/4\mu \\ 4\mu^2(1 - x) & \text{for } 1 - 1/4\mu < x \le 1 \end{cases}$$

The graph of $T_8^{[2]}$ appears in Figure 1.18, and suggests that for $1/2 < \mu < 1$, $T_\mu^{[2]}$ has four fixed points that can be found by solving the four equations $x = T_\mu^{[2]}(x)$ arising from the definition of $T_\mu^{[2]}$. We find that the fixed points are

$$0, \quad \frac{2\mu}{1 + 4\mu^2}, \quad \frac{2\mu}{1 + 2\mu}, \quad \text{and} \quad \frac{4\mu^2}{1 + 4\mu^2}$$

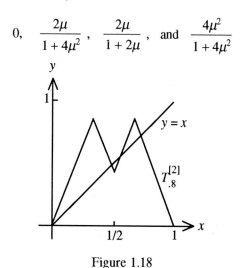

Figure 1.18

The first and third are the two fixed points of T_μ, so it follows that

$$\left\{ \frac{2\mu}{1 + 4\mu^2}, \quad \frac{4\mu^2}{1 + 4\mu^2} \right\}$$

is a 2-cycle for T_μ. This 2-cycle is repelling, because $|(T_\mu^{[2]})'(x)| = 4\mu^2 > 1$ wherever the derivative is defined. Because the graph of $T_\mu^{[n]}$ is linear on the 2^n subintervals $[0, 1/2^n], \ldots, [1 - 1/2^n, 1]$, it is possible (though tedious) to describe the various n-cycles of T_μ — all of which are repelling!

Case 4. $\mu = 1$

If $\mu = 1$ then $T_\mu = T$, the tent function given by

$$T(x) = \begin{cases} 2x & \text{for } 0 \le x \le 1/2 \\ 2(1 - x) & \text{for } 1/2 < x \le 1 \end{cases}$$

The graph of T appears in Figure 1.19(a). The major difference between the graph of T and the graph of T_μ when $\mu < 1$ is the fact that the range of T fills out the whole interval $[0, 1]$. The function T stretches the interval $[0, 1/2]$ over the entire interval $[0, 1]$, and folds the interval $[1/2, 1]$ back over the interval $[0, 1]$. It is this stretching and folding that is characteristic of many functions we will examine in this book, and which leads to the notion of chaos that we will discuss later.

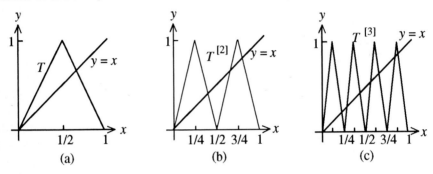

Figure 1.19

As with all members of $\{T_\mu\}$, 0 is a fixed point of T. Since $x = T(x) = 2(1 - x)$ if $x = 2/3$, we know that $2/3$ is the second fixed point of T. Figures 1.19(b)–(c) indicate that $T^{[2]}$ and $T^{[3]}$ have, respectively, four and eight fixed points. Thus T has two period-2 points and six period-3 points, which we could evaluate by solving the equations $x = T^{[2]}(x)$ and $x = T^{[3]}(x)$ for x.

Rather than determine the values of the period-n points for T, we turn to the number of period-n points of T for $n \ge 1$. The graph of $T^{[n]}$ has 2^{n-1} congruent spikes. Consequently there are 2^n fixed points for $T^{[n]}$, two in each of the subintervals $[0, 1/2^{n-1}]$, ..., $[(2^{n-1}- 1)/2^{n-1}, 1]$. Some of these fixed points of $T^{[n]}$ are fixed points for $T^{[k]}$ with $k < n$; the remaining fixed points of $T^{[n]}$ join together to form n-cycles for T. The breakdown is given in the following table:

n	1	2	3	4	5
# of fixed points for $T^{[n]}$	2	4	8	16	32
# of period-n points for T	2	2	6	12	30
# of n-cycles for T	2	1	2	3	6

To obtain the number of period-n points of T, we take the number, 2^n, of fixed points of $T^{\{n\}}$, and then subtract the total number of period-k points for all values of k for which $k < n$ and k divides n. For example, if $n = 4$, then there are $2^4 = 16$ fixed points for $T^{\{4\}}$. Two of these fixed points are fixed points of T, and 2 others are fixed points of $T^{\{2\}}$. Therefore the remaining twelve fixed points of $T^{\{4\}}$ are necessarily period-4 points, and form three 4-cycles for T.

Eventually Periodic and Periodic Points of T

In this subsection we will determine the eventually periodic points and periodic points for T. This analysis is more technical than what has preceded it, and the results are independent from what follows. In order to facilitate notation in the discussion that follows, we will allow the expression "eventually periodic point" to include periodic, eventually fixed, and fixed points.

THEOREM 1.14. Let x be in the interval $(0, 1)$. Then x is eventually periodic for T if and only if x is rational.

Proof. Assume that x is eventually periodic. Either $T(x) = 2x$ or $T(x) = 2 - 2x$, so that $T(x) = $ integer $\pm 2x$. Similarly, $T^{\{2\}}(x) = $ integer $\pm 2^2 x$, and in general,

$$T^{\{n\}}(x) = \text{integer} \pm 2^n x$$

For each n, let a_n and i_n be integers such that $a_n = 2^n$ or $a_n = -2^n$, and such that $T^{\{n\}}(x) = i_n + a_n x$. Since x is eventually periodic, there are positive integers k and m with $k \neq m$ and such that $T^{\{k\}}(x) = T^{\{m\}}(x)$. Thus $i_k + a_k x = i_m + a_m x$. Since $k \neq m$, it follows that $a_k \neq a_m$, so that

$$x = \frac{i_m - i_k}{a_k - a_m}$$

which means that x is a rational number.

To prove the converse, we will assume for the moment that $x = k/p$ with p an odd integer. Then $T(k/p) = 2k/p$ or $T(x) = 2(p - k)/p$, which means that $T(x) = $ (even integer)$/p$. There are only finitely many distinct numbers in $(0, 1)$ of the form (even integer)$/p$, so x is eventually periodic. Next, assume that $x = k/p$ with p even. In this case,

$$T(x) = \frac{2k}{p} = \frac{k}{p/2} \qquad \text{or} \qquad T(x) = 2\left(1 - \frac{k}{p}\right) = \frac{p - k}{p/2}$$

so that the denominator has become $p/2$. Continuing the process with $T(x)$ substituted for x, we find that for some positive integer i (depending on x), either $T^{\{i\}}(x) = $ (integer)/(odd integer), or $T^{\{i\}}(x) = 1$. The first possibility means that x

is eventually periodic by the preceding argument, and the second possibility means that $T^{[i+1]}(x) = 0$, so that x is eventually fixed (and hence eventually periodic). We conclude that if x is rational, then x is eventually periodic. ∎

Theorem 1.14 implies that all rational numbers in $(0, 1)$ are eventually periodic for T. To determine which rational numbers are actually periodic, we first have two lemmas (that is, pre-theorems).

LEMMA 1. Suppose that p is odd, and let $x = k/p$ be in $(0, 1)$. Then x is periodic for T if and only if k is even.

Proof. Suppose that x is periodic with period n. Since p is odd by hypothesis, and since $T(x) = 2k/p$ or $T(x) = 2(p - k)/p$, it follows that $T(x) = $ (even integer)/p. The same is true for all iterates of x, so in particular, $x = T^{[n]}(x) = $ (even integer)/p. Thus if x is periodic, then k must be even.

To prove the converse, assume that k is even. We will show that x is periodic. For any positive integer i, $T^{[i-1]}(x) = $ (even integer)/p. Thus

$$T^{[i]}(x) = \begin{cases} \dfrac{4\,(\text{integer})}{p} & \text{if } T^{[i-1]}(x) \leq 1/2 \\[3mm] 2\left(1 - \dfrac{\text{even integer}}{p}\right) = \dfrac{4\,(\text{integer}) + 2}{p} & \text{if } T^{[i-1]}(x) > 1/2 \end{cases}$$

Therefore as soon as we see the form of $T^{[i]}(x)$, we know whether $T^{[i-1]}(x)$ is in $[0, 1/2]$ or in $(1/2, 1]$.

Now recall from Theorem 1.14 that x is eventually periodic, so there are a least nonnegative integer i and a least positive integer n such that $n > i$ and $T^{[i]}(x) = T^{[n]}(x)$. If $i = 0$, then x is periodic with period n. Next, we will show that i cannot be positive (so that i must be 0). To obtain a contradiction, assume that $i > 0$. Then by the discussion in the preceding paragraph, both $T^{[i-1]}(x)$ and $T^{[n-1]}(x)$ lie in $[0, 1/2]$ or both lie in $(1/2, 1]$. Since T is strictly increasing on $[0, 1/2]$ and strictly decreasing on $(1/2, 1]$, we conclude that $T^{[i]}(x) = T^{[n]}(x)$ only if $T^{[i-1]}(x) = T^{[n-1]}(x)$. But that contradicts the minimality of i and n. Therefore $i = 0$, so that x is periodic. This completes the proof. ∎

LEMMA 2. Suppose that p is even, and let $x = k/p$ be in $(0, 1)$. Then x is not periodic for T.

Proof. Since x is assumed to be in reduced form, with p even, it follows that k must be odd, so that $x = $ (odd integer)/p. But then $T(x) = 2k/p$ or $T(x) = 2(p - k)/p$, so that in any case, $T(x) = $ (integer)/$(p/2)$. Thus the reduced form for $T(x)$, like $T^{[n]}(x)$ for any $n > 1$, cannot be (odd integer)/p. Thus x is not periodic.   ■

THEOREM 1.15. The rational number x in $(0, 1)$ is periodic for T if and only if x has the form (even integer)/(odd integer).

Proof. Because we assume that x is in reduced form, a moment's reflection tells us that Theorem 1.14, Lemma 1 and Lemma 2 together imply the result.   ■

An analogous result for eventually fixed points is the following: a number x in $[0, 1]$ is eventually fixed if and only if x has the form $k/2^m$ or $k/(3 \cdot 2^m)$ for appropriate nonnegative integers k and m (see Exercise 4(a)).

Now we list eventually fixed, periodic, and eventually periodic points for T.

<u>Iterates</u>

$$x = \frac{3}{16} : \quad \frac{3}{16} \quad \frac{3}{8} \quad \frac{3}{4} \quad \frac{1}{2} \quad 1 \quad 0 \quad 0 \quad \text{eventually fixed}$$

$$x = \frac{6}{13} : \quad \frac{6}{13} \quad \frac{12}{13} \quad \frac{2}{13} \quad \frac{4}{13} \quad \frac{8}{13} \quad \frac{10}{13} \quad \frac{6}{13} \quad \text{periodic}$$

$$x = \frac{7}{10} : \quad \frac{7}{10} \quad \frac{3}{5} \quad \frac{4}{5} \quad \frac{2}{5} \quad \frac{4}{5} \quad \frac{2}{5} \quad \frac{4}{5} \quad \text{eventually periodic}$$

There is much more to say about the tent function T. It will reappear in Section 2.2 when we discuss the concept of chaos.

EXERCISES 1.4

1. Determine which of the following are eventually fixed, which are eventually periodic, and which are periodic points for T.
    a. 3/11   b. 10/33   c. 5/18   d. 6/23   e. 3/16

2. Find as many 5-cycles of T as you can.

3. For T find the total number of
    a.   8-cycles   b.   15-cycles

4. a.   Let x be in $[0, 1]$. Show that x is eventually fixed for T if and only if x has the form $k/2^m$ or $k/(3 \cdot 2^m)$ for nonnegative integers k and m.

b.   Show that .3 is eventually periodic but not periodic for T.

c.   Use the computer program ITERATE to compute the first 100 iterates of .3 for T. Do you notice anything strange in the behavior of the iterates the computer provides? If so, give an explanation.

5.   Show that for each x in $(0, 1)$ that is not an eventually fixed point, and for each positive integer N, there is an $n > N$ such that $T^{[n]}(x) < 1/2$.

6.   Show that $1/2$ is an eventually fixed point of $T_{\sqrt{2}/2}$.

7.   Let n be an arbitrary positive integer, and let μ be fixed. Find the maximum value of $T_\mu^{[n]}$, under the condition that μ is in
a. $(0, 1/2)$ b. $(1/2 , 1)$

8.   Assume that $1/2 \le \mu \le 1$, and let $h(\mu)$ = relative minimum value of $T_\mu^{[2]}$ on the interval $(0, 1)$.
a.   Find a formula for $h(\mu)$.
b.   Find the maximum value of $h(\mu)$.

9.   Let $1/2 < \mu < 1$. Prove that the iterates under T_μ of each x in $(0, 1)$ are eventually "trapped" in the interval $[2\mu(1 - \mu), \mu]$, in the sense that there is an n (depending on x) such that if $k \ge n$, then $2\mu(1 - \mu) \le T_\mu^{[k]}(x) \le \mu$.

10.   Consider the function g_μ, where $0 < \mu < 1/4$. Show that the basin of attraction of the fixed point p_μ is the open interval $(-q_\mu, q_\mu)$.

Exercises 11–15 concern the family $\{E_\mu\}$, where $E_\mu(x) = \mu e^x$ for all x and $\mu > 0$.

11.   a.   Show that $E_{1/e}(x) \ge x$ for all x, and that $E_{1/e}$ has a single fixed point. Find the fixed point.
b.   Find $\lim_{n \to \infty} E_{1/e}^{[n]}(x)$ for all x.

12.   Let $\mu > 1/e$. Find $\lim_{n \to \infty} E_\mu^{[n]}(x)$ for all x, and thereby show that E_μ has no periodic points.

13.   Let $0 < \mu < 1/e$. Show that E_μ has two fixed points. Denote them by p_μ and q_μ, with $q_\mu < p_\mu$. Determine which fixed point is attracting and which is repelling.

14.   Let $0 < \mu < 1/e$. Show that E_μ has no periodic points that are not fixed points.

15.   Show that $\lim_{\mu \to 0+} q_\mu = 0$ and $\lim_{\mu \to 0+} p_\mu = \infty$.

Exercises 16–21 relate $\{E_\mu\}$ where $\mu < 0$.

16.   a.   Show that E_μ has a unique fixed point p_μ for each $\mu < 0$.
      b.   Find the fixed point of E_{-e}.

17.   Show that p_μ is repelling if $\mu < -e$ and is attracting if $\mu > -e$.

18.   Show that the maximum value of $(E_\mu^{[2]})'$ occurs for $x = -\ln(-\mu)$.

19.   Let $-e < \mu < 0$. Show that the single fixed point of E_μ has basin of attraction $(-\infty, \infty)$, and hence that there are no period-n points for $n > 1$.

20.   a.   Show that $E_{-e}^{[2]}$ has the unique fixed point p_μ found in Exercise 16(a).
      b.   Show that the graph of $E_{-e}^{[2]}$ is tangent to the line $y = x$ at (p_μ, p_μ).

21.   Let $\mu < -e$. Show that E_μ has one fixed point and one 2-cycle, and no other cycles.

1.5 THE QUADRATIC FAMILY

Consider a population of organisms for which there is a constant supply of food and limited space, and no predators. Many insect populations in the temperate zones fit this description at certain times in their history. In order to model the populations in successive generations, let N_n denote the population of the nth generation, and adjust the numbers so that the capacity of the environment equals 1, which means that $0 \le N_n \le 1$. One formula that has gained widespread fame is

$$N_{n+1} = \mu N_n(1 - N_n), \text{ for } 0 \le N_n \le 1$$

Sometimes the equation in (1) is called a "logistic equation," after a differential equation studied by the Belgian mathematician P. F. Verhulst 150 years ago. The parameter μ indicates the rate at which the population grows when it is very small.

Two properties of the equation in (1) are relevant to the study of population dynamics:

   i.   If the population is 0 at generation n, then the population remains 0.
   ii.   The population grows when N_n is small, and declines when N_n is large.

Property (ii) is reasonable, because when the population is small there is ample food and space, so the population can grow without hindrance. However, when the population is sufficiently large (that is, close to 1), the new generations are smaller because of food shortage and overcrowding.

The continuous version of (1) is the quadratic function given by

$$Q_\mu(x) = \mu x(1 - x) = \mu x - \mu x^2, \text{ for } 0 \le x \le 1$$

which appeared in Section 1.2. The present section is devoted to a detailed analysis of the quadratic family $\{Q_\mu\}$, not only because of its importance to the study of population dynamics, but also because its members, which are very simple polynomials, can have very complicated dynamics and can exhibit many of the characteristics that are associated with the study of chaotic dynamics.

In our study of the family $\{Q_\mu\}$, we will restrict the values of the parameter μ to be in $(0, 4]$. To see why we make this restriction, note that $Q_\mu'(x) = \mu - 2\mu x$, so that $Q_\mu'(x) = 0$ if $x = 1/2$. Since $Q_\mu''(x) = -2\mu$, we know that $Q_\mu(1/2) = \mu/4$ is an extreme value of Q_μ if $\mu \ne 0$. In order that $Q_\mu(1/2)$ lie in the domain $[0, 1]$ when $\mu \ne 0$, we must have $0 < \mu/4 \le 1$, that is, $0 < \mu \le 4$. It follows that if $0 < \mu \le 4$, then the range of Q_μ is contained in the domain of Q_μ. The graph of a representative Q_μ appears in Figure 1.20.

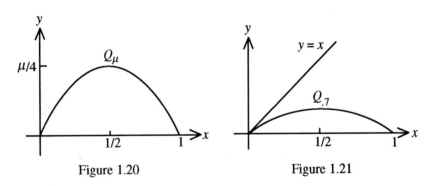

Figure 1.20   Figure 1.21

Although we know from Section 1.2 that Q_μ has one fixed point when $0 < \mu \le 1$ and two fixed points when $1 < \mu \le 4$, we have not discussed possible periodic points for Q_μ that are not fixed points. We will address this issue carefully below. Since it turns out that the dynamics of Q_μ change noticeably as μ passes through each of the integers 1, 2, and 3, we will split our discussion into four cases: $0 < \mu \le 1$, $1 < \mu \le 2$, $2 < \mu \le 3$, and $3 < \mu \le 4$.

Case 1. $0 < \mu \le 1$

This first case is the easiest. Since

$$0 < Q_\mu(x) = \mu x (1 - x) < \mu x \le x \text{ for } 0 < x < 1$$

it follows that $\{Q_\mu^{[n]}(x)\}_{n=0}^\infty$ is a positive, decreasing sequence, which by Corollary 1.4 converges to the fixed point 0. We conclude that the basin of attraction of 0 is the interval $[0, 1]$, so that there are no periodic points other than the fixed point 0.

The same result could be achieved by noticing that the graph of Q_μ lies below the line $y = x$ because $Q_\mu(0) = 0$ and $Q'_\mu(x) = \mu - 2\mu x \leq 1$ for all x in $[0, 1]$. Figure 1.21 supports these conclusions.

Case 2. $1 < \mu \leq 2$

By Example 3 of Section 1.2, we know that Q_μ has the two fixed points, 0 (which is repelling) and $1 - 1/\mu$ (which is attracting). If you apply graphical analysis to the graphs in Figure 1.22, you could convince yourself that the basin of attraction of $1 - 1/\mu$ is the whole interval $(0, 1)$. To prove it rigorously, we will denote $1 - 1/\mu$ by p_μ, and let $0 < x < p_\mu = 1 - 1/\mu$. Then

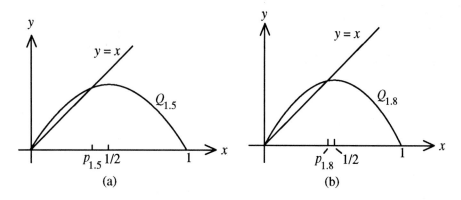

Figure 1.22

$$\frac{1}{\mu} < 1 - x, \text{ so that } 1 < \mu(1 - x), \text{ and thus } x < \mu x (1 - x) = Q_\mu(x)$$

Since Q_μ is increasing on $[0, p_\mu]$, this means that if $0 < x < p_\mu$, then

$$x < Q_\mu(x) < Q_\mu(p_\mu) = p_\mu$$

Consequently $\{Q_\mu^{[n]}(x)\}_{n=0}^\infty$ is a bounded increasing sequence when $0 < x < p_\mu$, so by Corollary 1.4 converges to a fixed point, which in this case must be p_μ. In an analogous fashion one can show that if $p_\mu < x < 1/2$, then $\{Q_\mu^{[n]}(x)\}_{n=0}^\infty$ is a bounded decreasing sequence, which also converges to p_μ (see Exercise 1). Finally, if $1/2 < x < 1$, then $0 < Q_\mu(x) < 1/2$, so by the above analysis, $\{Q_\mu^{[n]}(x)\}_{n=0}^\infty$ converges to p_μ. Therefore when $1 < \mu \leq 2$, the basin of attraction of p_μ is the open interval $(0, 1)$, and there are no periodic points other than the two fixed points.

Case 3. $2 < \mu \leq 3$

As μ increases from 1 toward 2, $p_\mu = 1 - 1/\mu$ increases from 0 toward

1/2. By contrast, if $\mu > 2$, then $p_\mu > 1/2$ (Figure 1.23(a)), which makes the analysis more difficult. Nevertheless, we will be able to show that if $2 < \mu \leq 3$, then again the basin of attraction of p_μ is $(0, 1)$.

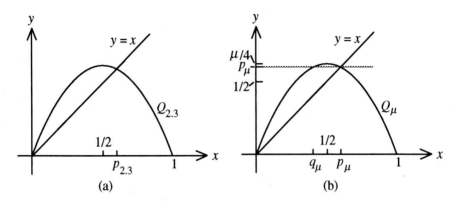

Figure 1.23

In our discussion, let us assume that $\mu \neq 3$, and denote by q_μ the unique number in $(0, 1/2)$ such that $Q_\mu(q_\mu) = Q_\mu(p_\mu) = p_\mu$ (Figure 1.23(b)). Notice that q_μ and p_μ are symmetric with respect to the line $x = 1/2$. Our first goal is to show that any x in the interval $(0, 1)$ has an iterate in the interval $(q_\mu, p_\mu]$.

Toward that end, fix x in $(0, q_\mu)$. Notice that on the interval $(0, q_\mu)$ the graph of Q_μ lies above the line $y = x$ (Figure 1.23(b)), so that $x < Q_\mu(x)$. If $\{Q_\mu^{[n]}(x)\}_{n=0}^{\infty}$ were an increasing sequence contained in $(0, q_\mu)$, then by Corollary 1.4 the sequence would need to converge to a fixed point, of which there is none in the interval $(0, q_\mu]$. It follows that

$$\text{if } 0 < x < q_\mu, \text{ then } x \text{ has an iterate } > q_\mu \qquad (2)$$

Next, by noting that $Q_\mu(q_\mu) = p_\mu = Q_\mu(p_\mu)$ and glancing at Figure 1.23(b), we find that

$$\text{if } q_\mu < x \leq p_\mu, \text{ then } p_\mu \leq Q_\mu(x) \leq \mu/4 \qquad (3)$$

In addition, by using the fact that $q_\mu < Q_\mu(\mu/4)$ (see Exercise 3), and the fact that Q_μ is decreasing on $[1/2, 1]$, we conclude that

$$\text{if } p_\mu < x \leq \mu/4, \text{ then } q_\mu < Q_\mu(\mu/4) \leq Q_\mu(x) < p_\mu \qquad (4)$$

Finally, since $p_\mu < \mu/4$ and Q_μ is decreasing on $[p_\mu, 1]$, we know that

$$\text{if } \mu/4 < x < 1, \text{ then } 0 < Q_\mu(x) < p_\mu \qquad (5)$$

From (2)–(5) it follows that if $0 < x < 1$, then x has an iterate in the interval $(q_\mu, p_\mu]$. We can actually deduce more from (2)–(5). In fact, since $q_\mu < 1/2$, it follows from (3) and (4) that the iterates of x oscillate between the intervals $(q_\mu, p_\mu]$ and $[p_\mu, \mu/4]$. Thus

$$\text{if } x \text{ is in } (q_\mu, p_\mu], \text{ then so is the sequence } \{Q_\mu^{[2n]}(x)\}_{n=0}^\infty \tag{6}$$

$$\text{if } x \text{ is in } [p_\mu, \mu/4], \text{ then so is the sequence } \{Q_\mu^{[2n]}(x)\}_{n=0}^\infty$$

Our next goal is to prove that $Q_\mu^{[2]}$ has only two fixed points: 0 and p_μ. This is equivalent to showing that $Q_\mu^{[2]}(x) - x = 0$ only if $x = 0$ or $x = p_\mu$. However, notice that

$$Q_\mu^{[2]}(x) - x = \mu Q_\mu(x)\, [1 - Q_\mu(x)] - x = \mu[\mu x\,(1-x)\,][1 - \mu x\,(1-x)\,] - x$$

Since p_μ is a fixed point of Q_μ and hence of $Q_\mu^{[2]}$, we know that $x - p_\mu$ is a divisor of $Q_\mu^{[2]}(x) - x$. By synthetic division we find that

$$Q_\mu^{[2]}(x) - x = \mu x\,(x - p_\mu)[-\mu^2 x^2 + (\mu^2 + \mu)x - \mu - 1]$$

By the quadratic formula, the expression inside the brackets has no real roots when $2 < \mu < 3$ (see Exercise 2). Therefore if $2 < \mu < 3$, then the only roots of $Q_\mu^{[2]}(x) - x$ are 0 and p_μ. This means that the only fixed points of $Q_\mu^{[2]}$ are 0 and p_μ .

Now we will show that if $q_\mu < x \le p_\mu$, then $\{Q_\mu^{[2n]}(x)\}_{n=0}^\infty$ converges to p_μ. By the preceding paragraph, $Q_\mu^{[2]}(x) - x$ has no roots in the interval (q_μ, p_μ). Thus $Q_\mu^{[2]}(x) - x$ has the same sign throughout that interval. To determine the sign, we calculate $Q_\mu^{[2]}(1/2) - 1/2$:

$$Q_\mu^{[2]}(\frac{1}{2}) - \frac{1}{2} = \frac{1}{2}\,\mu\,(\frac{1}{2} - p_\mu)\,(\frac{\mu^2}{4} - \frac{\mu}{2} - 1)$$

$$= \frac{1}{8}\,\mu\,[\frac{1}{2} - p_\mu]\,[(\mu - 1)^2 - 5]$$

Since $q_\mu < 1/2 < p_\mu$ and $2 < \mu < 3$, each bracketed expression is negative, so that $Q_\mu^{[2]}(x) - x$ is positive throughout the interval (q_μ, p_μ). Consequently $x < Q_\mu^{[2]}(x)$ for all x in (q_μ, p_μ). By (6) this means that if $q_\mu < x \le p_\mu$, then the sequence $\{Q_\mu^{[2n]}(x)\}_{n=0}^\infty$ is increasing and lies in $(q_\mu, p_\mu]$. Corollary 1.4 then implies that the sequence converges to the only positive fixed point of $Q_\mu^{[2]}$, namely p_μ.

We have just shown that $Q_\mu^{[2n]}(x) \to p_\mu$ as n increases without bound, for every x in $(q_\mu, p_\mu]$. Using the continuity of Q_μ, we find that

$$Q_\mu^{[2n+1]}(x) = Q_\mu(Q_\mu^{[2n]}(x)) \to Q_\mu(p_\mu) = p_\mu$$

as n increases without bound. Therefore $Q^{[n]}(x) \to p_\mu$ whenever x is in $(q_\mu, p_\mu]$.
Since every x in $(0, 1)$ has an iterate in $(q_\mu, p_\mu]$ by the comment following (5),
we conclude that $Q_\mu^{[n]}(x) \to p_\mu$ as n increases without bound, for all x in $(0, 1)$.
In other words, when $2 < \mu < 3$, the basin of attraction of p_μ is the interval $(0, 1)$.
A consequence of this result is the fact that if $2 < \mu \leq 3$, then there are no periodic
points for Q_μ other than fixed points. The same conclusions are valid when $\mu = 3$,
but the argument is more subtle.

Case 4. $3 < \mu \leq 4$

We have analyzed the dynamics of Q_μ when $0 < \mu \leq 3$. Now we turn to our
last case, $3 < \mu \leq 4$. As before, 0 and $p_\mu = 1 - 1/\mu$ are fixed points of Q_μ.
Since

$$Q_\mu'(p_\mu) = Q_\mu'(1 - 1/\mu) = \mu - 2\mu(1 - 1/\mu) = 2 - \mu$$

it follows that as μ increases to 3, $Q_\mu'(p_\mu)$ decreases to -1 (Figures 1.24(a)–(b)).
When μ increases further, $Q_\mu'(p_\mu) < -1$, so that p_μ is a repelling fixed point,
signifying that the quadratic family bifurcates when $\mu = 3$.

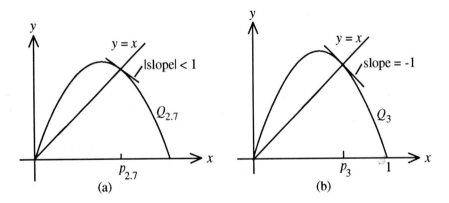

Figure 1.24

From the preceding discussion, if $\mu > 3$, then both fixed points 0 and p_μ
are repelling. So if $3 < \mu \leq 4$, do the iterates of other points in $(0, 1)$ converge, or
oscillate, or have no pattern? Are there periodic points different from 0 and p_μ?
The analysis of $\{Q_\mu\}$ becomes more and more complicated as μ increases from 3
toward 4. The information we will obtain concerning the dynamics of the family
when $3 < \mu \leq 4$ will come from an analysis of the dynamics of $Q_\mu^{[2]}$.
    Figures 1.25(a)–(c) display the graphs of $Q_\mu^{[2]}$ for $\mu = 2.7$, 3, and 3.3. From
the graphs it appears that as μ increases from 2.7 to 3.3, the middle trough descends
and pierces the line $y = x$. In particular, when $\mu = 3$ the graph is tangent to the line

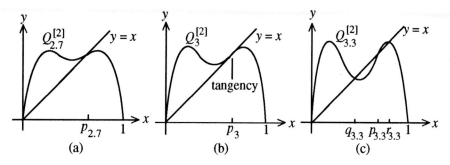

Figure 1.25

$y = x$ at the point (p_μ, p_μ), and when $\mu = 3.3$ the graph intersects the line $y = x$ four times. We conclude that in addition to two fixed points it has when $\mu < 3$, $Q_\mu^{[2]}$ appears to be endowed with two new fixed points, q_μ and r_μ, when $\mu > 3$. Since Q_μ has only two fixed points, $\{q_\mu, r_\mu\}$ would need to be a 2-cycle for Q_μ. Could there be more than one such 2-cycle for Q_μ? To answer this question, we notice that if Q_μ had two 2-cycles, then including the two fixed points of Q_μ and two points each for the two cycles, $Q_\mu^{[2]}$ would have at least six fixed points. But that would mean that the polynomial $Q_\mu^{[2]}(x) - x$ would have at least six distinct roots. However, since this polynomial has degree four, it is impossible for it to have more than four roots. Therefore Q_μ can have at most one 2-cycle.

Next we will show formally that if $\mu > 3$, then Q_μ does indeed have a 2-cycle $\{q_\mu, r_\mu\}$. We will also derive formulas for numerical values of q_μ and r_μ.

EXAMPLE 1. Let $\mu > 3$. Show that Q_μ has a 2-cycle $\{q_\mu, r_\mu\}$, and find formulas for the numerical values of q_μ and r_μ.

Solution. Suppose that Q_μ has a 2-cycle, which we will write as $\{q, r\}$. Then

$$r = Q_\mu(q) = \mu q (1 - q) \quad \text{and} \quad q = Q_\mu(r) = \mu r (1 - r) \qquad (7)$$

It follows that

$$r - q = \mu q (1 - q) - \mu r (1 - r) = \mu(q - r) - \mu(q^2 - r^2)$$

Since $q \neq r$, we can divide through by $r - q$, which yields

$$r + q = \frac{1}{\mu} + 1, \text{ or equivalently, } r = \frac{1}{\mu} + 1 - q \qquad (8)$$

From (7) we obtain

$$r^2 = \mu q r (1 - q) \quad \text{and} \quad q^2 = \mu q r (1 - r) \qquad (9)$$

Then (9) yields

$$r^2 - q^2 = \mu q r\, (r - q)$$

so that

$$r + q = \mu q r \tag{10}$$

Applications of (8), (10), and (8), respectively, yield

$$\frac{1}{\mu} + 1 = r + q = \mu q r = \mu q \left(\frac{1}{\mu} + 1 - q\right)$$

which simplifies to

$$\mu^2 q^2 - \mu^2 q - \mu q + \mu + 1 = 0$$

Solving this equation for q by means of the quadratic formula, we find that the roots are

$$\frac{1}{2} + \frac{1}{2\mu} \pm \frac{1}{2\mu} \sqrt{(\mu - 3)(\mu + 1)} \tag{11}$$

If we let q_μ be the smaller and r_μ the larger of the two values, then

$$q_\mu = \frac{1}{2} + \frac{1}{2\mu} - \frac{1}{2\mu} \sqrt{(\mu - 3)(\mu + 1)} \quad \text{and} \quad r_\mu = \frac{1}{2} + \frac{1}{2\mu} + \frac{1}{2\mu} \sqrt{(\mu - 3)(\mu + 1)}$$

It is easy to check that $0 < q_\mu < 1$ and $0 < r_\mu < 1$, and that $\{q_\mu, r_\mu\}$ is a 2-cycle for Q_μ. Consequently we have completed the proof. ❑

We are ready to determine the values of μ for which $\{q_\mu, r_\mu\}$ is attracting.

THEOREM 1.16. Let $3 < \mu < 4$. The 2-cycle $\{q_\mu, r_\mu\}$ is attracting for Q_μ if $3 < \mu < 1 + \sqrt{6}$.

Proof. For simplicity we will use q and r for q_μ and r_μ, respectively. Since $\{q, r\}$ is a 2-cycle and $Q'_\mu(x) = \mu - 2\mu x$, it follows from Theorem 1.11 that

$$(Q_\mu^{[2]})'(q) = Q'_\mu(q)\, Q'_\mu(r) = (\mu - 2\mu q)(\mu - 2\mu r) = \mu^2 - 2\mu^2(q + r) + 4\mu^2 q r$$

In order to be able to write the right side as a function of μ alone, we use (8) to substitute for $q + r$; after that we use (10) and then (8) to substitute for $\mu q r$. We obtain

$$(Q_\mu^{[2]})'(q) = \mu^2 - 2\mu^2 (\frac{1}{\mu} + 1) + 4\mu (\frac{1}{\mu} + 1) = -\mu^2 + 2\mu + 4$$

Therefore $|(Q_\mu^{[2]})'(q)| < 1$ if and only if $|\mu^2 - 2\mu - 4| < 1$. This inequality is equivalent to $-1 < (\mu - 1)^2 - 5 < 1$, which yields $3 < \mu < 1 + \sqrt{6}$. Thus the 2-cycle $\{q_\mu, r_\mu\}$ is attracting if $3 < \mu < 1 + \sqrt{6}$. ∎

The fact that the 2-cycle $\{q_\mu, r_\mu\}$ is attracting if $\mu = 1 + \sqrt{6}$ is harder to prove, as is the fact that if $3 < \mu \leq 1 + \sqrt{6}$, then the basin of attraction of the 2-cycle $\{q_\mu, r_\mu\}$ consists of all x in (0, 1) except the fixed point p_μ and the points whose iterates are eventually p_μ. We will not prove these results. We mention, however, that they imply that for $3 < \mu \leq 1 + \sqrt{6}$ the only periodic points of Q_μ are the repelling fixed points 0 and p_μ, and the period-2 points q_μ and r_μ that form an attracting 2-cycle.

Since

$$|(Q_\mu^{[2]})'(q_\mu)| = |\mu^2 - 2\mu - 4| < 1 \text{ if and only if } 3 < \mu < 1 + \sqrt{6}$$

it follows that if $\mu > 1 + \sqrt{6}$, then $|(Q_\mu^{[2]})'(q_\mu)| > 1$, so that $\{q_\mu, r_\mu\}$ becomes a repelling 2-cycle. As you might suspect, a new, attracting 4-cycle is born as μ increases beyond $1 + \sqrt{6}$. As μ increases still further, the attracting 4-cycle becomes repelling, and a new 8-cycle is born. The process continues indefinitely as μ increases.

One might imagine that the values of μ at which new 2^k-cycles emerge would march unboundedly toward ∞. However, this turns out to be not true. Let μ_k be the bifurcation point defined by

μ_k = maximum value of μ for which Q_μ has an attracting 2^k-cycle, for $k = 0, 1, 2, ...$

From our previous results, we know that Q_μ has an attracting fixed point for $0 \leq \mu \leq 3$ and an attracting 2-cycle for $3 < \mu \leq 1 + \sqrt{6}$. Therefore

$$\mu_0 = 3 \quad \text{and} \quad \mu_1 = 1 + \sqrt{6}$$

However, numerical values of μ_k for $k > 1$ are not so easy to determine. What we can say, though, is that

if $\mu_0 < \mu \leq \mu_1$, then Q_μ has 2 fixed points and a 2-cycle

if $\mu_1 < \mu \leq \mu_2$, then Q_μ has 2 fixed points, a 2-cycle and a 2^2-cycle

if $\mu_2 < \mu \leq \mu_3$, then Q_μ has 2 fixed points, a 2-cycle, a 2^2-cycle, and a 2^3-cycle

In general,

if $\mu_{n-1} < \mu \leq \mu_n$, then Q_μ has a 2^k-cycle for $k = 0, 1, 2, \ldots, n$

It is known, but difficult to prove, that $\mu_{k+1} \approx 1 + \sqrt{3 + \mu_k}$ for $k = 2, 3, \ldots$, and that the sequence $\{\mu_k\}_{k=1}^\infty$ has a limit μ_∞ given by

$$\mu_\infty = 3.61547\cdots$$

The number μ_∞ is sometimes called the **Feigenbaum number** for the quadratic family $\{Q_\mu\}$, named after the physicist Mitchell Feigenbaum, who in the mid-1970's conjectured that the bifurcation points had a limit and found a very precise value for it. (See Feigenbaum, 1978 or 1983, for accounts of his discovery.)

The surprising part of the story is yet to come. Let

$$d_k = \frac{\mu_k - \mu_{k-1}}{\mu_{k+1} - \mu_k}, \quad \text{for } k = 2, 3, 4, \ldots$$

Since $\mu_k - \mu_{k-1}$ represents the distance between μ_k and μ_{k-1}, it follows that d_k compares distances between successive pairs of μ_k's. Feigenbaum found that the sequence $\{d_k\}_{k=1}^\infty$ converges to a number we will denote d_∞, where

$$d_\infty = 4.669202\cdots$$

What is astonishing is that this constant d_∞ seems to be universal. That is, for many families of one-humped functions like the family of quadratic functions, bifurcations occur in such a regular fashion that the distances between successive pairs of bifurcation points approach the very same value d_∞! It is for this reason that d_∞ is called a **universal constant**. More particularly, it is referred to as the **Feigenbaum constant**, because Feigenbaum was the first to discover it and its universality.

We conclude by noting that the quadratic family $\{Q_\mu\}$ is one of the most illustrious parametrized families. Its functions are easy to describe, and have properties many more complicated functions have. Moreover, there is an enormous wealth of information concerning the family, spurred in part by the captivating article in the magazine *Nature* by Robert May (1975). Books by Pierre Collet and Jean-Pierre Eckmann (1980) and by Chris Preston (1983) give detailed analysis of functions like quadratic functions. We will once again study properties of the family when we investigate properties of chaos in Chapter 2.

EXERCISES 1.5

1.   Let $1 < \mu \leq 2$. Prove that if $p_\mu < x < 1/2$, then $\{Q_\mu^{[n]}(x)\}_{n=0}^\infty$ is a bounded decreasing sequence that converges to p_μ.

2. Let $2 < \mu < 3$. Show that $-\mu^2 x^2 + (\mu^2 + \mu)x - \mu - 1$ has no (real) roots.

3. Let $2 < \mu < 3$. Show that $q_\mu < 1/2 < Q_\mu(\mu/4)$. (*Hint*: Let $h(\mu) = Q_\mu(\mu/4)$, and find the range of h for μ in the interval $(2, 3)$.)

4. a. Show that $(Q_\mu^{[2]})'(1/2) = 0$ for $0 < \mu < 4$.
    b. Since $1/2$ is a zero of $(Q_\mu^{[2]})'$ by part (a), we can write $(Q_\mu^{[2]})'(x)$ in the form $\mu^2 (x - 1/2)(ax^2 + bx + c)$. Find the values of a, b, and c in terms of μ.
    c. Use part (b) to show that if $0 < \mu < 2$, then $Q_\mu^{[2]}$ has a unique relative extreme (maximum) value for x in $(0, 1)$.

5. Let $2 < \mu \leq 4$. Find the relative minimum value of $Q_\mu^{[2]}$ in the interval $(0, 1)$.

6. a. Let $2 \leq \mu \leq 4$. Find the maximum value of $Q_\mu^{[n]}$ for any positive integer n.
    b. Let $0 < \mu < 2$. Describe the difficulties, if any, that you encounter when trying to find the maximum value of $Q_\mu^{[n]}$.

7. Let $3 < \mu \leq 4$.
    a. Show that $0 < q_\mu < 1$.
    b. Show that $Q_\mu(q_\mu) = r_\mu$ and $Q_\mu(r_\mu) = q_\mu$.

8. Let $0 < \mu_0$, and let $\varepsilon > 0$. Show that if $|\mu - \mu_0| < \varepsilon$, then

$$|Q_\mu(x) - Q_{\mu_0}(x)| < \varepsilon$$

for all x in $[0, 1]$.

9. a. Let $0 < \mu < 1$. Find c in $(0, 1)$ such that $Q_\mu^{[n]}(x) < c^n x$ for all x in $(0, 1)$.
    b. Let $\mu = 1$. Show that no such c in $(0, 1)$ exists.

10. Let $1 < \mu < 2$. Find c in $(0, 1)$ such that

$$|Q_\mu^{[n]}(x) - p_\mu| < c^n |x - p_\mu|$$

for all x in $(0, 1)$ with $x \neq p_\mu$. (*Hint*: You only need to consider $0 < x \leq 1/2$. Analyze the cases $0 < x < p_\mu$ and $p_\mu < x < 1/2$.)

11. Let $M_0 = 3$, and let $M_{k+1} = 1 + \sqrt{3 + M_k}$ for $k = 1, 2, 3, \ldots$. Also let

$$M_\infty = \lim_{k \to \infty} M_k$$

It is known, but hard to prove, that $M_k \approx \mu_k$ for all k, and that $M_\infty \approx \mu_\infty$.

    a.   Use induction to find M_∞.

    b.   Compare the numerical value of M_∞ found in (a) to that of μ_∞.

12.   Using the definition of M_k in Exercise 11, let

$$D_k = \frac{M_k - M_{k-1}}{M_{k+1} - M_k}$$

Show that $\lim_{k \to \infty} D_k = 1 + \sqrt{17}$.

1.6 BIFURCATIONS

In the study of a family of functions such as $\{Q_\mu\}$ and $\{T_\mu\}$, values of μ at which the family bifurcates play a prominent role. After all, these values indicate where periodic points arise or disappear, as well as where periodic points become or cease to be attracting. This section is devoted to bifurcations. First we consider a method of displaying bifurcation points on a graph. Then we discuss two basic kinds of bifurcations: period-doubling bifurcations and tangent bifurcations.

Bifurcation Diagrams

One method of displaying the points at which a parametrized family of functions $\{f_\mu\}$ bifurcates is called a bifurcation diagram, and is designed to give information about the behavior of higher iterates of arbitrary members of the domain of f_μ for all values of the parameter μ.

The **bifurcation diagram** of $\{f_\mu\}$ is a graph for which the horizontal axis represents values of μ and the vertical axis represents higher iterates of the variable (normally x). For each value of μ, the diagram includes (in theory) all points of the form $(\mu, f_\mu^{[n]}(x))$, for values of n larger than, say, 50 or 100. The reason we only use the higher iterates of x is that the diagram is designed to show eventual behavior of iterates, such as convergence or periodicity or unpredictability.

Now we will study the bifurcation diagram of the quadratic family $\{Q_\mu\}$. In order to give as much detail as we can in the bifurcation diagram, we have split the diagram into two parts, $0 \le \mu \le 1 + \sqrt{6}$ in Figure 1.26 and $1 + \sqrt{6} \le \mu \le 4$ in

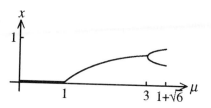

partial bifurcation diagram for $\{Q_\mu\}$

Figure 1.26

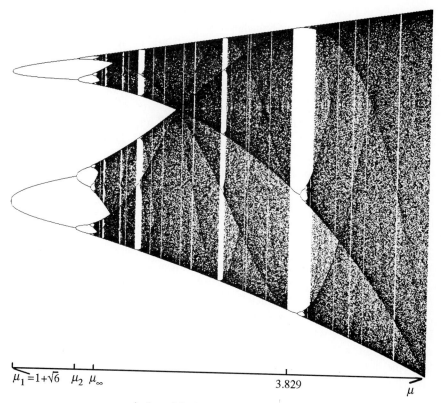

remainder of the bifurcation diagram for $\{Q_\mu\}$

Figure 1.27

Figure 1.27. The bifurcation diagram was obtained on the computer by letting x be 1/2, taking increments of 1/1000 for μ in the interval [0, 4], and plotting all the points of the form $(\mu, Q_\mu^{[n]}(x))$, for $201 \leq n \leq 700$. You can check that if another value of x were chosen, or if the increments remained small but were altered, or if the range of n were changed, then the corresponding bifurcation diagram would have been indistinguishable from the one pictured in Figures 1.26 and 1.27.

To analyze the diagram, notice in Figure 1.26 that for $0 < \mu \le 1$, the points in the diagram lie on the x axis, because iterates of all x in the domain $[0, 1]$ are attracted to 0. Next, for $1 < \mu \le 3$, the curve represents the points of the form $(\mu, p_\mu) = (\mu, 1 - 1/\mu)$, since p_μ attracts the iterates of all x in $(0, 1)$. When $3 < \mu \le 1 + \sqrt{6}$, iterates of all x in $(0, 1)$ that are not eventually fixed are attracted to the 2-cycle $\{q_\mu, r_\mu\}$, so for such values of μ there are two curves, one with points of the form (μ, q_μ) and the other with points of the form (μ, r_μ). We observe that at the bifurcation point $\mu = 3$ a single curve splits into two curves representing the attracting 2-cycle that emerges at $\mu = 3$.

Turning to Figure 1.27, which commences with $\mu = 1 + \sqrt{6}$, we see that to the right of $\mu = 1 + \sqrt{6}$, four branches appear (corresponding to an attracting 4-cycle). The four branches extend until $\mu = \mu_2$, at which point eight branches start to appear (corresponding to an attracting 8-cycle). In general these branches represent the various attracting 2^k-cycles, and split or fork at points corresponding to the various bifurcation points of $\{Q_\mu\}$. The branches represent the various attracting 2^k-cycles that appear in sequence.

The heavily shaded vertical strips in Figure 1.27 represent values of μ for which the iterates $\{Q_\mu^{[n]}(x)\}_{n=0}^{\infty}$ of "most" points x in $(0, 1)$ are not eventually periodic, but spread out over a subinterval or collection of subintervals of $[0, 1]$. These darkened patches represent unpredictable, chaotic-like patterns for iterates of such values of x. The dark curves that give interesting patterns to the heavily shaded regions represent extensions (for larger values of μ) of attracting cycles, as you can verify by following the curves to the left.

In the diagram we also observe "windows" containing isolated curves that represent attracting cycles. For example, the window that appears at $\mu \approx 3.829$ corresponds to an attracting 3-cycle. Notice that toward the right-hand side of that window, the three curves divide into six curves, then twelve curves, and so forth, each set in turn representing an attracting $3 \cdot 2^k$-cycle for appropriate values of μ. As before, the μ-coordinate of each point where a fork occurs is a bifurcation point. In addition to the 3-cycle window, there are 5-cycle, 6-cycle, and 8-cycle windows that should be visible in the diagram. At the right end of each window the branches undergo the same succession of splitting that occurs in the 3-cycle window. What is not so obvious is that there are infinitely many windows representing attracting cycles; these windows are in general too narrow to detect without a strong zoom feature on the computer screen.

Having described the bifurcation diagram for $\{Q_\mu\}$, we turn to the bifurcation diagram for the tent family $\{T_\mu\}$, shown in Figure 1.28. This diagram was obtained from a computer in the same way as the bifurcation diagram for $\{Q_\mu\}$ was. Values of μ in $[0, 1]$ were selected in increments of $1/1000$, and plotted the points $(\mu, T_\mu^{[n]}(x))$ for $1001 \le n \le 2000$. Other choices of x, increments for μ, and number of plotted points can lead to virtually identical bifurcation diagrams for $\{T_\mu\}$.

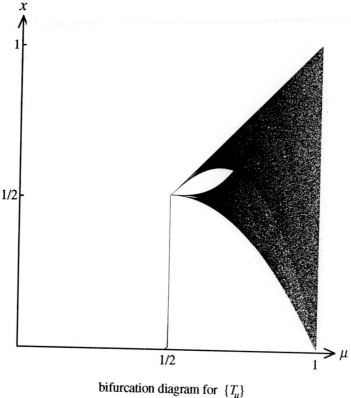

bifurcation diagram for $\{T_\mu\}$

Figure 1.28

In the diagram, the line segment [0, 1/2] on the μ-axis appears because when $0 \le \mu < 1/2$, iterates of each x in the domain approach 0. If $\mu = 1/2$, then each x in the interval [0, 1/2] is a fixed point, so the diagram contains a vertical line above 1/2 on the μ axis. The diagram becomes more complicated for $\mu > 1/2$, because for such values of μ, the iterates of "most" values of x in (0, 1) spread out over intervals that widen as μ approaches 1. These darkened patches suggest an increasingly chaotic type of behavior for the iterates of corresponding tent functions. The apparent "eye" in the diagram can be justified by careful analysis of the iterates of T_μ for μ in the interval [1/2, 7/10].

An alternative method of displaying bifurcation points is the **orbit diagram**, which includes solid curves for attracting fixed points and attracting cycles, and dashed curves for repelling fixed points and repelling cycles. Unlike the bifurcation diagram, it does not indicate the various iterates of points in the domain. (In fact, it would perhaps make better sense to switch the names of orbit diagram and bifurcation diagram, but that would counter common usage.)   Orbit diagrams for portions of $\{Q_\mu\}$ and $\{E_\mu\}$ appear in Figure 1.29.

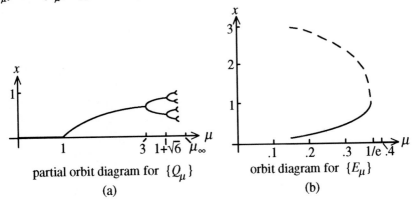

partial orbit diagram for $\{Q_\mu\}$

(a)

orbit diagram for $\{E_\mu\}$

(b)

Figure 1.29

Period-Doubling Bifurcations

A bifurcation at which an attracting period-n cycle becomes repelling and gives birth to an attracting $2n$-cycle is a **period-doubling bifurcation**.   The bifurcations of $\{Q_\mu\}$ represented in Figure 1.29(a) are period-doubling bifurcations. Because the graph resembles a pitchfork near the point on the orbit diagram corresponding to the bifurcation, this kind of bifurcation is often called a **pitchfork bifurcation** (Figure 1.30).

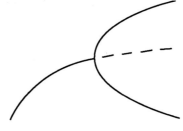

"pitchfork" bifurcation

Figure 1.30

Period-doubling bifurcations form one important class of bifurcations.   To understand better the properties of families of functions at period-doubling bifurcation points, we will focus on Q_μ and $Q_\mu^{[2]}$ when μ is near 3.

EXAMPLE 1. Show that the quadratic family $\{Q_\mu\}$ has the following properties relating to the bifurcation point $\mu = 3$:

  i. $Q_3(2/3) = 2/3$, so that $2/3$ is a fixed point of Q_3.
  ii. $Q_3'(2/3) = -1$.
  iii. $Q_\mu'(2/3)$ decreases as μ increases through 3.
  iv. The graph of $Q_3^{[2]}$ has an inflection point at $(2/3, 2/3)$.

Solution.   First we recall that $Q_\mu(x) = \mu x(1 - x)$, so that $Q_3(2/3) = 2/3$. Thus (i) is proved. Since $Q_\mu'(x) = \mu - 2\mu x$, it follows that $Q_\mu'(2/3) = -\mu/3$. Therefore $Q_3'(2/3) = -1$, and $Q_\mu'(2/3)$ decreases as μ increases through 3. Consequently (ii) and (iii) are both verified. To prove (iv) we calculate that

$$(Q_3^{[2]})'(x) = 9(1 - 8x + 18x^2 - 12x^3) \quad \text{and} \quad (Q_3^{[2]})''(x) = -36(3x - 2)(3x - 1)$$

As a result, $(Q_3^{[2]})''$ changes sign at $2/3$, so that $(2/3, 2/3)$ is an inflection point. That finishes the proof of (iv).   ❑

Using Example 1 as a model, we can describe general properties of families of functions that accompany period-doubling bifurcations. To that end, let $\{f_\mu\}$ be a parametrized family, and assume that the family has the following properties:

  i. f_{μ^*} has a fixed point p.
  ii. $f_{\mu^*}''(p)(p) = -1$.
  iii. The graph of $f_\mu^{[2]}$ crosses the line $y = x$ when $\mu < \mu^*$, is tangent to the line $y = x$ when $\mu = \mu^*$, and snakes around the line $y = x$ when $\mu > \mu^*$. (See Figures 1.31(a)–(c), which are portions of Figures 1.25(a)–(c), respectively.)

Properties (i) and (ii) indicate that the graphs of f_μ are positioned correctly with respect to the line $y = x$ when μ is near to μ^*. However, it is the rotation of the graphs of $f_\mu^{[2]}$ around the point (p, p) as μ passes through μ^*, as described in (iii), that signals the period-doubling bifurcation of $\{f_\mu\}$ at μ^*. From Example 1 and Figure 1.31, we see that (i)–(iii) hold if $f_\mu = Q_\mu$, $\mu^* = 3$, and $p = 2/3$.

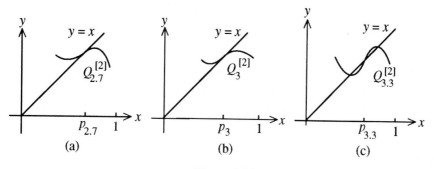

Figure 1.31

Tangent Bifurcations

A bifurcation with a far different character occurs at $\mu = 1/e$ for the exponential family $\{E_\mu\}$. To support this assertion, let us first recall that

$$E_\mu(x) = \mu e^x, \text{ where } \mu > 0$$

In Figures 1.32(a)–(c), when μ decreases through $1/e$, the graph of E_μ descends until it is tangent to the line $y = x$ and then breaks through the line $y = x$, giving rise to two fixed points that separate from one another as μ continues to decrease. The bifurcation at $\mu = 1/e$ occurs because the graph of E_μ breaks through the line $y = x$. Since the graph of E_μ is tangent to the line $y = x$ at the point corresponding to the bifurcation point, the bifurcation is called a tangent bifurcation.

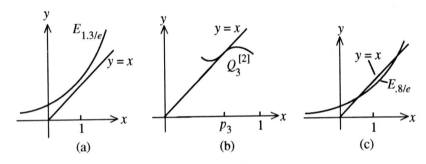

Figure 1.32

More generally, a family $\{f_\mu\}$ has a **tangent bifurcation** (or **saddle**, or **fold bifurcation**) at μ^* if a pair of fixed points are born as a curve in the graph of f_μ becomes tangent to and then crosses the line $y = x$ when μ passes through μ^*. It can be shown that a parametrized family $\{f_\mu\}$ has a tangent bifurcation at μ^* if

i. $f_{\mu^*}(p) = p$, so p is a fixed point of f_{μ^*}.
ii. $f'_{\mu^*}(p) = 1$, so the graph of f_{μ^*} is tangent to the line $y = x$ at (p, p).
iii. $f_\mu(p)$ is a monotone function of μ near μ^*.
iv. The graph of f_μ is concave upward (or downward) at (p, p).

A major difference between period-doubling and tangent bifurcations is the derivative of f_{μ^*} at p: -1 for period-doubling bifurcations, and 1 for tangent bifurcations.

In the following example we will show that the exponential family satisfies conditions (i)–(iv) at $\mu = 1/e$.

EXAMPLE 2. Show that $\{E_\mu\}$ satisfies (i)–(iv) relative to the bifurcation point $\mu = 1/e$.

Solution. Since

$$(E_{1/e})'(1) = E_{1/e}(1) = \frac{1}{e} e^1 = 1$$

(i) and (ii) are satisfied with $p = 1$. The fact that $E_\mu(1) = \mu e$ verifies (iii). Finally, the graph of E_μ is concave upward because $E_\mu''(x) = E_\mu(x) = \mu e^x > 0$ for all x and all μ. Thus (iv) is true as well. ❑

The bifurcations of $\{Q_\mu\}$ that we have discussed are period-doubling. Now we will show why the family $\{Q_\mu^{[3]}\}$ has a tangent bifurcation at $\mu \approx 3.83$. Two local minima and one local maximum in Figure 1.33(a) are nearly tangent to the line $y = x$ when $\mu = 3.81$, are tangent to the line $y = x$ when $\mu \approx 3.83$ (Figure 1.33(b)), and break through the line $y = x$ as μ increases through 3.83. At the bifurcation an attracting period-3 cycle and a repelling period-3 cycle are born.

In general a parametrized family $\{f_\mu\}$ *cannot* have a bifurcation at μ^* unless f_{μ^*} has a periodic point p such that $|f_{\mu^*}'(p)| = 1$. Thus if p is a fixed point of f_{μ^*}, if $|f_{\mu^*}'(p)| \neq 1$, and if both f_μ and f_μ' vary continuously as μ varies, then in effect the graph of f_μ will cross the line $y = x$ exactly once for all μ near enough to μ^*, so that μ^* is *not* a bifurcation point of $\{f_\mu\}$. (We have not defined what it means

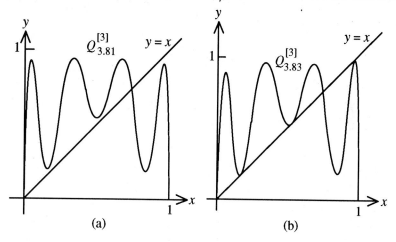

Figure 1.33

for f_μ or f_μ' to vary continuously as μ varies, but all the families we have considered have this property.) Figure 1.34 demonstrates this type of condition.

The implication is that in order for a parametrized family $\{f_\mu\}$ to bifurcate at μ^*, there must be a fixed point p such that $|f_{\mu^*}'(p)| = 1$. If the derivative is -1, then there may be a period-doubling bifurcation at μ^*; if the derivative is 1, then there may be a tangent bifurcation at μ^*. The books by Devaney (1989) and Guckenheimer and Holmes (1983) contain further information on other types of bifurcations.

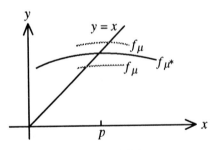

Figure 1.34

It is worth noting that scientists have encountered period-doubling bifurcations in many different experiments. Some thirty years ago a Russian chemist named Boris Belousov observed interesting oscillations in the concentration of bromide when he mixed together sulfuric acid, potassium bromate, cerium sulfate and malonic acid. At a later date, details of the reaction were confirmed by another Russian, Anatol Zhabotinskii. Nowadays, the reaction is normally called the B-Z reaction. During the reaction, bromate ions oxidize to form bromine, after which cerium oxidizes and makes the solution change from red to blue. Then autocatalysis makes the color switch back from blue to red. If the reaction is continued for a period of time by continually pumping in new reactants, the colors can flip back and forth between red and blue, displaying not only immense complexity but also period-doubling toward chaos. The entertaining article by Stephen Scott (1989) entitled "Clocks and Chaos in Chemistry," in the *New Scientist* magazine, discusses the B-Z reaction in more detail.

A bifurcation diagram that looks rather similar to the diagram for the quadratic family occurs in the study of periodically forced nonlinear circuits (Figure 1.35). In

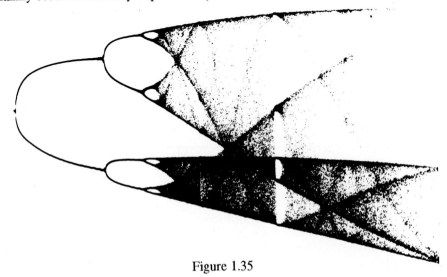

Figure 1.35

the figure, the junction voltage is shown as a function of the drive voltage, and shows the onset of chaos through period-doubling. This diagram appeared in an article by Robert Van Buskirk and Carson Jeffries (1985).

EXERCISES 1.6

1. Let $f_\mu(x) = x^2 + 1/4 + \mu$, for all x and all real μ. Show that there is a bifurcation at $\mu = 0$, and classify it as a period-doubling bifurcation, a tangent bifurcation, or neither.

In Exercises 2–3, determine a bifurcation for the given parametrized family, and classify it as period-doubling bifurcation, tangent bifurcation or neither.

2. The exponential family $\{E_\mu\}$, where $\mu < 0$

3. $\{f_\mu\}$, where $f_\mu(x) = \mu \arctan x$ for all x, with $\mu > 0$

4. Let x_1, x_2, ..., x_8 denote the eight fixed points of $Q_\mu^{[3]}$ when $3.83 < \mu < 3.84$. Assume that $x_1 < x_2 < \cdots < x_8$. Using Figure 1.33(b) as a guide, determine which of the x_k's form the attracting 3-cycle for Q_μ.

5. Use the computer program BIFURCATION to approximate (to the nearest thousandth) the interval on the μ axis for which there is an attracting 5-cycle window for the quadratic family.

6. Use the computer program BIFURCATION to approximate (to the nearest thousandth) the interval on the μ axis for which there is an attracting period-10 window for the quadratic family.

7. Use the computer program BIFURCATION to make a bifurcation diagram for $\{Q_\mu\}$ in which the increments are $1/100$ and the initial point is $x = 1/2$. How does the portion between $1 + \sqrt{6}$ and 4 compare with Figure 1.27?

8. Alter the computer program PLOT to plot the graph of $Q_\mu^{[2]}$. Observe the behavior of the graph as μ increases from 1 to 4.

9. Alter the computer program PLOT to plot the graph of $Q_\mu^{[8]}$. Observe the behavior of the graph as μ increases from 3 to 4.

10. Alter the computer program PLOT to plot the graphs of Q_μ and $Q_\mu^{[2]}$ simultaneously. Determine a value of μ in $[3, 4]$ such that the middle portion of the graph of $Q_\mu^{[2]}$ looks like an inverted copy of Q_μ.

11.   Consider the function T_6.   Alter the computer program ITERATE to approximate (to within one thousandth) the largest subinterval J of   [.5, .6] such that the higher iterates of $T_6(x)$ lie outside J for all x in J.

1.7   PERIOD-3 POINTS

In Section 1.5 we discovered values of μ for which Q_μ has fixed points and 2^k-cycles for all positive integers k less than any given positive integer n,   but no other cycles. In other words, there are values of μ such that Q_μ has points of certain periods but no points of other periods. In the present section we will describe what the presence of a period-3 point, or any period-n point, implies about the existence of other periodic points.   The answer will derive from two famous theorems, those of Li and Yorke (1975) and of Sharkovsky (1964).

Our first goal will be to prove a wonderful theorem due to James Yorke and his student Tien-Yien Li. It tells us that if Q_μ has a period-3 point, then Q_μ has a period-n point for every $n \geq 1!$ Their ingenious proof relies on two powerful theorems from calculus: the Maximum-Minimum Theorem and the Intermediate Value Theorem. For reference we state them here without proof.

THEOREM 1.17 (Maximum-Minimum Theorem).   Suppose that f is continuous on the interval $[a, b]$.   Then f has a maximum value and a minimum value.

Concretely, the Maximum-Minimum Theorem says that if f is continuous on $[a, b]$,   then there are numbers u and v in $[a, b]$ such that if x is any number in $[a, b]$,   then $f(u) \leq f(x) \leq f(v)$ (Figure 1.36). Thus

$$f(u) = \text{minimum value of } f \text{ on } [a, b]$$
$$f(v) = \text{maximum value of } f \text{ on } [a, b]$$

THEOREM 1.18 (Intermediate Value Theorem).   Suppose that f is continuous on the interval $[a, b]$, and let p be any number between $f(a)$ and $f(b)$. Then

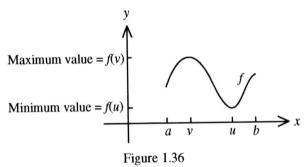

Figure 1.36

there is a number c in $[a, b]$ such that $f(c) = p$.

The Intermediate Value Theorem says in effect that if f is continuous on the closed interval $[a, b]$, then it cannot skip over any values between $f(a)$ and $f(b)$; in other words, the range of f contains all values between $f(a)$ and $f(b)$.

In order to make the proof of the Li-Yorke Theorem more accessible, we will prepare for it with four lemmas. The first of the lemmas is a direct consequence of the Intermediate Value Theorem.

LEMMA 1. Let f be continuous on an interval J. Let $f(J)$ denote the collection of all values $f(x)$ for x in J. Then $f(J)$ is also an interval.

Proof. Suppose that $f(J)$ were not an interval. Then there would exist two numbers y and z in $f(J)$, with $y < z$, and a number p in (y, z) such that p is not in the range of f. By the Intermediate Value Theorem applied to $[y, z]$, the range of f must contain the entire interval $[y, z]$, and in particular must contain p. This contradiction implies that $f(J)$ is an interval. ∎

The next lemma will be used repeatedly in the proof of the Li-Yorke Theorem.

LEMMA 2. Let f be continuous on a closed interval J, and assume that $f(J) \supseteq [a, b]$. Then there is a closed interval K such that $J \supseteq K$ and $f(K) = [a, b]$.

Proof. Since $f(J) \supseteq [a, b]$, there are numbers x and z in J such that $f(x) = a$ and $f(z) = b$. Of all such numbers z, there is one closest to x. Call it y (Figure 1.37, where $J = [r, s]$). That y exists is assured by the continuity of f. Similarly, there is a w between x and y that is the closest to y for which $f(w) = a$. By Lemma 1, the closed interval K determined by w and y (which is either $[w, y]$ or $[y, w]$) has the property that $f(K) = [a, b]$. ∎

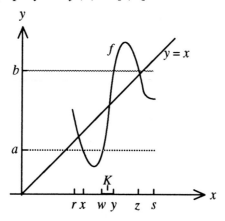

Figure 1.37

The proof of the following lemma uses both the Maximum-Minimum Theorem and the Intermediate Value Theorem.

LEMMA 3. Suppose that J is a closed interval, and assume that f is continuous on J and $f(J) \supseteq J$. Then f has a fixed point in J.

Proof. By the Maximum-Minimum Theorem, f has a minimum value r and a maximum value s on J, so for suitable y and z in J,

$$r = f(y) = \text{minimum value of } f \text{ on } J, \qquad s = f(z) = \text{maximum value of } f \text{ on } J$$

Since r is the minimum value of $f(J)$ and s is the maximum value of $f(J)$, and since $f(J) \supseteq J$, it follows that $r \leq y \leq s$ and $r \leq z \leq s$. Now let $g(x) = f(x) - x$. Then g is continuous on J since f is, and furthermore,

$$g(y) = f(y) - y = r - y \leq 0 \quad \text{and} \quad g(z) = f(z) - z = s - z \geq 0$$

By the Intermediate Value Theorem there is an x between y and z such that $g(x) = 0$, or equivalently, $f(x) = x$. Since x is in J, f has a fixed point in J. ∎

Our final lemma tells us that if a continuous function f has a period-3 point, then it also has a fixed point and a period-2 point.

LEMMA 4. Let f be continuous and suppose that $f(a) = b$, $f(b) = c$ and $f(c) = a$. Then f has a fixed point and a period-2 point.

Proof. Without loss of generality we may suppose that $a < b < c$ (Figure 1.38(a)). Since $f(b) = c$ and $f(c) = a$, we know that $f[b, c] \supseteq [a, c] \supseteq [b, c]$, so by Lemma 3, f has a fixed point in $[b, c]$.

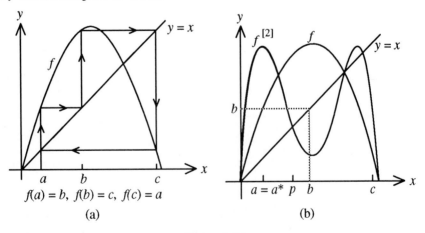

$f(a) = b,\ f(b) = c,\ f(c) = a$

(a)   (b)

Figure 1.38

To show that f has a period-2 point, let

$a^* = $ the largest number such that $a \le a^* < b$ and $f(a^*) = b$

(Since f is continuous on $[a, b]$ and since $f(b) = c > b$, such an a^* exists.)
Then $f[a^*, b] \supseteq [b, c]$, so that

$$f^{[2]}[a^*, b] \supseteq f[b, c] \supseteq [a, c] \supseteq [a^*, b]$$

By Lemma 3, with $J = [a^*, b]$, there is a fixed point p of $f^{[2]}$ in $[a^*, b]$ (Figure 1.38(b)). Since $a^* < b < c$, $f(a^*) = b$ and $f(b) = c$, we know that $p \ne a^*$ and $p \ne b$. It follows from the definition of a^* that if $a^* < x < b$, then $f(x) > b$, so that $f(p) > b > p$. Therefore p is not a fixed point of f, so p is a period-2 point of f. (You can check in Figure 1.38(b) that p is indeed a period-2 point of f.) ∎

Now we are ready to state and prove the theorem of Li and Yorke.

THEOREM 1.19 (Li-Yorke Theorem). Suppose that f is continuous on the closed interval J, with $J \supseteq f(J)$. If f has a period-3 point, then f has points of all other periods.

Proof. Assume that $a < b < c$ and that $f(a) = b$, $f(b) = c$, and $f(c) = a$. By Lemma 4, f has points of period 1 and 2, and by assumption, f has points of period 3. Now let $n > 3$. We will show that f has points of period n. The idea of the proof is to show that there is a point p in $[b, c]$ such that

  i. $f^{[k]}(p)$ lies in $[b, c]$ for $k = 1, 2, 3, ..., n - 2$.

  ii. $f^{[n-1]}(p)$ lies in (a, b).

  iii. $f^{[n]}(p) = p$, and lies in $[b, c]$.

Then automatically p will have period n.
  The proof that f has a period-n point proceeds as follows. Let $J_0 = [b, c]$. Since

$$f(J_0) = f[b, c] \supseteq [a, c] \supseteq [b, c] = J_0$$

Lemma 2 assures the existence of a closed interval J_1 such that $J_0 \supseteq J_1$ and $f(J_1) = [b, c] = J_0$. Next,

$$f^{[2]}(J_1) = f(J_0) \supseteq J_0$$

so by Lemma 2 there is a closed interval J_2 such that $J_1 \supseteq J_2$ and $f^{[2]}(J_2) = J_0$. Now

$$f^{[3]}(J_2) \; = \; f(f^{[2]}(J_2)) \; = \; f(J_0) \; \supseteq \; J_0$$

so that again by Lemma 2 there is a closed interval J_3 such that $J_2 \supseteq J_3$ and $f^{[3]}(J_3)$ $= J_0$. Inductively we obtain a nested sequence of closed intervals $J_0, \; J_1, \; J_2, \; ..., \; J_{n-2}$ with

$$[b, c] = J_0 \supseteq J_1 \supseteq J_2 \supseteq \cdots \supseteq J_{n-2} \text{ and } f^{[k]}(J_k) = J_0 = [b, c], \text{ for } k = 1, 2, ..., n-2 \qquad (1)$$

In particular, $f^{[n-2]}(J_{n-2}) = [b, c]$. Therefore

$$f^{[n-1]}(J_{n-2}) \; = \; f(f^{[n-2]}(J_{n-2})) \; = \; f[b, c] \; \supseteq \; f[a, c] \; \supseteq \; [a, b]$$

so that by Lemma 2 there is a closed interval J_{n-1} such that $J_{n-2} \supseteq J_{n-1}$ and

$$f^{[n-1]}(J_{n-1}) \; = \; [a, b] \qquad (2)$$

Consequently by (1) and (2),

$$f^{[n]}(J_{n-1}) \; = \; f(f^{[n-1]}(J_{n-1})) \; = \; f[a, b] \; \supseteq \; [b, c] \; \supseteq \; J_{n-2} \; \supseteq \; J_{n-1}$$

It follows from Lemma 3 that there is a point p in J_{n-1} (and hence in $[b, c]$) that is a fixed point of $f^{[n]}$. To show that p has period n, we first observe that $f^{[k]}(p)$ is in $[b, c]$ for $k = 0, 1, 2, ..., n - 2$ because for each such k, p is in J_k and $[b, c] = f^{[k]}(J_k)$ by (1). We will complete the proof that p has period n by showing that $f^{[n-1]}(p)$ is in $[a, b)$. To that end, we recall from (2) that $f^{[n-1]}(J_{n-1}) = [a, b]$. Since p is in J_{n-1}, we know that $f^{[n-1]}(p)$ is in $[a, b]$. If it were true that $f^{[n-1]}(p)$ $= b$, then

$$p \; = \; f^{[n]}(p) \; = \; f(f^{[n-1]}(p)) \; = \; f(b) \; = \; c$$

so that $f(p) = f(c) = a$. However, since $J_1 \supseteq J_{n-2} \supseteq J_{n-1}$ and p is in J_{n-1}, it follows that $f(p)$ is in $f(J_1) = [b, c]$, so that $f(p) \neq a$. This contradiction implies that $f^{[n-1]}(p) \neq b$, so that $f^{[n-1]}(p)$ is in $[a, b)$. Therefore all the first $n - 2$ iterates of p lie in $[b, c]$, the $(n - 1)$st iterate lies in $[a, b)$, and the nth iterate lies again in $[b, c]$. Consequently p really does have period n. This completes the proof in case $f(a) = b$, $f(b) = c$ and $f(c) = a$. Because the case in which $f(a) = c$, $f(b) = a$, and $f(c) = b$ is entirely similar, it is left as an exercise (see Exercise 9). ∎

If $3.829 \leq \mu \leq 3.840$, then the quadratic function Q_μ has period-3 points, so the Li-Yorke Theorem implies that such a function has points (and hence cycles) with every possible period. Finding such cycles is another story. For example, you might try to locate a point of period 11 for, say, $\mu = 3.83$. In Section 2.3 we will be able to show that the value of μ where the 3-cycles emerge is actually $1 + 2\sqrt{2}$. (See Exercise 3 in Section 2.3.)

In the same vein, in Section 1.4 we noted that $\{2/7, 4/7, 6/7\}$ is a 3-cycle for the tent function T, so that as a direct consequence of the Li-Yorke Theorem, T has cycles of all possible periods. You might look back at the discussion of T and see whether this fact is implied by any of our results in Section 1.4.

Why does the bifurcation diagram for $\{Q_\mu\}$ in Figure 1.39 basically show only three curves in the period-3 window, since for such μ the function Q_μ has points of all possible periods? The reason is that the 3-cycle is attracting (so that the orbits of almost all other points converge to it), whereas the other cycles are repelling. Nevertheless, Li and Yorke have shown that there are uncountably many numbers in $[0, 1]$ that are not in the basin of attraction of the attracting 3-cycle.

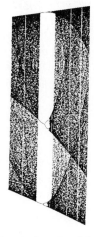

3-cycle window for $\{Q_\mu\}$

Figure 1.39

The Li-Yorke Theorem says that if f has a period-3 point, then it has points of all other periods. But suppose that we can only show that f has, say, a period-5 point. Then must f have points of all periods? A remarkable theorem by the Russian mathematician A. N. Sharkovsky provides a complete answer. In order to present Sharkovsky's result, we need to define the **Sharkovsky ordering** of the positive integers:

$$3 \dashv 5 \dashv 7 \dashv \cdots 2\cdot3 \dashv 2\cdot5 \dashv 2\cdot7 \dashv \cdots 2^2\cdot3 \dashv 2^2\cdot5 \dashv 2^2\cdot7 \dashv \cdots \dashv \cdots \dashv 2^3 \dashv 2^2 \dashv 2 \dashv 1$$

odd integers $2 \cdot$ (odd integers) $2^2 \cdot$ (odd integers)   powers of 2

Here $m \dashv n$ signifies that m appears before n in the Sharkovsky ordering. Thus $17 \dashv 14$ (because $14 = 2 \cdot 7$) and $40 \dashv 64$ (because $40 = 2^3 \cdot 5$ and $64 = 2^6$). Since every positive integer can be written as $2^k \cdot$ (odd integer) for a suitable nonnegative integer k and a suitable odd integer, the Sharkovsky ordering is an ordering of the collection of *all* positive integers. Now we are ready for the theorem.

THEOREM 1.20 (Sharkovsky Theorem). Let f be a continuous function defined on the interval J, and suppose that $J \supseteq f(J)$. If f has a point with period m, then f has a point with period n for all n such that $m \dashv n$.

The original proof of this theorem was long and technical. Even though accessible proofs involving concepts from graph theory have been given (see the papers by Straffin, 1978, and Ho and Morris, 1981), we must omit the proof.

By letting $m = 3$ in Sharkovsky's Theorem, we see that the Li-Yorke Theorem (as stated above) is an immediate corollary of Sharkovsky's Theorem. Moreover, from the Sharkovsky ordering we can imagine why the period-5 window visible in the bifurcation diagram for $\{Q_\mu\}$ lies to the left of the large period-3 window.

By Sharkovsky's Theorem, if a continuous function on a closed interval has a period-5 point, then it has points of all periods except possibly 3, since 5 precedes all positive integers except 3 in the Sharkovsky ordering. That such a function need not have a period-3 point is illustrated in the following example.

EXAMPLE 1. Show that the function f defined in Figure 1.40 has a period-5 point but no period-3 point.

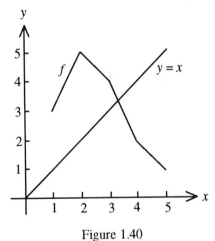

Figure 1.40

Solution. The function f in Figure 1.40 is linear on each of the intervals $[1, 2]$, $[2, 3]$, $[3, 4]$, and $[4, 5]$. It is easy to check that $\{1, 3, 4, 2, 5\}$ is a 5-cycle of f, so that 1 is a period-5 point. To verify that f has no period-3 point, let $f(m, n)$ denote the image of the open interval (m, n). We obtain

$$f(1, 2) = (3, 5), \quad f(3, 5) = (1, 4), \quad f(1, 4) = (2, 5)$$

so that $f^{[3]}(1, 2) = (2, 5)$. Similarly,

$$f^{[3]}(2, 3) = (3, 5), \quad f^{[3]}(3, 4) = (1, 5), \quad f^{[3]}(4, 5) = (1, 4)$$

Therefore $f^{[3]}$ has no fixed points on $(1, 2)$, $(2, 3)$ or $(4, 5)$, and hence f has no period-3 points in these intervals. By contrast, $f(3, 4) = (2, 4) \supseteq [3, 4]$, so Lemma 3 implies that f has a fixed point p in $(3, 4)$. Next, we notice that

> f is decreasing on $(3, 4)$, with image $(2, 4)$
> f is decreasing on $(2, 4)$, with image $(2, 5)$, so $f^{[2]}$ is increasing on $[3, 4]$
> f is decreasing on $(2, 5)$, with image $(1, 5)$, so $f^{[3]}$ is decreasing on $[3, 4]$

It follows that $f^{[3]}$ can cross the line $y = x$ at most once in $(3, 4)$, and that crossing must occur at (p, p) corresponding to the fixed point p. Therefore the only fixed point of $f^{[3]}$ in the interval $(3, 4)$ is the fixed point of f. Consequently in $(3, 4)$ there is no period-3 point of f, so f has no period-3 point at all. ❑

More generally, it can be shown that for any positive integer n, there is a continuous function with a period-n point but no period-m points for any integer m such that $m \dashv n$.

Sharkovsky's Theorem does not tell us how many period-n points a function must have for those positive integers to the right of a given integer m in the Sharkovsky ordering. The minimum number of such period-n points guaranteed to exist is known. (See, for example, the article by Bau-Sen Du, 1985.)

Finally, we mention that Sharkovsky's Theorem was published in a Russian journal in 1964, and apparently was unknown to mathematicians in the western world until after Li and Yorke's celebrated paper "Period Three Implies Chaos" appeared in 1975. Their major theorem included what we have stated as the Li-Yorke Theorem, as well as information about the orbits of various points. It is the information about the orbits that gave rise to the notion of "chaos," a topic we will discuss in Chapter 2.

EXERCISES 1.7

1. a. Find an approximate value of μ such that

$$\{.149407, .488004, .959447\}$$

is (approximately) a 3-cycle for Q_μ.

 b. Show that the 3-cycle in part (a) is attracting.

2. Draw the graph of $f^{[3]}$ for the function f in Example 1.

3.   Suppose that f is defined on the interval $[1, 7]$, passes through the points $(1, 4)$, $(2, 7)$, $(3, 6)$, $(4, 5)$, $(5, 3)$, $(6, 2)$, and $(7, 1)$, and is linear in between. Show that f has a period-7 point but no period-5 point.

4.   a.   Define a function f on the interval $[1, 3]$ that is linear on $[1, 2]$ and on $[2, 3]$, and such that the point 1 has period 3.
     b.   Find the numerical value of a period-2 point for your function f.

5.   a.   Define a function that has a 1-cycle, 2-cycle and 3-cycle, but does not have any n-cycles for $n \geq 4$.
     b.   Can a function such as that described in (a) be continuous? Explain why or why not.

6.   Suppose that $|f'(x)| < 1$ for all x. Show that f cannot have any periodic points other than a unique fixed point.

7.   Use the Intermediate Value Theorem to prove that every real number has a cube root.

8.   Use the Intermediate Value Theorem to prove that every real number is the tangent of a number in the interval $(-\pi/2, \pi/2)$.

9.   Prove the case in which $f(a) = c$, $f(b) = a$, and $f(c) = b$ in the Li-Yorke Theorem.

10. Assume that f is differentiable on the closed interval $[a, b]$, and let $f'(a) < A < f'(b)$. Prove that there exists a number c in the open interval (a, b) such that $f'(c) = A$. (*Hint*: Let $g(x) = f(x) - Ax$ for $a \leq x \leq b$, and use the Maximum-Minimum Theorem.) This result is called **Darboux's Theorem**, and is a kind of Intermediate Value Theorem for the derivative of a differentiable function.

11. Use the computer program ITERATE to show that if $\mu = 1 + 2\sqrt{2}$, then Q_μ has a period-3 point.

1.8 THE SCHWARZIAN DERIVATIVE

The question we address in this section is the following: How many attracting cycles can a differentiable function have? To understand why the question might be relevant, we need only look at the window in the bifurcation diagram for $\{Q_\mu\}$ that appears in Figure 1.41. For a given μ in the interval, the six horizontal curves that dominate the window could, in theory, represent an attracting 6-cycle, two attracting

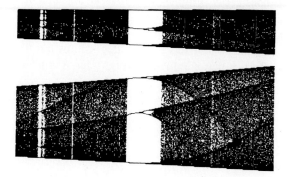

window in the bifurcation diagram of $\{Q_\mu\}$

Figure 1.41

3-cycles, three attracting 2-cycles, or even six attracting fixed points. We will be able to resolve this issue by Theorem 1.22.

Recall that if f is a differentiable function defined on an interval J, then x is a critical point in the interior of J if $f'(x) = 0$. The goal of this section is to prove a wonderful theorem by the American mathematician David Singer (1978) to the effect that under certain conditions, if f has n critical points, then it has at most $n + 2$ attracting cycles. The proof of this result is long (and entails seven lemmas). The hypothesis that is crucial involves what is called the Schwarzian derivative.

DEFINITION 1.21. Let f be defined on the interval J, and assume that the third derivative f''' is continuous on J. Define Sf by

$$(Sf)(x) = \frac{f'''(x)}{f'(x)} - \frac{3}{2}\left(\frac{f''(x)}{f'(x)}\right)^2$$

Then $(Sf)(x)$ is the **Schwarzian derivative of f at x** whenever it exists as a number or as $-\infty$ or ∞.

The Schwarzian derivative is named for the German mathematician Hermann Schwarz, who in 1869 defined it and used it in the study of complex-valued functions. For our purposes the Schwarzian derivative has two important features:

i. Composites of functions with negative Schwarzian derivatives also have negative Schwarzian derivatives (Lemma 4).
ii. If a function f has a negative Schwarzian derivative, together with enough fixed points, then f has a critical point (Lemma 3).

To illustrate (ii), consider a continuously differentiable function f having four isolated fixed points a, b, c, and d with $a < b < c < d$ (Figure 1.42). Notice that $f'(x) > 1$ and $f'(y) < 1$ for appropriate values of x and y in $[a, d]$. If f has a negative Schwarzian derivative on $[a, d]$, then it turns out that there must be a number z in $[a, d]$ such that $f'(z) = 0$, that is, f has a critical point.

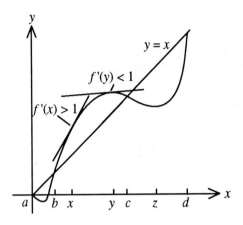

Figure 1.42

One can show that any polynomial the zeros of whose derivative are real and distinct has a negative Schwarzian derivative. (For the general result, see page 69 of Devaney, 1989; for third degree polynomials, see Exercise 21 of this section.) We will show, more modestly, that the quadratic function Q_μ has a negative Schwarzian derivative for all values of μ.

EXAMPLE 1. Let $Q_\mu(x) = \mu x(1 - x)$. Show that $SQ_\mu(x) < 0$ for $0 < x < 1$.

Solution. Notice that $Q_\mu'(x) = \mu(1 - 2x)$, $Q_\mu''(x) = -2\mu$, and $(Q_\mu)'''(x) = 0$. Therefore if $x \neq 1/2$, then by the definition of SQ_μ,

$$(SQ_\mu)(x) = -\frac{3}{2}\left(\frac{-2\mu}{\mu(1 - 2x)}\right)^2 = \frac{-6}{(1 - 2x)^2} < 0$$

Finally, $\lim_{x \to 1/2}(Sf)(x) = -\infty$, so we write $(Sf)(1/2) = -\infty$. ❑

Having defined the Schwarzian derivative, we are ready to direct our attention toward Singer's Theorem. Throughout we will assume that the third derivative of each function under discussion exists and is continuous. The goal of the first three lemmas is to show that any function with negative Schwarzian derivative and four isolated fixed points has a critical point. We will write "$Sg < 0$" for the more complete "$Sg(x) < 0$ for all x in the domain of g."

LEMMA 1. Let $Sg < 0$. If g' has a relative minimum value at x^*, then $g'(x^*) < 0$.

Proof. Suppose that g' has a relative minimum value at x^*. Then $g''(x^*) = 0$, so that

$$\frac{g'''(x^*)}{g'(x^*)} = (Sg)(x^*) < 0$$

However, since $g'(x^*)$ is a relative minimum value, the Second Derivative Test from calculus tells us that $g'''(x^*) \geq 0$. Consequently $g'(x^*) < 0$. ∎

The same reasoning as we used in Lemma 1 implies that if $Sg < 0$, then any relative maximum value of g' must be positive. Lemma 1 and the preceding comment imply that if $Sg < 0$, then the graph of g cannot appear as in Figure 1.43(a) because g' has a positive relative minimum value at b and a negative relative maximum value at c. By contrast, the graph appearing in Figure 1.43(b) is allowed because g' has a negative relative minimum value at b and a positive relative maximum value at c.

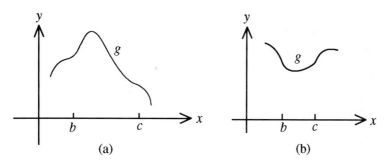

Figure 1.43

Before we turn to Lemmas 2 and 3, we remark that if $Sg < 0$ on an open interval, then there must be an x in the interval such that $g(x) \neq x$.

LEMMA 2. Let a, b, and c be fixed points of g, with $a < b < c$. Assume also that $Sg < 0$ on (a, c). If $g'(b) \leq 1$, then g has a critical point in (a, c).

Proof. Since $g(a) = a$, $g(b) = b$, and $g(c) = c$, it follows from the Mean Value Theorem that there exist an r in (a, b) and s in (b, c) such that $g'(r) = 1 = g'(s)$

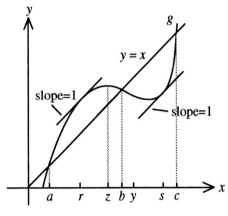

Figure 1.44

(Figure 1.44). Since $g'(r) = 1 = g'(s)$ and $g'(b) \leq 1$, and since g' is continuous on $[r, s]$, the Maximum-Minimum Theorem implies that g' has a minimum value $g'(y)$ on $[r, s]$. Since $Sg < 0$ by hypothesis, g' cannot be constant on $[r, s]$, so $g'(y) < 1$. It then follows from Lemma 1 that $g'(y) < 0$. Since $g'(s) = 1$ and $g'(y) < 0$, the Intermediate Value Theorem guarantees a z in (r, s) such that $g'(z) = 0$. This z is the critical point of g that we seek. ■

LEMMA 3. Suppose that g has fixed points $a, b, c,$ and $d,$ with $a < b < c < d.$ Assume also that $Sg < 0$ on $[a, d].$ Then g has a critical point in $(a, d).$

Proof. If $g'(b) \leq 1,$ then Lemma 2 implies that g has a critical point in $(a, c).$ Similarly, if $g'(c) \leq 1,$ then g has a critical point in $(b, d).$ So let us assume that $g'(b) > 1$ and $g'(c) > 1$ (Figure 1.45). Then there are r and t such that $b < r < t < c$ and such that $g(r) > r$ and $g(t) < t$ (Figure 1.45). Hence by the Mean Value Theorem there is an s in (r, t) such that $g'(s) < 1.$ Since g' is continuous on $[b, c],$ g' must have a relative minimum value $g'(y)$ on $(b, c).$ Lemma 1 implies that $g'(y) < 0.$ Then the Intermediate Value Theorem yields a z in (y, c) such that $g'(z) = 0.$ Thus z is a critical point of $g.$ ■

In the proof we assumed that $g'(b) > 1$ and $g'(c) > 1.$ Since b and c were fixed points, it followed that b and c were *not* adjacent fixed points of $g.$

With Lemma 3 we have shown that a negative Schwarzian derivative and the existence of four fixed points together imply the existence of a critical point. The next lemma analyzes the Schwarzian derivative of a composite of functions.

LEMMA 4. Suppose that $Sf < 0$ and $Sg < 0.$ Then $S(f \circ g) < 0.$

Proof. First we use the Chain Rule to calculate that

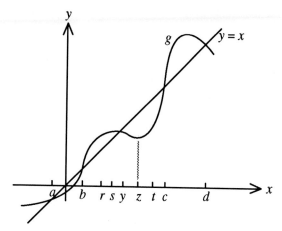

Figure 1.45

$(f \circ g)'(x) = [f'(g(x))] [g'(x)]$

$(f \circ g)''(x) = [f''(g(x))] [g'(x)]^2 + [f'(g(x))] [g''(x)]$

$(f \circ g)'''(x) = [f'''(g(x))] [g'(x)]^3 + 3[f''(g(x))] [g'(x)] [g''(x)] + [f'(g(x))] [g'''(x)]$

Then

$$[S(f \circ g)](x) = \frac{(f \circ g)'''(x)}{(f \circ g)'(x)} - \frac{3}{2} \left(\frac{(f \circ g)''(x)}{(f \circ g)'(x)} \right)^2$$

$$= \frac{[f'''(g(x))][g'(x)]^3 + 3[f''(g(x))][g'(x)][g''(x)] + [f'(g(x))][g'''(x)]}{[f'(g(x))][g'(x)]}$$

$$- \frac{3}{2} \left(\frac{[f''(g(x))][g'(x)]^2 + [f'(g(x))][g''(x)]}{[f'(g(x))][g'(x)]} \right)^2$$

$$= \underset{<0}{[(Sf)(g(x))]} \; \underset{\geq 0}{[g'(x)]^2} + \underset{<0}{(Sg)(x)}$$

$$< \quad 0$$

This completes the proof. ∎

LEMMA 5. Suppose that $Sf < 0$. Then $Sf^{[n]} < 0$ for any positive integer n.

Proof. We will use induction. We know that $Sf^{[2]} = S(f \circ f) < 0$ by letting $g = f$ in Lemma 4. Therefore the result is valid for $n = 2$. Next, assume that $Sf^{[n]} < 0$ for a given positive integer n. Then $Sf^{[n+1]} = S(f \circ f^{[n]}) < 0$ by letting $g = f^{[n]}$ in Lemma 4. By the Law of Induction, $Sf^{[n]} < 0$ for each positive integer n. ∎

Since period-n points of f are fixed points of $f^{[n]}$, Lemma 5 implies that arguments involving periodic points of f can reduce to arguments about fixed points of $f^{[n]}$.

Lemmas 6 and 7 assure us that if f has a finite number of critical points, then f has a finite number of period-m points, for each positive integer m.

LEMMA 6. Let f be differentiable, and suppose that f has a finite number of critical points. Then $f^{[m]}$ has a finite number of critical points for each $m > 1$.

Proof. To start the proof, suppose that $x < y$ and $f(x) = f(y)$. Then the Mean Value Theorem implies that there is a z in the interval (x, y) such that $f'(z) = 0$. This means that z is a critical point of f. Since f has only a finite number of critical points by hypothesis, it follows that the collection of points x such that x or $f(x)$ is a critical point of f is a finite set of points. We will use this fact later in the proof.

Our proof proceeds by induction. Let x be a critical point of $f^{[2]}$. Then

$$0 = (f^{[2]})'(x) = [f'(f(x))] [f'(x)]$$

so that x or $f(x)$ is a critical point of f. Since there are only a finite number of such points by our preceding comment, we deduce that $f^{[2]}$ has but a finite set of critical points. Thus the proof is complete for $m = 2$. Now we let $m > 2$, and assume that x is critical point of $f^{[m]}$. In this case,

$$0 = (f^{[m]})'(x) = [f'(f^{[m-1]}(x))] [f'(f^{[m-2]}(x))] \cdots [f'(f(x))] [f'(x)]$$

so that x, $f(x)$, . . ., or $f^{[m-1]}(x)$ is a critical point of f. Using the same reasoning as for $m = 2$, we conclude that $f^{[m]}$ can have only a finite number of critical points. ∎

LEMMA 7. Let f have a finite number of critical points, and assume that $Sf < 0$. Then for any positive integer m there is a finite number of period-m points of f.

Proof. Let $g = f^{[m]}$. Then $Sg < 0$ by Lemma 5, and g has finitely many critical points by Lemma 6. If g were to have an increasing (or decreasing) infinite sequence of fixed points $\{p_n\}$, then Lemma 3 tells us that there would be a critical point between p_1 and p_4, between p_5 and p_8, etc. Thus g would have infinitely many critical points. This contradiction proves that g has but a finite number of fixed points. Since $g = f^{[m]}$, this is tantamount to f having only finitely many points of period m. ∎

We are prepared to state and prove Singer's Theorem. For convenience, when we refer to cycles, we will include fixed points.

THEOREM 1.22 (Singer's Theorem). Let f be defined on a closed interval J, and suppose that $J \supseteq f(J)$. Assume that $Sf < 0$, and that f has n critical points. Then f has at most $n + 2$ attracting cycles.

Proof. Let $J = [A, B]$, where $-\infty \le A$ and $B \le \infty$. Suppose that p is an attracting period-m point of f. This means that if $g = f^{[m]}$, then p is an attracting fixed point of g. We will focus on g. Let (L, R) be the largest open interval about p all of whose points are attracted to p. (We allow the possibility that $L = -\infty$ or $R = \infty$ or both.) If L and R are both interior to J, so that L and R are finite, then since g is continuous and the interval (L, R) is maximal, it follows that $g(L) = L$ and $g(R) = R$, or $g(L) = R$ and $g(R) = L$, or $g(L) = g(R)$. We will show that in each case there is a critical point of g that is attracted to p, and hence a critical point of f that is attracted to the orbit of p.

Case 1. $g(L) = L$ and $g(R) = R$

Since p is an attracting fixed point of g, we know by Theorem 1.6 that $|g'(p)| \le 1$, and therefore $g'(p) \le 1$. Next, $Sg < 0$ by Lemma 5. If we let $L = a$, $p = b$, and $R = c$ in Lemma 2, we find that g has a critical point x^* in the interval (L, R). Now

$$0 = g'(x^*) = (f^{[m]})'(x) = [f'(f^{[m-1]}(x^*))]\,[f'(f^{[m-2]}(x^*))] \cdots [f'(x^*)]$$

Therefore one of x^*, $f(x^*)$, $f^{[2]}(x^*)$, . . . , $f^{[m-1]}(x^*)$ is a critical point of f. Since x^* is in (L, R) and hence is attracted (by the iterates of g) to p, all of the iterates of x^* are similarly attracted to p. Consequently f has a critical point that is attracted (by the iterates of f) to the orbit of p.

Case 2. $g(L) = R$ and $g(R) = L$

In this case we first consider $g^{[2]}$. Notice that $g^{[2]}(L) = L$ and $g^{[2]}(R) = R$, and that p is an attracting fixed point of $g^{[2]}$. By Case 1 this means that $g^{[2]}$ has a critical point x^* in the interval (L, R) that is attracted (by the iterates of g) to p. However,

$$0 = (g^{[2]})'(x^*) = [g'(g(x^*))]\,[g'(x^*)]$$

so that x^* or $g(x^*)$ is a critical point of g. Therefore by the argument in Case 1, there is an iterate (with respect to f) of x^* that is a critical point of f. Since x^* is in (L, R) and hence is attracted (by the iterates of g) to p, all of the iterates of x^* are similarly attracted to p. Consequently f has a critical point that is attracted (by the iterates of f) to the orbit of p.

Case 3. $g(L) = g(R)$

If $g(L) = g(R)$, then by the Mean Value Theorem there is an x^* in the interval (L, R) such that $g'(x^*) = 0$, meaning that x^* is a critical point of g. Once again some iterate of x^* not only must be a critical point of f, but also must be attracted to the orbit of p.

In each of Cases 1–3, there is a critical point of f that is attracted to the orbit of p. Since a critical point can be attracted to at most one orbit, and since by hypothesis f has only n critical points, we conclude that there are at most n attracting periodic orbits that are associated with intervals of the form (L, R) that are interior to J. Orbits associated with intervals of the form (A, R) or (L, B) add a maximum of 2 more possible attracting cycles, making the maximum possible number of attracting cycles be $n + 2$. ∎

Singer's Theorem yields the following corollary, which was known to the French mathematician Gaston Julia nearly a century ago.

COROLLARY 1.23.   Let $0 < \mu < 4$. Each function in the quadratic family $\{Q_\mu\}$ has at most one attracting cycle.

Proof. If $0 < \mu \leq 1$, then the basin of attraction of 0 is $[0, 1]$, so 0 is the only attracting periodic point. For the remainder of the proof, assume that $1 < \mu < 4$. Since Q_μ has the unique critical point $1/2$, Singer's Theorem implies that there can be at most 3 attracting cycles, one each associated with intervals of the form $[0, L)$, (L, R), and $(R, 1]$, where $0 < L < R < 1$. Since 0 is a repelling fixed point and since $Q_\mu(1) = 0$, neither $[0, L)$ nor $(R, 1]$ appears as a basin of attraction for cycles of Q_μ. Consequently Q_μ has at most one attracting cycle.   ∎

Corollary 1.23 implies that Q_μ can have only one attracting cycle. Therefore the six horizontal curves in the window featured in Figure 1.41 represent an attracting 6-cycle (rather than multiple 3-cycles or 2-cycles or fixed points). Thus we have answered the question posed at the outset of the section.

Corollary 1.23 also yields a method for detecting potential attracting cycles of a function Q_μ. The method involves calculating the iterates of the lone critical point $1/2$ of Q_μ. If the iterates appear to converge to a periodic orbit, then that orbit would be the candidate for the unique attracting periodic orbit of Q_μ. You might try this technique to search for the attracting 2-cycle of, say, the function $Q_{3.2}$. By contrast, if the iterates of $1/2$ do not seem to converge to a periodic orbit, then it may well happen that no attracting orbit exists (although it can be hard to tell because orbits can have arbitrarily many points). This is the case with Q_4.

Singer's Theorem indicates that a function with but one critical point and a negative Schwarzian derivative can have no more than three attracting cycles. Thus in theory there could be zero, one, two, or three attracting cycles (see Exercise 14).

We should mention that the result in Singer's Theorem need not hold if the hypothesis that $Sf < 0$ is ignored.

EXAMPLE 2. Let $f(x) = 1 - x^2/2 - x^{14}/2$ for $0 \le x \le 1$ (see Figure 1.46). Verify that f has not only an attracting fixed point but also an attracting 2-cycle in the interval $[0, 1]$. Show also that $Sf(x) > 0$ for some x in $(0, 1)$.

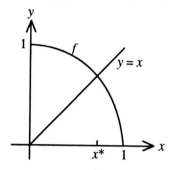

Figure 1.46

Solution. If $x^* \approx -0.72861$, then a routine calculation shows that x^* is an attracting fixed point for f, and that $Sf(x^*) \approx 18.77 > 0$. In addition, $\{0, 1\}$ is a 2-cycle that is attracting because $(f^{[2]})'(0) = [f'(1)] [f'(0)] = 0.$ ❏

We close the discussion of attracting cycles with an observation noted by Singer. Suppose a population is in stable equilibrium that could be constant (represented by a fixed point) or could involve oscillations (hence cycles). Singer's Theorem would imply that if the stable equilibrium is, say, constant, then by changing the population size (that is, the value of x), then the type of stable equilibrium cannot change.

EXERCISES 1.8

1. Let $E_\mu(x) = \mu e^x$ for all x. Show that whatever nonzero number μ is, SE_μ is a constant function. Determine that constant.

In Exercises 2–4, show that $Sf < 0$.

2. $f(x) = \sin x$ for $0 \le x < \pi/2$ 3. $f(x) = \cos x$ for $0 < x \le \pi/2$

4. $f(x) = x(1 + \mu(1 - x))$

In Exercises 5–6, determine whether the Schwarzian derivative of f is positive, or is negative.

5. $f(x) = x^2$ 6. $f(x) = x^2(x - 1)$

7. Let $f(x) = x^a$, where $0 < a$. Determine the values of a for which $Sf < 0$.

8. Let $f_\mu(x) = x^3 + \mu x$. Show that if $\mu > 0$, then $Sf(x) > 0$ for x near to 0.

9. Let $f(x) = (x^3/3) + x$. Determine the interval on which $Sf > 0$.

10. Let $f_\mu(x) = x - \mu x^3$. Show that $Sf_\mu(0)$ changes sign as μ passes through 0.

11. Let $f(x) = 7.86x - 23.31x^2 + 28.75x^3 - 13.30x^4$. (This function appeared in Singer's article.)
    a. Show that if $x^* \approx 0.7263986$, then x^* is an attracting fixed point.
    b. Show that if $y^* \approx 0.3217591$, then $\{y^*, f(y^*)\}$ is an attracting 2-cycle.
    c. How many critical points does f have? Explain your answer.
    d. Do the results of (a) and (b) contradict Singer's Theorem? Explain why or why not.
    e. Show that if $c \approx 0.3239799$, then c is a critical point.
    f. Determine whether c is attracted to the fixed point, the 2-cycle, or neither.

12. Let $f_\mu(x) = \mu(7.86x - 23.31x^2 + 28.75x^3 - 13.30x^4)$.
    a. Show that the attracting fixed point of f_μ becomes repelling as μ increases through 1.
    b. Using a computer, draw the graphs of f_μ for $\mu = .95$, $\mu = 1$ and $\mu = 1.05$, and show that as μ increases through the interval $[.95, 1]$, an attracting 2-cycle *and* a repelling 2-cycle are born.
    c. Using the graphs in part (b), show that as μ increases through the interval $[1, 1.05]$, the repelling 2–cycle coalesces with the fixed point, which is repelling.
    d. Show that the family $\{f_\mu\}$ has a bifurcation point at 1.

13. Let $f_\mu(x) = \mu \sin x$ for $0 \le x \le \pi$, where $0 < \mu < \pi$. Determine the maximum possible number of attracting periodic orbits of f.

14.   a.   Find a value of μ such that Q_μ has no attracting cycles.
      b.   Find a value of μ such that Q_μ has one attracting fixed point and no other cycles.
      c.   Let $g(x) = 1.2 \sin x$ on $[-p, \pi]$, where p is a fixed point of g that lies in $[0, \pi]$. Show that g has one critical point and two attracting fixed points, but no other cycles.

15.   Suppose that the differentiable function f is defined on $[0, 1]$, and that f satisfies the following conditions:

      i. $f(0) = 0 = f(1)$.
      ii. $1/2$ is the only critical point of f and $f(1/2) = 1$.
      iii. $Sf < 0$ on $[0, 1]$.
      iv.   The graph of f is concave downward on $(0, 1)$.

      Determine the maximum possible number of attracting cycles of f.

16.   Suppose that Sf exists.
      a.   Show that $S(af) = Sf$ for any nonzero constant a.
      b.   Show that $S(f + b) = Sf$ for any constant b.
      c.   Show that $S(1/f) = Sf$.
      d.   Let a, b, c, and d be nonzero constants, and define the function g by

$$g = \frac{af + b}{cf + d}$$

      Use parts (a)–(c) to show that $Sg = Sf$.

17.   Show that if $Sf > 0$ and $Sg > 0$, then $S(f \circ g) > 0$.

18.   Show that if $Sg < 0$, then g' cannot have a negative relative maximum value.

19.   Suppose that $Sg(x) < 0$ for all x in the interval J. Show that for each x in J, either $g''(x) \neq 0$ or $g'''(x) \neq 0$.

20.   Suppose that f has a second derivative.
      a.   Find a formula for $(f^{[2]})'(x)$ in terms of x, f, and f'.
      b.   Find a formula for $(f^{[2]})''(x)$ in terms of x, f, f', and f''.
      c.   Suppose that $f(p) = p$ and $f'(p) = -1$. Show that $(f^{[2]})''(p) = 0$.

21.   Consider the general third-degree polynomial function defined by $f(x) = ax^3 + bx^2 + cx + d$, with $a > 0$.

a.   Let $x = z - b/3$. Show that $ax^3 + bx^2 + cx + d = az^3 + (c - b^2/3)z + r$ for an appropriate constant r. (The substitution $x = z - b/3$ has been used in the process of identifying the solutions of the general cubic equation, and was first published in 1545 by the Italian mathematician Girolamo Cardano.)

b.   Let $g(z) = az^3 + (c - b^2/3)z + r$. Show that $g(z) = 0$ has three real roots if and only if $c < b^2/3$.

c.   Let $c < b^2/3$. Show that $Sg(z) < 0$ for all z.

d.   Show that $Sf(x) < 0$ for all x if and only if $Sg(z) < 0$ for all z.

CHAPTER
2

ONE-DIMENSIONAL CHAOS

In Chapter 1 the focus was mainly on periodic points, and more particularly, attracting periodic points. Attracting periodic points indicate a regularity, predictability, and stability in the dynamics of a function of a parametrized family of functions.

Chapter 2 is devoted to a contrasting dynamical action – points whose iterates separate from one another. This kind of behavior is symptomatic of what we call chaotic dynamics, or just plain chaos. It was only after the advent of the high-speed computer that such dynamics could be investigated and analyzed effectively.

In Section 2.1 we define the most illustrious concept in the study of chaotic dynamics: sensitive dependence on initial conditions. Sensitive dependence on initial conditions, along with the closely related notion of the Lyapunov exponent, serve as the ingredients in the definition of chaos. In Section 2.2 we turn to points whose orbits virtually fill up the whole domain space. If a function is chaotic and has this added property and enough periodic points, then it is strongly chaotic. Section 2.3 is devoted to the notion of conjugacy. If two functions are conjugate to one another, then they share many properties pertaining to chaos and strong chaos. We use conjugacy to prove the main result of the section: Q_4 is strongly chaotic. The final section concerns Q_μ for $\mu > 4$. We establish that such a Q_μ is strongly chaotic on a subset of $[0, 1]$ that is a so-called Cantor set. It is interesting that Cantor sets, which play a central role in analysis, play a like role in chaotic dynamics. Chapter 2 completes the study of chaos for functions of one variables.

2.1 CHAOS

In this section we study two methods of describing the way in which iterates of neighboring points separate from one another: sensitive dependence on initial conditions and the Lyapunov exponent. These notions are fundamental to the concept of chaos, which also will appear in the present section.

Sensitive Dependence on Initial Conditions

Before defining sensitive dependence on initial conditions, we adopt a notation that henceforth will facilitate our discussion. We will write $f : A \to B$ to indicate that the domain of the function f is A and the range of f is contained in B. Thus $f : J \to J$ signifies that the domain of f is J and the range is contained in J.

DEFINITION 2.1. Let J be an interval, and suppose that $f : J \to J$. Then f has **sensitive dependence on initial conditions at** x, or just **sensitive dependence at** x if there is an $\varepsilon > 0$ such that for each $\delta > 0$, there is a y in J and a positive integer n such that

$$|x - y| < \delta \quad \text{and} \quad |f^{[n]}(x) - f^{[n]}(y)| > \varepsilon$$

If f has sensitive dependence on initial conditions at each x in J, we say that f **has sensitive dependence on initial conditions on** J, or that f **has sensitive dependence on** J, or that f **has sensitive dependence**.

The "initial conditions" in the definition refer to the given, or initial, points x and y. The definition says that f has sensitive dependence if arbitrarily close to any given point x in the domain of f there is a point and an nth iterate that is farther from the nth iterate of x than a distance ε. This has practical significance, because in such instances higher iterates of an approximate value of x may not resemble the true iterates of x. Thus computer calculations may be misleading.

To illustrate sensitive dependence on initial conditions, we turn to the baker's function.

EXAMPLE 1.   Consider the baker's function B, given by

$$B(x) = \begin{cases} 2x & \text{for } 0 \leq x \leq 1/2 \\ 2x - 1 & \text{for } 1/2 < x \leq 1 \end{cases}$$

Show that after 10 iterates, the iterates of $1/3$ and $.333$ are farther than $1/2$ apart.

Solution.   Notice that $B(1/3) = 2/3$ and $B^{[2]}(1/3) = 1/3$, so that the iterates of $1/3$ alternate between $1/3$ and $2/3$. To compare the iterates of $1/3$ and $.333$ we make the following table (where we use 3-place approximations for the iterates of $.333$):

iterates	1	2	3	4	5	6	7	8	9	10
	$\frac{1}{3}$	$\frac{2}{3}$	$\frac{1}{3}$	$\frac{2}{3}$	$\frac{1}{3}$	$\frac{2}{3}$	$\frac{1}{3}$	$\frac{2}{3}$	$\frac{1}{3}$	$\frac{2}{3}$	$\frac{1}{3}$
	.333	.666	.332	.664	.328	.656	.312	.624	.248	.496	.992

Therefore the tenth iterates of 1/3 and .333 are, respectively, 1/3 and .992, which are farther apart than a distance 1/2. ❑

Letting

$$x_n = n\text{th iterate of } 1/3 \qquad \text{and} \qquad y_n = n\text{th iterate of } .333$$

we display the separation of the iterates of 1/3 and .333 in Figure 2.1.

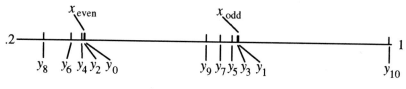

Figure 2.1

The implication of Example 1 is that higher iterates of .333 can have little relationship to corresponding iterates of 1/3. In fact, although we can tell precisely what the 20th or 35th iterate of 1/3 is, there is absolutely no way of determining the 20th or 35th iterate of .333 without actually calculating those iterates.

EXAMPLE 2. Show that the tent function T has sensitive dependence on initial conditions on [0, 1].

Solution. Let x be any number in [0, 1]. First we will show that if v is any dyadic rational number (of the form $j/2^m$, in lowest terms) in [0, 1] and w is any irrational number in [0, 1], then there is a positive integer n such that

$$|T^{[n]}(v) - T^{[n]}(w)| > \frac{1}{2} \tag{1}$$

Toward that goal, we recall from Section 1.4 that if $v = j/2^m$, then $T^{[m]}(v) = 1$ and $T^{[m+k]}(v) = 0$ for all $k > 0$. By contrast, if w is any irrational number in [0, 1], then since T doubles each number in (0, 1/2), there exists an $n > m$ such that $T^{[n]}(w) > 1/2$. Since $n > m$, it follows that $T^{[n]}(v) = 0$, so that (1) is valid. Next, let $\delta > 0$. Then there exist a dyadic rational v and an irrational number w in [0, 1] such that $|x - v| < \delta$ and $|x - w| < \delta$. Therefore (1) implies that

$$\text{either} \quad |T^{[n]}(x) - T^{[n]}(v)| > \frac{1}{4} \quad \text{or} \quad |T^{[n]}(x) - T^{[n]}(w)| > \frac{1}{4}$$

Thus if we let $\varepsilon = 1/4$, then sensitive dependence on initial conditions at the arbitrary number x, and hence on [0, 1], is proved. ❑

Basically the reason that T has sensitive dependence on initial conditions is that if $x \neq 1/2$, then $|T'(x)| = 2$, so that distances between pairs of numbers in $(0, 1/2)$ or in $(1/2, 1)$ are doubled by T.

If p is period-n point of f such that $|(f^{[n]})'(p)| < 1$, then f cannot have sensitive dependence on initial conditions at p (Exercise 5). Because this condition holds for any attracting periodic point of Q_μ whenever $0 < \mu < \mu_\infty$, it follows that Q_μ does not have sensitive dependence on initial conditions for such values of μ. However, for $\mu > \mu_\infty$, the story can be quite different. In fact, in Section 2.3 we will show that Q_4 has sensitive dependence on initial conditions.

Lyapunov Exponents

Although the concept of sensitive dependence on initial conditions is easy to visualize, actually determining that a given function has sensitive dependence is usually not so simple as it is for the tent function.

Before we indicate a second method of describing the separation of iterates of neighboring points, let us return to the tent function. Suppose that x is in $(0, 1)$ and is not a dyadic rational. If $\varepsilon > 0$ and is small enough, then

$$|T(x + \varepsilon) - T(x)| = 2\varepsilon \quad \text{and} \quad |T^{[2]}(x + \varepsilon) - T^{[2]}(x)| = 2^2\varepsilon$$

Now let n be an arbitrary positive integer. Again, if ε is small enough, then

$$|T^{[n]}(x + \varepsilon) - T^{[n]}(x)| = 2^n\varepsilon \qquad (2)$$

Next we divide both sides by ε and let ε approach 0. We obtain

$$|(T^{[n]})'(x)| = \lim_{\varepsilon \to 0} \left| \frac{T^{[n]}(x + \varepsilon) - T^{[n]}(x)}{\varepsilon} \right| = 2^n \qquad (3)$$

Thus for the tent function the derivative is related to separation of iterates of nearby points. We will now explore this idea for other functions.

Let J be a bounded interval, and consider a function $f : J \to J$ having a continuous derivative. By analogy with (2), we assume that for each x in the interior of J and each small enough $\varepsilon > 0$, there is a number $\lambda(x)$ such that for each positive integer n,

$$|f^{[n]}(x+\varepsilon) - f^{[n]}(x)| \approx [e^{\lambda(x)}]^n \, \varepsilon$$

(For the tent function T, the number corresponding to $e^{\lambda(x)}$ would be 2.) This implies that

$$e^{n\lambda(x)} \approx \left| \frac{f^{[n]}(x+\varepsilon) - f^{[n]}(x)}{\varepsilon} \right|$$

so that

$$e^{n\lambda(x)} = \lim_{\varepsilon \to 0} \left| \frac{f^{[n]}(x+\varepsilon) - f^{[n]}(x)}{\varepsilon} \right| = |(f^{[n]})'(x)| \tag{4}$$

If $(f^{[n]})'(x) \neq 0$, then by taking logarithms and dividing by n in (4), we obtain

$$\lambda(x) = \frac{1}{n} \ln |(f^{[n]})'(x)| \tag{5}$$

This leads us to make the following definition.

DEFINITION 2.2. Let J be a bounded interval, and $f : J \to J$ continuously differentiable on J. Fix x in J, and let $\lambda(x)$ be defined by

$$\lambda(x) = \lim_{n \to \infty} \frac{1}{n} \ln |(f^{[n]})'(x)| \tag{6}$$

provided that the limit exists. In that case, $\lambda(x)$ is the **Lyapunov exponent of** f **at** x. If $\lambda(x)$ is independent of x wherever $\lambda(x)$ is defined, then the common value of $\lambda(x)$ is denoted by λ, and is the **Lyapunov exponent** of f.

The definition honors the 20th century Russian mathematician A. M. Lyapunov. To apply the definition to the tent function, we use (3), which yields

$$\lim_{n \to \infty} \frac{1}{n} \ln |(T^{[n]})'(x)| = \lim_{n \to \infty} \frac{1}{n} \ln 2^n = \ln 2$$

Therefore $\lambda(x) = \ln 2$ for the tent function, whenever x is *not* a dyadic rational.
The number $\lambda(x)$ can be considered to measure "the average loss of information" of successive iterates of points near x. If y is near x, and if the iterates of x and y remain close together, then $\lambda(x)$ is apt to be negative because of the presence of the logarithm. By contrast, if the iterates separate from one another, then $\lambda(x)$ is apt to be positive. Thus the larger $\lambda(x)$ is, the greater the loss of information. For the baker's function B discussed in Example 1, we found that the iterates of 1/3 and .333 separate so that the 10th iterate of 1/3 sheds *no* information on the 10th iterate of .333. Thus for the baker's function, successive iterates of a number such as 1/3 provide less and less information about the corresponding iterates of nearby points. For B, $\lambda = \ln 2 > 0$ (Exercise 2).
To make general calculations of $\lambda(x)$ simpler, let $x_0 = x$ and $x_k = f^{[k]}(x)$ for $k = 1, 2, \ldots$. By (3) in Section 1.3 and the Law of Logarithms,

$$\ln |(f^{[n]})'(x)| = \ln |f'(x_{n-1}) f'(x_{n-2}) \cdots f'(x_1) f'(x_0)|$$

$$= \ln [|f'(x_{n-1})| \, |f'(x_{n-2})| \cdots |f'(x_1)| \, |f'(x_0)|]$$

$$= \sum_{k=0}^{n-1} \ln |f'(x_k)|$$

Therefore (6) can be rewritten as

$$\lambda(x) = \lim_{n \to \infty} \frac{1}{n} \sum_{k=0}^{n-1} \ln |f'(x_k)| \tag{7}$$

We will use (7) in order to determine the Lyapunov exponent for quadratic functions.

EXAMPLE 3. Let $Q_\mu(x) = \mu x(1 - x)$, for $0 \leq x \leq 1$, where $1 < \mu < 3$ and $\mu \neq 2$. Show that $\lambda = \ln |2 - \mu|$.

Solution. Let x be arbitrary in $(0, 1)$. Recall from Section 1.5 that Q_μ has the fixed point $p_\mu = 1 - 1/\mu$ that attracts all points in $(0, 1)$. Therefore

$$x_k = Q_\mu^{[k]}(x) \to p_\mu$$

as k increases without bound. Since $Q_\mu'(x) = \mu - 2\mu x$, it follows that

$$Q_\mu'(x_k) \to Q_\mu'(p_\mu) = \mu - 2\mu \left(1 - \frac{1}{\mu} \right) = 2 - \mu$$

By hypothesis, $\mu \neq 2$, so that

$$\ln |Q_\mu'(x_k)| \to \ln |2 - \mu|$$

as k increases without bound. Now let $\varepsilon > 0$. Because the natural logarithm is continuous and x_k approaches p_μ as k increases, there is a positive integer N such that if $k > N$, then

$$\ln |2 - \mu| - \varepsilon < \ln |Q_\mu'(x_k)| < \ln |2 - \mu| + \varepsilon$$

Consequently for $n > N$,

$$\frac{n-N}{n} (\ln |2 - \mu| - \varepsilon) = \frac{1}{n} \sum_{k=N+1}^{n} (\ln |2 - \mu| - \varepsilon) < \frac{1}{n} \sum_{k=N+1}^{n} \ln |Q_{\mu}'(x_k)|$$

$$< \frac{1}{n} \sum_{k=N+1}^{n} (\ln |2 - \mu| + \varepsilon) = \frac{n-N}{n} (\ln |2 - \mu| + \varepsilon)$$

If n is sufficiently large and much larger than N, then $(n - N)/n \approx 1$, so that the preceding inequalities reduce to

$$\ln |2 - \mu| - \varepsilon < \frac{1}{n} \sum_{k=N+1}^{n} \ln |Q_{\mu}'(x_k)| < \ln |2 - \mu| + \varepsilon$$

Moreover, for large enough n, we have

$$\left| \frac{1}{n} \sum_{k=0}^{N} \ln |Q_{\mu}'(x_k)| \right| < \varepsilon$$

Therefore

$$\lim_{n \to \infty} \frac{1}{n} \sum_{k=0}^{n} \ln |Q_{\mu}'(x_k)| = \lim_{n \to \infty} \left(\underbrace{\frac{1}{n} \sum_{k=0}^{N} \ln |Q_{\mu}'(x_k)|}_{\approx 0} + \underbrace{\frac{1}{n} \sum_{k=N+1}^{n} \ln |Q_{\mu}'(x_k)|}_{\approx \ln |2 - \mu|} \right)$$

$$= \ln |2 - \mu|$$

Since the final expression is independent of the number x in $(0, 1)$, we conclude that $\lambda = \ln |2 - \mu|$. This completes the solution. ❏

Because $\ln |2 - \mu| < 0$ whenever $1 < \mu < 2$ or $2 < \mu < 3$, it follows that $\lambda < 0$ for such values of μ. As μ approaches 2, $\ln |2 - \mu|$ approaches $-\infty$. Consequently if we consider $-\infty$ as a legitimate value of λ, then we can conclude that $\lambda < 0$ for all μ in the interval $(1, 3)$.

Using the same kind of analysis as in the solution of Example 3, one can show that if $3 < \mu < 1 + \sqrt{6}$ and if x is not eventually periodic, then $\lambda(x) < 0$ (Exercise 7). It turns out that if $3 < \mu < \mu_\infty$, then $\lambda(x) < 0$ for all x that are not eventually periodic. However, as μ increases toward 4, λ oscillates more and more wildly between positive and negative values, as Figure 2.2 shows. Thus λ seems to have more positive values as μ increases, and Q_μ is increasingly sensitive to initial conditions. Finally, for $\mu = 4$ it is known that the Lyapunov exponent $\lambda(x) = \ln 2$ whenever $0 < x < 1$ and x is not eventually periodic.

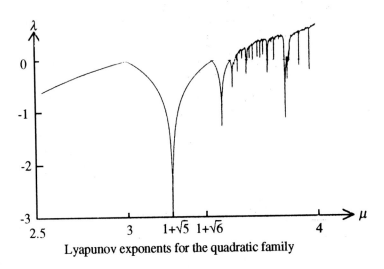

Lyapunov exponents for the quadratic family

Figure 2.2

The formula in (7) condenses considerably when x is eventually periodic. Indeed, if an iterate of x for a function f is eventually the fixed point p, then

$$\lambda(x) = \ln |f'(p)|$$

(Exercise 6). Similarly, if the iterates of x eventually join the 2-cycle $\{q, r\}$, then

$$\lambda(x) = \frac{1}{2} \ln |f'(q) f'(r)|$$

Chaos

The word "chaos" is familiar in everyday speech. It normally means a lack of order or predictability. Thus one says that the weather is chaotic, or that rising particles of smoke are chaotic, or that the stock market is chaotic. It is the lack of predictability that lies behind the mathematical notion of chaos. Both sensitive dependence on initial conditions and the Lyapunov exponent qualify as measures of unpredictability. Thus we have the following definition of chaos.

DEFINITION 2.3.   A function f is **chaotic** if it satisfies at least one of the following conditions:

    i. f has a positive Lyapunov exponent at each point in its domain that is not eventually periodic

or

    ii. f has sensitive dependence on initial conditions on its domain.

The term "chaos" in reference to functions was first used in Li and Yorke's paper "Period Three Implies Chaos" (1975). An essential portion of their theorem that the existence of period-3 points implies the existence of points with periods of all orders was that in such a case the function had a kind of sensitive dependence on initial conditions.

From Example 2 we see that the tent function T is chaotic. Similarly the function B in Example 1 is chaotic. In Section 2.3 we will prove that because T is chaotic, so is Q_4.

Notions of unpredictability in mathematics have been acknowledged for a long time. A century ago one of the outstanding unsolved problems was the following: Is the solar system stable? The problem was very well known, and even excited King Oscar II of Sweden, who in 1887 offered a prize of 2500 crowns to anyone who might correctly resolve it. At the turn of the century the great French mathematician Henri Poincaré tried to answer the simpler 3-body problem: Can one characterize the motion of a sun and two planets revolving about it? He concluded not only that there was no general solution to the 3-body problem, but also that minute differences in initial conditions for the three bodies could result in wildly divergent positions after a period of time. In other words, he concluded that the three bodies were sensitive to initial conditions. He wrote:

> If we knew exactly the laws of nature and the situation of the universe at the initial moment, we could predict exactly the situation of that same universe at a succeeding moment. But even if it were the case that the natural laws had no longer any secret for us, we could still know the situation approximately. If that enabled us to predict the succeeding situation with the same approximation, that is all we require, and we should say that the phenomenon had been predicted, that it is governed by the laws. But it is not always so; it may happen that small differences in the initial conditions produce very great ones in the final phenomena. A small error in the former will produce an enormous error in the latter. Prediction becomes impossible. (p. 321 of Gleick, 1987)

More recently, sensitive dependence on initial conditions, as well as the related positive Lyapunov exponent, have been observed in a wide variety of experiments. Below we discuss two such areas: weather prediction and the asteroid belt.

The Butterfly Effect

Nearly thirty years ago the American meteorologist Edward Lorenz published a paper "Deterministic Turbulent Flow," in which he concluded from computer simulations that weather patterns are sensitive to initial conditions (see Lorenz, 1963).   At that time the idea that ever so slight a variation in initial conditions could have profound repercussions on future weather was a startling discovery.   It gave rise to the phrase "butterfly effect," which in effect says that the weather is so sensitive to initial conditions that the mere flapping of butterfly wings in Rio de Janiero at one instant could (in theory) bring on a tornado in Texas several weeks later.   An example to illustrate the sensitivity to initial conditions in weather prediction appeared recently in an expository article by Tim Palmer (1989).

Examples of forecasts with nearly identical initial conditions were given in the article. Two of them are reproduced as the upper pictures in Figure 2.3. Notice that those pictures are nearly identical, reflecting virtually identical initial conditions for weather prediction. However, one week later the computer model shows weather patterns that are dramatically different, as shown in the lower pictures in Figure 2.3. This is a vivid example of sensitive dependence on initial conditions! From it we can understand why forecasters would have a difficult task in long-range weather prediction. In fact, it appears that we may never be able to predict weather far into the future. We will return to Lorenz's example relating to weather prediction in Section 5.4.

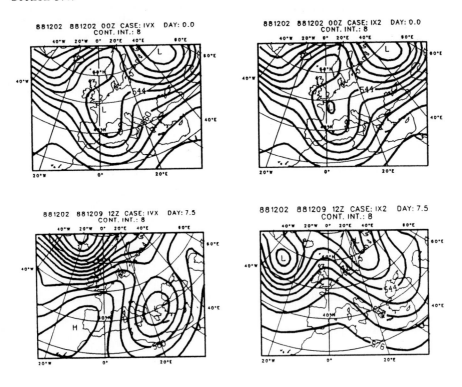

Figure 2.3

The Asteroid Belt

A quite different application of chaotic dynamics concerns the asteroid belt, which consists of several thousand objects that orbit around the sun and lie between Mars and Jupiter. The largest of the asteroids is Ceres, approximately 389 kilometers in radius. Most of the asteroids are much smaller, on the order of a kilometer in radius. Over four thousand asteroids have been catalogued.

Usually distances of such objects from the sun are given in terms of an **astronomical unit**, which by definition is the distance between the earth and the sun. Figure 2.4 is taken from an expository article of Carl Murray (1989), and shows the relative number of asteroids at various distances from the sun.

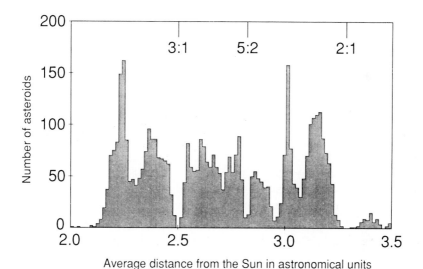

Figure 2.4

As you can see from the figure, the asteroid belt is not uniformly distributed. In 1867 the American astronomer, Daniel Kirkwood, observed gaps in the belt, now called **Kirkwood gaps**. In addition, he noted that those gaps correspond to asteroid orbits that are linked to the orbit of Jupiter. We say that an asteroid is in **resonance** with Jupiter if the ratio of the period of the asteroid to that of Jupiter is a simple fraction (such as 2:1 or 5:2 or 3:1). For example, if the ratio is 2:1, then the asteroid makes two revolutions around the sun every time Jupiter makes one (Figures 2.5(a)–(c)). When the sun, asteroid and Jupiter are aligned, then they are said to be in **conjunction**. Figures 2.5(a)–(d) show four possible conjunctions. It turns out that there is a maximum effect on the orbit of an asteroid if conjunction occurs when the asteroid is closest to Jupiter and simultaneously farthest from the sun, as is illustrated in Figure 2.5(d). The reason is that such an alignment gives a maximum

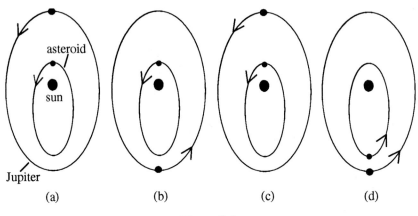

Figure 2.5

exchange of angular momentum between the asteroid and Jupiter, resulting in potential changes in the orbit, as well as the spin, of the asteroid.

For over a hundred years astronomers tried to understand why there were gaps in the asteroid belt. Finally in 1981 Jack Wisdom, then a graduate student at the California Institute of Technology, developed and demonstrated on a computer a theory that showed that asteroids moving in resonance with Jupiter could undergo large changes in their orbits that made them unpredictable and sensitive to initial conditions (see Wisdom, 1987). In particular, asteroids in a 3:1 resonance with Jupiter could receive such perturbations that their orbits could even cross the earth's orbit. This not only addressed the question of why there is a virtual absence of asteroids at resonances such as 3:1, 5:2, and 2:1 with Jupiter, as shown in Figure 2.4. This also finally explained how a special type of meteorite called chondrite, known to have come from the vicinity of the 3:1 resonance region, could have plunged into our atmosphere and struck earth. With Wisdom's theory, we now know why Kirkwood gaps exist in the asteroid belt. And it is comforting to know that only a relatively few asteroids remain in those gaps.

Conclusion

Chaotic motion is so common in natural and scientific phenomena that many scientists say that it should be considered to be the rule, rather than the exception, in the study of natural phenomena. Chaotic motion has been observed in such diverse areas as fluid dynamics, ecology, meteorology, optics, the dynamics of the heart and the brain, astrophysics, buckling beams, oceanography, and nonlinear electrical circuits. For this reason it is important to gain an understanding of chaotic motion.

EXERCISES 2.1

1.   Let $x = 1/3$, and let $y =$ the number that your calculator displays for 1/3.

Find the minimum positive integer n such that $|B^{[n]}(x) - B^{[n]}(y)| > 1/2$.

2. Show that the baker's function B has sensitive dependence.

3. a. Find the Lyapunov exponent associated with the baker's function B, and show that it is constant (where defined).
    b. Find the Lyapunov exponent of $B^{[2]}$.

4. Let $B_\mu(x) = \begin{cases} 2\mu x & \text{for } 0 \le x \le 1/2 \\ \mu(2x - 1) & \text{for } 1/2 < x \le 1 \end{cases}$, where $\mu > 0$.

    a. Determine the values of μ for which B_μ has sensitive dependence on initial conditions.
    b. Determine the Lyapunov exponent of B_μ.

5. Suppose that p is a period-n point of f such that $|(f^{[n]})'(p)| < 1$. Show that f cannot have sensitive dependence on initial conditions at p.

6. Suppose that f is differentiable on the interval J.
    a. Suppose that the iterates of x are eventually the fixed point p. Show that $\lambda(x) = \ln |f'(p)|$.
    b. Suppose that the iterates of x are eventually the cycle $\{p, q, r\}$. Find a formula for $\lambda(x)$.

7. Consider the quadratic function Q_μ, with $3 < \mu < 1 + \sqrt{6}$.
    a. Let x be in $(0, 1)$ but not eventually fixed. Show that if $\lambda(x)$ is finite, then

    $$\lambda(x) = \frac{1}{2} \ln |\mu^2 - 2\mu - 4| < 0$$

    b. Suppose that x is eventually fixed. Find $\lambda(x)$.

8. a. Find the value of $Q_4^{[n]}(1/2)$ for $n \ge 2$.
    b. Use the computer program ITERATE, with any necessary alterations, to find a value of y and a positive integer n such that

    $$\left| \frac{1}{2} - y \right| \le \frac{1}{100} \quad \text{and} \quad \left| Q_4^{[n]}(1/2) - Q_4^{[n]}(y) \right| > \frac{1}{2}$$

    (This should convince you that Q_4 has sensitive dependence at $1/2$.)

2.2 TRANSITIVITY AND STRONG CHAOS

The iterates of a fixed point do not wander at all; they remain the same point. At the other end of the spectrum are points whose iterates wander all over the domain of the function. Functions with such points are called transitive.

DEFINITION 2.4. Suppose that J is an interval and $f: J \to J$. Then f is **transitive** if for any pair of non-empty open intervals U and V that lie inside J, there is a positive integer n such that $f^{[n]}(U)$ and V have a common element.

It seems unlikely that we would be able to prove transitivity of a given function directly from the definition of transitivity. We need a manageable criterion for transitivity. The criterion we will present utilizes the notion of density of a subset of J.

A subset A of the interval J is **dense** in J if A intersects every non-empty open subinterval of J. Because every nonempty open interval contains rational numbers, it follows that the collection of rationals in the interval $[0, 1]$ is dense in $[0, 1]$ (Exercise 3). By contrast, the interval $[0, 1/2]$ is not dense in the interval $[0, 1]$ because $[0, 1/2]$ does not intersect open intervals like $(1/2, 1)$.

EXAMPLE 1. Show that the set P of periodic points of the tent function T is dense in $[0, 1]$.

Solution. Let U be an open subinterval in $[0, 1]$, with $U = (a, b)$. Let $d = b - a$, and let n be an odd integer such that $n > 2/d$. Since

$$\frac{k}{n} - \frac{k-1}{n} = \frac{1}{n} < \frac{d}{2}$$

and since U has length d, it follows that two successive numbers in the group $1/n, 2/n, ..., (n-2)/n$ lie in U. Of the two successive numbers, one of them must have the form (even integer)/(odd integer). By Theorem 1.15 such a number is periodic for T, so is in P. Therefore P is dense in $[0, 1]$. ❑

The definition of density leads to a criterion for transitivity.

THEOREM 2.5. Suppose that J is a closed interval and $f : J \to J$. Then f is transitive if and only if there is an x in J whose orbit is dense in J.

Partial Proof. We will prove that if there is a dense orbit, then f is transitive; the other half of the proof utilizes advanced mathematics and is omitted. Suppose that x has a dense orbit in J, and let U and V be arbitrary nonempty open subsets of J.

Because of the density of the orbit of x, there is a positive integer k such that $y = f^{[k]}(x)$ is in U. Using the density of the orbit of x a second time, we find a positive integer n such that $f^{[n]}(y) = f^{[k+n]}(x)$ is in V. Then y is in U and $f^{[n]}(y)$ is in V, which by the definition of transitivity implies that f is transitive.   ∎

Our next goal is to show that the tent function T is transitive. Even with the help of the criterion in Theorem 2.5, we have much work to accomplish before we can prove the transitivity of T in Theorem 2.9.

To begin the process, let us call a sequence of the form $x_0 x_1 x_2 \cdots \overline{0} \cdots$, in which all terms to the right of a given term are 0, a **finite sequence**. Let

A = the collection of all sequences of 0's and 1's that are not finite sequences

and define the function $h: [0, 1] \to A$ by

$$h(x) = \text{the sequence } x_0 x_1 x_2 \cdots, \text{ where } x_n = \begin{cases} 0 \text{ if } 0 \le T^{[n]}(x) \le 1/2 \\ 1 \text{ if } 1/2 < T^{[n]}(x) \le 1 \end{cases}$$

We will show that h defines a correspondence between the points of $[0, 1]$ and A.

The terms $x_0, x_1, x_2, \ldots$ of $h(x)$ indicate where in $[0, 1]$ the point x lies. In particular,

x_0 tells whether x lies in the left or right half of $[0, 1]$. Call the correct half J_0.
x_1 tells whether x lies in the left or right half of J_0. Call the correct half J_1.
x_2 tells whether x lies in the left or right half of J_1. Call the correct half J_2.

The sets $J_3, J_4, \ldots$ are defined analogously. It follows that $x_0 x_1$ indicates in which fourth of the interval $[0, 1]$ the point x lies, $x_0 x_1 x_2$ indicates in which eighth of the interval $[0, 1]$ the point lies, and so forth. In general, $x_0 x_1 x_2 \cdots x_n$ indicates in which subinterval of $[0, 1]$ of length $1/2^{n+1}$ the number x is located. Figure 2.6 gives the initial terms of the sequences with appropriate subintervals of $[0, 1]$.

EXAMPLE 2. Determine $h(3/11)$.

Solution. Let $x = 3/11$. We make the following table of values of $T^{[n]}(3/11)$ and corresponding x_n:

n	0	1	2	3	4	5	6
value of $T^{[n]}\left(\dfrac{3}{11}\right)$	$\dfrac{3}{11}$	$\dfrac{6}{11}$	$\dfrac{10}{11}$	$\dfrac{2}{11}$	$\dfrac{4}{11}$	$\dfrac{8}{11}$	$\dfrac{6}{11}$
x_n	0	1	1	0	0	1	1

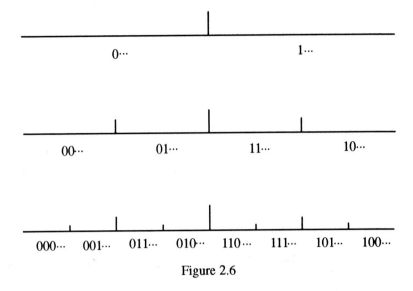

Figure 2.6

Since $T^{[6]}(3/11) = T(3/11)$, it follows that $3/11$ is eventually repeating, so that

$$h\left(\frac{3}{11}\right) = 01100\overline{11001}\cdots$$

where the digits with overbars indicate the group of digits that is repeated in the sequence $h(3/11)$. ❑

The sequence $h(3/11)$ is called a **repeating sequence** because of the group of digits 11001 that is repeated. By Theorem 1.14 every rational number x in the interval $[0, 1]$ is eventually periodic for T, which means that $h(x)$ is a repeating sequence the way $h(3/11)$ is.

By the definition of h,

$$\text{if } h(x) = x_0 x_1 x_2 \cdots, \text{ then } h(T(x)) = x_1 x_2 x_3 \cdots \tag{1}$$

By induction we deduce that for any positive integer,

$$\text{if } h(x) = x_0 x_1 x_2 \cdots, \text{ then } h(T^{[n]}(x)) = x_n x_{n+1} x_{n+2} \cdots \tag{2}$$

The next step in proving that T is transitive involves showing that h is **one-to-one**, that is, showing that if $h(x) = h(y)$, then $x = y$.

THEOREM 2.6. h is one-to-one.

Proof. Let

$$h(x) = x_0x_1x_2\cdots \quad \text{and} \quad h(y) = y_0y_1y_2\cdots$$

and suppose that $h(x) = h(y)$. We will prove that $x = y$. The hypothesis that $h(x) = h(y)$ means that $x_k = y_k$ for $k = 0, 1, 2, \ldots$. Consequently for any positive integer n we have $x_0x_1x_2\cdots x_n = y_0y_1y_2\cdots y_n$, so that x and y lie in the same subinterval of width $1/2^{n+1}$. Therefore $|x - y| \le 1/2^{n+1}$. Since n is arbitrary, we conclude that $|x - y| = 0$, that is, $x = y$. ∎

Although Theorem 2.6 associates every number in $[0, 1]$ with a sequence of 0's and 1's, it is *not* true that, conversely, every sequence of 0's and 1's corresponds to a number in $[0, 1]$ by means of h. (See Exercise 5.) The reason is that h is not defined symmetrically on $[0, 1/2]$ and on $[1/2, 1]$: the term x_n in the sequence $x_0x_1x_2\cdots x_n$ is 0 if $T^{[n]}(x)$ is in the *closed* interval $[0, 1/2]$, whereas x_n is 1 if $T^{[n]}(x)$ is in the *half-open* interval $(1/2, 1]$.

Despite the fact that not every sequence of 0's and 1's correponds to a number in $[0, 1]$, we will prove that any sequence that is not a finite sequence corresponds to a number in $[0, 1]$. But to prove this we need a simplified version of the famous theorem due to two nineteenth century European mathematicians, Eduard Heine and Emile Borel.

THEOREM 2.7   (Heine-Borel Theorem). Suppose that $B_0, B_1, B_2, \ldots$ is a sequence of closed, bounded intervals of reals such that $B_n \supseteq B_{n+1}$ for all n. Then there is a point common to *all* the B_n's.

The proof utilizes notions usually studied in advanced analysis, so we omit it. (A proof of a version of Theorem 2.7 can be found in the book by Kenneth Ross, 1980.) The sequence $B_0, B_1, B_2, \ldots$ appearing in the Heine-Borel Theorem is called a **nested sequence**, since B_{n+1} is "nested" inside B_n for each n. The theorem says that a nested sequence of closed, bounded intervals of reals has at least one common point. Thus if

$$J_n = [\frac{1}{2} - \frac{1}{2n} , \frac{1}{2} + \frac{1}{2n}], \text{ for } n = 0, 1, 2, \ldots$$

then each interval is closed and bounded, and the sequence of intervals is nested. Consequently the Heine-Borel Theorem indicates that there is a point common to all of the J_n's. Of course that point is $1/2$! We emphasize that a nested sequence of sets can contain infinitely many common points. However, if the sequence does not consist of closed intervals or does not consist of bounded intervals, then the intersection may well be empty. (See Exercises 1 and 2.)

The next result says that h is **onto** A, which means that if a is any element of A, then there is an x in $[0, 1]$ such that $h(x) = a$. In proving the result, we will use the notation $\{x \text{ in } [0, 1]: x_0 = z_0\}$, which is to be read "the set of all x in $[0, 1]$ such that $x_0 = z_0$."

THEOREM 2.8. h maps $[0, 1]$ onto A .

Proof. Let a be in A , and let $a = \{z_n\}_{n=0}^{\infty}$. We will show that $h(x) = a$. Next, let J_0 be the smallest closed interval containing $\{x$ in $[0, 1]: x_0 = z_0\}$, so that J_0 is either the interval $[0, 1/2]$ or the interval $[1/2, 1]$. In addition, let J_1 be the smallest closed interval containing $\{x$ in $[0, 1]: x_0 x_1 = z_0 z_1\}$, so that J_1 is $[0, 1/4]$, or $[1/4, 1/2]$, or $[1/2, 3/4]$, or $[3/4, 1]$. Then $J_0 \supseteq J_1$. In general, let

J_n be the smallest closed interval containing $\{x$ in $[0, 1]: x_0 x_1 x_2 \cdots x_n = z_0 z_1 z_2 \cdots z_n\}$

Then J_n is a closed, bounded subinterval of $[0, 1]$ with length $1/2^{n+1}$, and for all n we have $J_n \supseteq J_{n+1}$. By the Heine-Borel Theorem there is a number x common to all J_n . In addition, x is unique because the length of each J_n is $1/2^{n+1}$. Since x is in J_n for each n , it follows that $x_0 x_1 x_2 \cdots x_n = z_0 z_1 z_2 \cdots z_n$ for each n , and thus $h(x) = a$. ∎

Finally we are ready for the promised theorem about the transitivity of T .

THEOREM 2.9. T is transitive.

Proof. Consider the sequence

$$s \;=\; 0\,1 \qquad 00\;01\;10\;11 \qquad 000\;001\;010\;011\;100\;101\;110\;111 \;\cdots$$

$$|\,1\text{ block}\,|\quad 2\text{ block}\quad|\qquad\qquad 3\text{ block}\qquad\qquad\qquad|$$

which is composed of blocks consisting of all singles, all pairs, all triples, etc., of 0's and 1's. Because s is *not* a finite sequence, by Theorem 2.8 there is an x_t in $[0, 1]$ such that $h(x_t) = s$. Thus

$$h(x_t) \;=\; 0\,1\;\;00\;01\;10\;11\;\;000\;001\;010\;011\;100\;101\;110\;111\;\cdots$$

Using the blocks of 0's and 1's in the formula for $h(x_t)$, along with (1) and the definitions of h and T , we find that

x_t is in $[0, 1/2]$	$T(x_t)$ is in $[1/2, 1]$	$T^{[2]}(x_t)$ is in $[0, 1/4]$
$T^{[4]}(x_t)$ is in $[1/4, 1/2]$	$T^{[6]}(x_t)$ is in $[3/4, 1]$	$T^{[8]}(x_t)$ is in $[1/2, 3/4]$

and so forth. In general, for any interval of the form

$$L \;=\; [\,\frac{k}{2^n}, \frac{k+1}{2^n}\,]$$

there is a positive integer m such that $T^{[m]}(x_t)$ is in L. Thus the iterates of x_t are dense in $[0, 1]$, so that T is transitive by Theorem 2.5. ∎

Proving that T is transitive by means of dense orbits was made possible by means of the use of sequences of the "symbols" 0 and 1. Employing any other pair of symbols, such as ♦ and ♥, to define the sequences would have worked in the same manner. The use of such sequences in analyzing the dynamics of functions is referred to as **symbolic dynamics**. Thus we have proved that the tent function T is transitive by employing symbolic dynamics.

From Theorem 1.14 we know that the rationals in $[0, 1]$ are the numbers that are eventually periodic under the function T. However, eventually periodic points cannot have dense orbits, because they entail but a finite number of points. Consequently the number x_t that arose in the preceding proof is irrational. Given our previous results, one might expect that each irrational number in $[0, 1]$ would have a dense orbit for T. However, this is not true. (See Exercise 7.)

It would appear that one might be able to prove the transitivity of T directly, without employing symbolic dynamics. In fact, were there infinitely many numbers of the form $2/n$, where n and $(n-1)/2$ are prime and the orbit of $2/n$ is spread evenly throughout the interval $[0, 1]$, then we could prove directly that T is transitive. (See Exercise 13.) However we are not able to prove that there are infinitely many such numbers, so must leave it as a conjecture. The conjecture is apparently intimately related to several famous conjectures discussed in the delightful book *Solved and Unsolved Problems in Number Theory*, by Daniel Shanks (1978).

Strong Chaos

The function T exhibits three important features:

    i. T is chaotic (by Example 2 of Section 2.1)
   ii. T has a dense set of periodic points (by Example 1)
  iii. T is transitive (by Theorem 2.9)

In addition to the unpredictability of orbits that follows from (i), T has a certain regularity because of its dense set of periodic points. Finally, the transitivity tells us that the iterates of certain points spread themselves throughout the domain of T. Thus T is more than chaotic, and we say that it is strongly chaotic.

DEFINITION 2.10. A function f on an interval J is **strongly chaotic** if

    i. f is chaotic
   ii. f has a dense set of periodic points
  iii. f is transitive

By our comments above, T is strongly chaotic. As we indicated when we defined sensitive dependence and Lyapunov exponents in Section 2.1, it is not so easy to prove that Q_4 is chaotic. The same is true for a proof that Q_4 has a dense set of periodic points, or that it is transitive. By the results of Section 2.3, the fact that T is strongly chaotic will imply that Q_4 is also strongly chaotic.

EXERCISES 2.2

1.   Find a nested sequence $\{B_n\}_{n=1}^{\infty}$ of closed intervals whose intersection is void.

2.   Find a nested sequence $\{B_n\}_{n=1}^{\infty}$ of bounded intervals whose intersection is void.

3.   a.   Show that the set of rationals in $[0, 1]$ is dense in $[0, 1]$.
     b.   Show that the set of dyadic rationals in $[0, 1]$ is dense in $[0, 1]$.

Exercises 4–8 relate to the tent function T.

4.   a.   Find the number x in $[0, 1]$ such that $h(x) = \overline{1}\cdots$.
     b.   Find the number x in $[0, 1]$ such that $h(x) = \overline{10}\cdots$.

5.   a.   Show that $11\overline{0}\cdots$ is not the image under h of any x in $[0, 1]$.
     b.   Show that any sequence of the form $x_0 x_1 x_2 \cdots x_n 11\overline{0}\cdots$ is not the image under h of any x in $[0, 1]$.

6.   Let D consist of all irrationals in $[0, 1]$ with dense orbits for T. Show that D is dense in $[0, 1]$.

7.   a.   Find an irrational number x in $[0, 1]$ that does not have a dense orbit.
     b.   Let D^* consist of all irrationals in $[0, 1]$ whose orbits are not dense in $[0, 1]$. Show that D^* is dense in $[0, 1]$.

8.   Let $x_0 x_1 x_2 \cdots$ be a sequence of 0's and 1's. For each positive integer n, let
     $g(x_0, x_1, x_2, \ldots, x_n) = (x_0 + x_1 + x_2 + \cdots + x_n) \pmod 2$.
     a.   Show that if $h(x) = x_0 x_1 x_2 \cdots$, then

$$x = \frac{g(x_0)}{2} + \frac{g(x_0 x_1)}{2^2} + \frac{g(x_0 x_1 x_2)}{2^3} + \cdots$$

     b.   Use part (a) to find an approximation to within .01 of the irrational number x_t with dense orbit and appears in the proof of Theorem 2.9.

9. Let $x = .28$ and $z = .63$. Use the computer program NUMBER OF ITERATES in order to determine an integer n such that $|Q_4(x) - z| < .001$.

10. Let $x = .45$ and $z = .99$. Use the computer program NUMBER OF ITERATES in order to determine an integer n such that $|T(x) - z| < .001$.

11. Show that the baker's function B is strongly chaotic.

12. Let $f(x) = \begin{cases} 3x & \text{for } 0 \le x \le 1/3 \\ 3x - 1 & \text{for } 1/3 < x \le 2/3 \\ 3x - 2 & \text{for } 2/3 < x \le 1 \end{cases}$

    a. Let A_0 denote the collection of all sequences of 0's, 1's and 2's. Define a function h_0: $[0, 1] \to A_0$ in the same spirit as the function h defined in this section.
    b. Show that there is a dense set of periodic points in $[0, 1]$.
    c. Show that f is strongly chaotic.

13. Suppose that both n and $(n - 1)/2$ are prime numbers, with $n > 2$.
    a. Find the elements in the orbit of $2/n$.
    b. Suppose that there are infinitely many prime numbers n with the property that $(n - 1)/2$ is a prime number. Then show directly that the tent function T is transitive.

2.3   CONJUGACY

Among the many features of T already discussed are transitivity and the existence of a dense set of periodic points. We might ask whether these features apply to the quadratic function Q_4. In the present section we will define the concept of conjugacy. We will show that if two functions are conjugate to one another, then one function inherits such properties as transitivity and the existence of a dense set of periodic points from the other function. In particular, this information will apply to T and Q_4, and hence will yield valuable information about Q_4.

Before we define the notion of conjugacy we make an important definition.

DEFINITION 2.11. Let J and K be intervals. The function $f\colon J \to K$ is a **homeomorphism of J onto K** provided that h is one-to-one and onto, and provided that both f and f^{-1} are continuous.

Recall from calculus that if $f\colon J \to K$ is a continuous, one-to-one function that is onto K, then $f^{-1}\colon K \to J$ is a continuous function, so that f is automatically a homeomorphism. As a result, if $f(x) = x^2$ for $0 \leq x \leq 2$, then f is a homeomorphism of [0, 2] onto [0, 4] (Figure 2.7(a)). Similarly, if $g(x) =$ arctan x, then g is a homeomorphism of $(-\infty, \infty)$ onto $(-\pi/2, \pi/2)$ (Figure 2.7(b)). Finally, if $h(x) = \sin^2(\pi x/2)$ for $0 \leq x \leq 1$, then h is a homeomorphism of [0, 1] onto itself (Figure 2.7(c)). We will use h when we discuss the relationship between T and Q_4 later in this section.

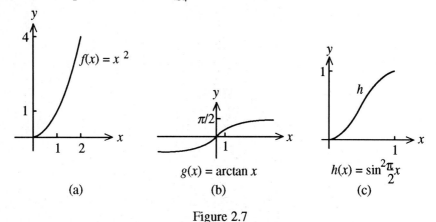

Figure 2.7

We remark that if f is a homeomorphism from J onto K, then f^{-1} is automatically a homeomorphism from K onto J. For example, since the function h defined above is a homeomorphism, it follows that h^{-1}, defined by

$$h^{-1}(x) = \frac{2}{\pi} \arcsin \sqrt{x}$$

is also a homeomorphism.

Suppose that $f: J \rightarrow K$ is a homeomorphism. It is an important consequence of the concept of homeomorphism that f maps open (closed) subintervals interior to J onto open (closed) subintervals interior to K. Moreover, if A is a dense subset of J, then $f(A)$ is a dense subset of K (Exercise 12).

We are ready to define the notion of conjugacy.

DEFINITION 2.12. Let J and K be intervals, and suppose that $f: J \rightarrow J$ and $g: K \rightarrow K$. Then f and g are **conjugate** (to one another) if there is a homeomorphism $h: J \rightarrow K$ such that $h \circ f = g \circ h$. In this case, we write $f \underset{h}{\approx} g$.

Using the fact that h^{-1} is a homeomorphism whenever h is, one can prove that if $f \underset{h}{\approx} g$, then $g \underset{h^{-1}}{\approx} f$ (Exercise 11). Thus g is conjugate to f whenever f is conjugate to g (although the homeomorphisms h and h^{-1} are normally quite different from one another!).

EXAMPLE 1. Let

$$g(x) = x^2 - \frac{3}{4} \text{ for } -\frac{3}{2} \le x \le \frac{3}{2} \text{ and } Q(x) = 3x(1-x) \text{ for } 0 \le x \le 1$$

Show that $g \underset{h}{\approx} Q$, where $h(x) = -\frac{1}{3}x + \frac{1}{2}$.

Solution. We will verify that $h \circ g = Q \circ h$. A routine check verifies that h is a homeomorphism from $[-3/2, 3/2]$ onto $[0, 1]$, and that

$$(h \circ g)(x) = h(g(x)) = h\left(x^2 - \frac{3}{4}\right) = -\frac{1}{3}\left(x^2 - \frac{3}{4}\right) + \frac{1}{2} = -\frac{1}{3}x^2 + \frac{3}{4}$$

and

$$(Q \circ h)(x) = Q(h(x)) = Q\left(-\frac{1}{3}x + \frac{1}{2}\right) = 3\left(-\frac{1}{3}x + \frac{1}{2}\right)\left(1 - \left(-\frac{1}{3}x + \frac{1}{2}\right)\right)$$

$$= -\frac{1}{3}x^2 + \frac{3}{4}$$

Therefore $h \circ g = Q \circ h$. ❑

You might recognize the functions g and Q in Example 1 as $g_{-3/4}$ and Q_3, respectively, from the parametrized families $\{g_\mu\}$ and $\{Q_\mu\}$.

EXAMPLE 2. Let

$$Q_4(x) = 4x(1-x) \text{ for } 0 \le x \le 1 \text{ and } T(x) = \begin{cases} 2x & \text{for } 0 \le x \le 1/2 \\ 2(1-x) & \text{for } 1/2 < x \le 1 \end{cases}$$

Show that $T \underset{h}{\approx} Q_4$, where $h(x) = \sin^2 \frac{\pi}{2} x$.

Solution. As mentioned earlier, h is a homeomorphism from $[0, 1]$ onto $[0, 1]$. On the one hand we observe that

$$(Q_4 \circ h)(x) = Q_4(h(x)) = Q_4\left(\sin^2 \frac{\pi}{2} x\right) = 4\left(\sin^2 \frac{\pi}{2} x\right)\left(1 - \sin^2 \frac{\pi}{2} x\right)$$

$$= 4\left(\sin^2 \frac{\pi}{2} x\right)\left(\cos^2 \frac{\pi}{2} x\right) = \left(2 \sin \frac{\pi}{2} x \cos \frac{\pi}{2} x\right)^2 = \sin^2 \pi x$$

where the last equality results from a trigonometric double-angle formula. On the other hand, we find that

if $0 \le x \le \dfrac{1}{2}$, then $(h \circ T)(x) = h(2x) = \sin^2 \pi x$

if $\dfrac{1}{2} < x \le 1$, then $(h \circ T)(x) = h(2 - 2x) = \left(\sin \frac{\pi}{2} (2 - 2x)\right)^2$

$$= [\sin (\pi - \pi x)]^2 = \sin^2 \pi x$$

Consequently $(h \circ T)(x) = (Q_4 \circ h)(x)$ for $0 \le x \le 1$, so that $h \circ T = Q_4 \circ h$. ❑

Example 2 shows that the tent function T and the quadratic function Q_4 are conjugates. This conjugacy will play a decisive role in our proof that Q_4 is strongly chaotic.

Periodic points are inherited through conjugacy, as Theorem 2.13 tells us.

THEOREM 2.13. Suppose that $f \underset{h}{\approx} g$. Then

    i. $h \circ f^{[n]} = g^{[n]} \circ h$ for $n = 1, 2, \dots$.
    ii.   If x^* is a period-n point of f, then $h(x^*)$ is a period-n point of g.
    iii.   If f has a dense set of periodic points, then so does g.

Proof. Since $f \underset{h}{\approx} g$, we know that $h \circ f = g \circ h$. For the purposes of an induction proof, let us assume that $h \circ f^{[n-1]} = g^{[n-1]} \circ h$. Then

$$h \circ f^{[n]} = (h \circ f) \circ f^{[n-1]} = (g \circ h) \circ f^{[n-1]} = g \circ (h \circ f^{[n-1]}) = g \circ (g^{[n-1]} \circ h) = g^{[n]} \circ h$$

so that

$$h \circ f^{[n]} = g^{[n]} \circ h \quad \text{for} \quad n = 1, 2, \dots$$

By the Law of Induction, (i) is proved. To prove (ii), assume that x^* is a period-n point of f. Together with the fact that $f^{[n]}(x^*) = x^*$, (i) implies that

$$g^{[n]}(h(x^*)) = (g^{[n]} \circ h)(x^*) = (h \circ f^{[n]})(x^*) = h(f^{[n]}(x^*)) = h(x^*)$$

Consequently $h(x^*)$ is a period-n point of g, which proves (ii). Finally, the image $f(A)$ of the dense set A of periodic points of f contains only periodic points of g by (ii), and is dense in the range of g by a remark following the definition of homeomorphism. ∎

Theorem 2.13 has immediate application to the periodic points of Q_4.

EXAMPLE 3. Find a 3-cycle for Q_4.

Solution. A simple calculation shows that $\{2/7, 4/7, 6/7\}$ is a 3-cycle for T. By Example 2, $T \underset{h}{\approx} Q_4$, where $h(x) = \sin^2 \dfrac{\pi}{2} x$. Therefore by Theorem 2.13,

$$\{\sin^2 \frac{\pi}{7}, \ \sin^2 \frac{2\pi}{7}, \ \sin^2 \frac{3\pi}{7}\}$$

is a 3-cycle for Q_4. By calculator we find that the 3-cycle is approximately

$$\{.1882550991, \ .611260467, \ .950484434\} \quad \square$$

As you can see, it would be hard to guess this 3-cycle of Q_4 by trial and error. Our next result shows that transitivity is inherited through conjugacy.

THEOREM 2.14.　Assume that $f \underset{h}{\approx} g$, and that $f: J \to K$ is transitive. Then g is also transitive.

Proof. Recall from Theorem 2.5 that a function is transitive if and only if it has a dense orbit. To prove our result, we suppose that the orbit of x for f is dense in the domain J. We will show that the orbit of $h(x)$ for g is dense in K. To that end, let U be a nonempty open subinterval of K. Because h is a homeomorphism, it follows that $h^{-1}(U)$ is an open subinterval of J. Since the orbit of x is dense in J, there is a positive integer n such that $f^{[n]}(x)$ is in $h^{-1}(U)$, so $h(f^{[n]}(x))$ is in $h(h^{-1}(U)) = U$. By (i) of Theorem 2.13,

$$h(f^{[n]}(x)) = (h \circ f^{[n]})(x) = (g^{[n]} \circ h)(x) = g^{[n]}(h(x))$$

Therefore $g^{[n]}(h(x))$ is in U. Since U is an arbitrary open interval in K, we have succeeded in proving that the orbit of $h(x)$ for g is dense in K, so that by Theorem 2.5, g is transitive. ∎

We are now equipped to prove that Q_4 is strongly chaotic.

THEOREM 2.15. The function Q_4 is strongly chaotic.

Proof. The function Q_4 has sensitive dependence on initial conditions (mentioned following Definition 2.3) and a positive Lyapunov exponent (mentioned after Example 3 of Section 2.1). Either way, Q_4 is chaotic. Since T is transitive and has a dense set of periodic points, the same properties hold for Q_4, because of the conjugacy of T and Q_4, and because of Theorems 2.13 and 2.14. These observations together imply that Q_4 is strongly chaotic. ∎

In our discussion of $f \underset{h}{\approx} g$, we have previously assumed that h is a homeomorphism. There are important cases in which the homeomorphism is also a linear function. We say that f and g are **linearly conjugate** if there is a linear homeomorphism h such that $f \underset{h}{\approx} g$.

THEOREM 2.16. Let

$$f(x) = ax^2 + bx + c \quad \text{and} \quad g(x) = rx^2 + sx + t$$

where $a \neq 0$ and $r \neq 0$, and where

$$c = \frac{b^2 - s^2 + 2s - 2b + 4rt}{4a} \tag{1}$$

Then f and g are linearly conjugate to one another, with associated homeomorphism given by

$$h(x) = \frac{a}{r}x + \frac{b - s}{2r} \tag{2}$$

Proof. Let $h(x) = dx + e$. We will prove that there are constants $d \neq 0$ and e such that $h \circ f = g \circ h$. Now $(h \circ f)(x) = (g \circ h)(x)$ if $h(f(x)) = g(h(x))$, that is, if

$$d(ax^2 + bx + c) + e = r(dx + e)^2 + s(dx + e) + t$$

or equivalently, if

$$dax^2 + dbx + (dc + e) = rd^2x^2 + (2rde + sd)x + (re^2 + se + t)$$

Collecting coefficients of like powers of x, and using the hypothesis that $r \neq 0$, we find that

x^2 terms: $da = rd^2$, so that if $d \neq 0$, then $d = \dfrac{a}{r}$

x terms: $db = 2rde + sd$, so that if $d \neq 0$, then $e = \dfrac{b - s}{2r}$

constant terms: $\quad dc + e = re^2 + se + t$

Substituting for d and e in the last equation, we obtain

$$\frac{a}{r}c + \frac{b-s}{2r} = r\left(\frac{b-s}{2r}\right)^2 + s\frac{b-s}{2r} + t$$

which yields

$$4ac + 2b - 2s = b^2 - 2bs + s^2 + 2bs - 2s^2 + 4rt$$

Solving for c and using the fact that $a \neq 0$, we conclude that

$$c = \frac{b^2 - s^2 + 2s - 2b + 4rt}{4a}$$

which is (1). We conclude that for this value of c, f and g are linearly conjugate, with h as given in (2). ∎

In effect, Theorem 2.16 says that each quadratic function g is linearly conjugate to a suitable vertical shift of any other quadratic function f. The shift of f must be such that the constant term c of f satisfies (1).

EXAMPLE 4. Let $g_c(x) = x^2 + c$ and $Q_\mu(x) = \mu x(1 - x)$, with $0 < \mu < 4$. Find c and h so that $g_c \underset{h}{\approx} Q_\mu$ and h is linear.

Solution. We use Theorem 2.16 with the following substitutions:

$$a = 1, \quad b = 0, \quad c = c, \quad r = -\mu, \quad s = \mu, \quad t = 0$$

Using (1), we find that

$$c = \frac{b^2 - s^2 + 2s - 2b + 4rt}{4a} = -\frac{\mu^2}{4} + \frac{\mu}{2}$$

Then $g_c \underset{h}{\approx} Q_\mu$, where by (2), h satisfies

$$h(x) = \frac{a}{r}x + \frac{b-s}{2r} = -\frac{1}{\mu}x + \frac{1}{2}$$

This completes the solution. ❑

In a similar fashion, one can show that the following are pairwise conjugate (with appropriate domains and values of μ, r, and c):

$$Q_\mu(x) = \mu x(1-x), \quad f_\mu(x) = (2-\mu)x - \mu x^2, \quad K_c(x) = c - x^2, \quad \text{and} \quad F_C(x) = 1 - Cx^2$$

EXERCISES 2.3

1.   Let $f_\mu(x) = (2-\mu)x - \mu x^2$ for x in some interval L. Determine L so that $f_\mu \underset{h}{\approx} Q_\mu$. (Notice that μ is the same for both functions.)

2.   Let $K_c(x) = c - x^2$ for x in an interval L. Find L and c so that Q_μ and K_c are linearly conjugate.

3.   Let $F_C(x) = 1 - Cx^2$ for all real x.
     a.   Assume that Q_μ is defined for all real x. Show that for a given μ, there is a constant C such that F_C and Q_μ are linearly conjugate.
     b.   By altering the computer program BIFURCATION, show that F_C has a 3-cycle if $C > 7/4$, and does not appear to have a 3-cycle if $C < 7/4$. (Thus $7/4$ is a special bifurcation point of F_C.)
     c.   Use (b) to show that Q_μ has a 3-cycle if $\mu > 1 + 2\sqrt{2} \approx 3.828427...$.

4.   Show that for any given c, there is an appropriate C such that the functions K_c and F_C appearing in Exercises 2 and 3 are linearly conjugate.

5.   Using the fact that Q_4 and T are conjugates, approximate a period-n point of Q_4, where
     a. $n = 4$ b. $n = 5$

6.   Let $f_\mu(x) = (2-\mu)x - \mu x^2$, for $0 \le x \le 1$, where $3 < \mu < 3.25$. Use Exercise 1 to show that a 2-cycle exists for f_μ, and find it.

7.   Let $f(x) = x^3$, and let $g_\mu(x) = x - \mu x^3$, where μ is a positive constant. Show that there is no nontrivial polynomial h such that $f \underset{h}{\approx} g_\mu$.

8.   Is there a linear function h on $[0, 1]$ such that Q_4 is linearly conjugate to the tent function T? Explain why or why not.

9.   Suppose that $0 < \mu < 4$. Does there exist a λ in $(0, 1)$ such that Q_μ and T_λ are linearly conjugate? Explain why or why not.

10. Suppose that $f \underset{h}{\approx} g$ and $g \underset{H}{\approx} k$, where h and H are linear. Show that f and k are linearly conjugate.

11.   Suppose that $f \underset{h}{\approx} g$. Show that $g \underset{h^{-1}}{\approx} f$.

12.   Let $f: J \to K$ be a homeomorphism.
       a.   Let U be an open interval interior to J. Prove that $f(U)$ is an open interval interior to K.
       b.   Prove that if A is dense in J, then $f(A)$ is dense in K.

13.   Let $\{f_\mu\}$ be a parametrized family of functions, and assume that x_μ is a fixed point of f_μ, for all μ. Define a parametrized family $\{g_\mu\}$ such that for each μ, $f_\mu \approx g_\mu$ and 0 is a fixed point of g_μ.

14.   Let $g: [0, 1] \to [0, 1]$ be continuous, and let $f(x) = \dfrac{1}{3} g(3x)$ for $0 \le x \le 1/3$.
       a.   Find a linear function h such that $f \underset{h}{\approx} g$.
       b.   Let F be defined by

$$F(x) = \begin{cases} 2/3 + f(x) \text{ for } 0 \le x \le 1/3 \\ [f(1/3)](2 - 3x) \text{ for } 1/3 < x \le 2/3 \\ x - 2/3 \text{ for } 2/3 < x \le 1 \end{cases}$$

(Figure 2.8). Show that F has one fixed point, and that all other periodic points of F have double the periods of the corresponding periodic points of g. (*Hint*: Use part (a) to show that $f \underset{h}{\approx} g$, and then show that $F^{[2n]}(x) = f^{[n]}(x)$ for all x in $[0, 1/3]$.)

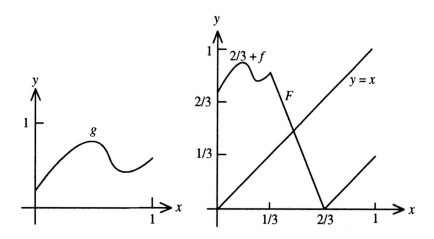

Figure 2.8

2.4   CANTOR SETS

In Sections 1.5 and 2.3 we studied the functions Q_μ for $0 < \mu \leq 4$. Now in this final section of Chapter 2 we will study the behavior of Q_μ for $\mu > 4$. The first substantial difference we notice is that if $\mu > 4$, then the range of Q_μ is not totally contained in $[0, 1]$ (Figure 2.9(a)).   In our study we will introduce several kinds of sets, such as closed, perfect, uncountable, and Cantor sets. As a result, the section is more technical and theoretical than those that have preceded it.

Throughout the section we will assume that $\mu > 4$.   The graph of Q_μ extends above the line $y = 1$, so that in order to be able to describe the iterates of all x in the domain of Q_μ, we need to enlarge the domain. This we do by taking the domain of Q_μ to be $(-\infty, \infty)$ whenever $\mu > 4$ (Figure 2.9(b)). Then surely the range of Q_μ is contained in the domain! If $x < 0$, then

$$Q_\mu(x) = \mu x(1 - x) < x < 0 \tag{1}$$

so that $\{Q_\mu^{[n]}(x)\}_{n=1}^\infty$ is a negative, decreasing sequence. If it converged, its limit would need to be a fixed point of Q_μ by Corollary 1.4. However, by (1) there are no negative fixed points, so that we conclude that

$$\lim_{n \to \infty} Q_\mu^{[n]}(x) = -\infty \text{ for all } x < 0 \tag{2}$$

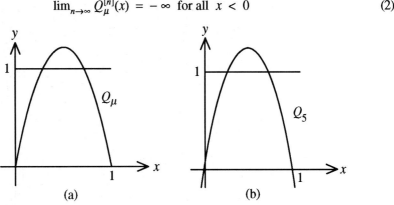

Figure 2.9

If $Q_\mu(x) > 1$, then $Q_\mu^{[2]}(x) = Q_\mu(Q_\mu(x)) < 0$, so that by (2), $\lim_{n \to \infty} Q_\mu^{[n]}(x) = -\infty$.   Thus if any iterate of x is greater than 1, then the successive iterates approach $-\infty$. Figures 2.10(a)–(b) show the behavior of $Q_\mu^{[2]}$ and $Q_\mu^{[3]}$ when $\mu = 5$, and give an indication of the unboundedness of iterates.

Although many iterates converge to $-\infty$, it is those x in $[0, 1]$ whose iterates remain in $[0, 1]$ that we will study. The set of numbers in $[0, 1]$ for which $0 \leq Q_\mu(x) \leq 1$ consists of two closed subintervals which we will call J_{11} and J_{12} (Figure 2.11(a)). The set on which $0 \leq Q_\mu^{[2]} \leq 1$ consists of the four closed

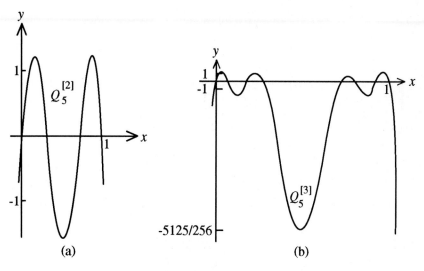

Figure 2.10

subintervals J_{21}, J_{22}, J_{23}, and J_{24} (Figure 2.11(b)). In general, at the nth stage the set on which $0 \leq Q_\mu^{[n]} \leq 1$ consists of the 2^n closed subintervals J_{n1}, J_{n2}, ..., J_{n2^n}. We observe that $J_{nk} \supseteq J_{(n+1)m}$ for all n, and all k and m in $[1, 2^n]$. Next, let

$$K_n = J_{n1} \cup J_{n2} \cup \cdots \cup J_{n2^n}$$

Therefore K_n is the collection of all numbers in J_{nk} for various k. Alternatively, K_n consists of all points in $[0, 1]$ whose iterates through the nth iterate lie in $[0, 1]$.

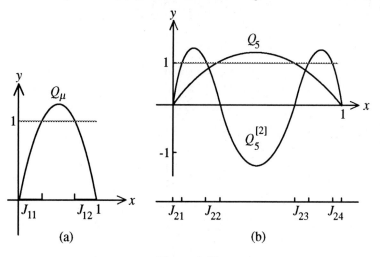

Figure 2.11

Moreover, $K_n \supseteq K_{n+1}$ for all n . Finally, let

C_μ = the collection of all points in $[0, 1]$ that are in each K_n , for $n = 1, 2, \ldots$

By its definition, C_μ consists of those points in $[0, 1]$ *all* of whose iterates lie in $[0, 1]$. Does C_μ contain any points? It does, for it contains not only the fixed points 0 and $1 - 1/\mu$, but all periodic points of Q_μ (Exercise 9). We will discover later in the section that C_μ contains points that are not periodic.

The Cantor Ternary Set

You may have noticed that the set C_μ is described by the same process, for every value of $\mu > 4$. In order to study the special features of the set C_μ , we will focus on one single set C in the interval $[0, 1]$ that not only is described in a way analogous to the description of the sets C_μ , but whose defining subintervals all have the same size at each stage.

Specifically, we define C by first selecting the two closed subintervals P_{11} and P_{12} of length $1/3$ shown in Figure 2.12, then selecting the four closed subintervals P_{21} , P_{22} , P_{23} , and P_{24} of length $1/3^2$ (Figure 2.12). The process continues indefinitely (as with the construction of C_μ). At the *n*th stage we select 2^n closed subintervals P_{n1} , P_{n2} ..., P_{n2^n} of length $1/3^n$, two each from each of the subintervals in the $(n-1)$ st stage. If

$$L_n = P_{n1} \cup P_{n2} \cup \cdots \cup P_{n2^n} \tag{3}$$

then we define C as follows:

C = the collection of all points in $[0, 1]$ that are in each L_n , for $n = 1, 2, \ldots$

The set C is called the **Cantor ternary set**. The word "ternary" means third, and indicates that at each stage the middle third of each remaining open interval is deleted. The name Cantor honors the Russian-born German mathematician Georg Cantor; he

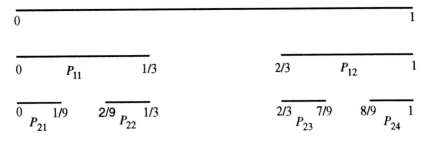

Figure 2.12

developed the theory of sets a century ago. His work has had a profound effect on the logical basis for theorems of calculus and higher mathematics.

In the discussion below we will identify several basic features of C. Because of the similar ways that C and C_μ are constructed, with a minimum of effort we will then be able to associate features of C with corresponding features of C_μ for $\mu > 4$.

The first property concerns closed sets. A set is said to be **closed** if it contains all its limit points. In other words, a set D is closed if whenever $\{x_i\}_{i=1}^{\infty}$ is a sequence of points in D such that $\lim_{i \to \infty} x_i = x$, then x is also in D. It is evident that any closed interval is a closed set. In addition, any finite point set is closed, as is the union of a finite number of closed intervals (Exercise 1). Also the set consisting of the sequence $\{1/n\}_{n=1}^{\infty}$ and 0 is closed. By contrast, the set of rationals in the interval $[0, 1]$ is not closed because, for example, the irrational number $.50550555055550555550\cdots$ is the limit of the rational numbers $.5$, $.505$, $.5055$, $.505505$, ...

Now we will show that C is closed.

LEMMA 1. C is closed.

Proof. By our remarks above, for each n the set L_n is closed since it is the union of a finite collection of closed intervals. Next, let $\{x_i\}_{i=1}^{\infty}$ be a sequence in C with $\lim_{i \to \infty} x_i = x$. By the definition of C, the sequence is in L_n for each n. Since L_n is closed, it follows that x is also in L_n. Consequently x is in C. Therefore every limit point of a sequence in C is in C, so that C is closed. ∎

The next property is total disconnectedness. A set D is **totally disconnected** if it contains *no* nonempty open intervals. Thus finite point sets, the rationals in the interval $(0, 1)$, and the sequence $\{1/n\}_{n=1}^{\infty}$ are totally disconnected. In the proof of Lemma 2 we will use the fact that every nonempty open interval has positive length, and hence its length is larger than $1/3^n$ for an appropriate positive integer n.

LEMMA 2. C is totally disconnected.

Proof. Since L_n consists of intervals of length $1/3^n$ and since $C \subset L_n$ for each n, we conclude from the remark preceding the lemma that C cannot contain any nonempty open intervals. Thus C is totally disconnected. ∎

The third property involves perfect sets. A set D of real numbers is **perfect** provided that it is closed and each of its points is a limit of other points in D. Any closed interval with more than one point is perfect. By contrast, any closed set with isolated points is not perfect.

LEMMA 3. C is perfect.

Proof. We know that C is closed, by Lemma 1. Let x be in C. Then for each positive integer n, there is a positive integer $k(n)$ in $[1, 2^n]$ depending on n and such that x is in $P_{nk(n)}$. By the definition of $P_{nk(n)}$,

$$P_{nk(n)} = P_{(n+1)k(n+1)} \cup P_{(n+1)m(n+1)}$$

where the two sets on the right-hand side are closed and disjoint, and where x is in $P_{(n+1)k(n+1)}$. For each n let y_n be in $P_{(n+1)m(n+1)}$. Since $P_{(n+1)k(n+1)}$ and $P_{(n+1)m(n+1)}$ are disjoint, we know that $x \neq y_n$. However, since x and y_n are both in $P_{nk(n)}$ and since the length of $P_{nk(n)}$ is $1/3^n$, we know that $\lim_{n\to\infty} y_n = x$. Therefore C is perfect. ∎

Lemmas 1–3 together yield the following theorem.

THEOREM 2.17. C is a closed, bounded, totally disconnected and perfect subset of $[0, 1]$.

Sets of real numbers that are simultaneously closed, totally disconnected and perfect play a significant role in analysis, and are accorded a special name.

DEFINITION 2.18. A set D of real numbers that is closed, totally disconnected and perfect is called a **Cantor set**.

By the definition of a Cantor set, C is a Cantor set. Another property of C is that it has a "large" infinite number of points. To tell what we mean by a "large" infinite number of points, we will define countable and uncountable sets. We say that a subset D of real numbers is **countable** if we can list all of its members in the fashion d_1, d_2 , ..., where the sequence is either finite or infinite.   In mathematical language this means that we can find a one-to-one function F from D into the set of positive integers. It follows that finite sets, as well as the sequence $\{1/n\}_{n=1}^{\infty}$, are countable.   Any set that is not countable is **uncountable**. We mention that uncountable sets are larger than countable ones, since uncountable sets are so large that they cannot be identified with subsets of the positive integers. As a result, if D is uncountable and $D \subseteq E$, then E is also uncountable. Below we will show that the collection of numbers in the interval $[0, 1]$ is uncountable.

THEOREM 2.19. The set of positive rational numbers is countable.

Proof. Let F be the function whose domain is the set of positive rational numbers and whose values are positive integers, defined in the following way. For any rational number p/q in reduced form, let

$$F(p/q) = 2^p\, 3^q$$

To show that F is one-to-one, we suppose that $F(p/q) = F(r/s)$, so that

$$2^p \, 3^q \; = \; 2^r \, 3^s \, , \quad \text{or equivalently,} \quad 2^{p-r} \; = \; 3^{s-q}$$

Let us assume without loss of generality that $p > r$, so that 2^{p-r} is an integer. The only way that a power of 2 can equal a power of 3 is for the powers to be 0. Therefore $p - r = 0 = s - q$, so that $p = r$ and $q = s$. Thus $p/q = r/s$. Consequently F is one-to-one, so that the set of positive rational numbers is a countable set. ∎

Using Theorem 2.19, one can show that the collection of all rationals is also countable (Exercise 2(b)). In addition, since the set of rationals in $(0, 1)$ is a subset of the positive rationals, and since subsets of countable sets are again countable, we have the following corollary.

COROLLARY 2.20. The set of rational numbers in $(0, 1)$ is countable.

Next, we will prove that the set D of irrational numbers in $(0, 1)$ is uncountable by showing that there is *no* one-to-one correspondence between D and the set of positive integers. Before proving the result, we note that each number in $(0, 1)$ has a decimal expansion (which is not necessarily unique). The expansion is repeating (like $.23\overline{495}\cdots$) if and only if the number is rational (Exercise 2(a)).

THEOREM 2.21. The set of irrational numbers in $(0, 1)$ is uncountable.

Proof. We will prove the result by contradiction. Suppose that D were countable, so we could list its members as $\{d_n\}_{n=1}^{\infty}$, as decimals:

$$
\begin{aligned}
d_1 &= x_{11}x_{12}x_{13}x_{14}\cdots \\
d_2 &= x_{21}x_{22}x_{23}x_{24}\cdots \\
d_3 &= x_{31}x_{32}x_{33}x_{34}\cdots \\
d_4 &= x_{41}x_{42}x_{43}x_{44}\cdots
\end{aligned}
$$

In general,

$$d_n \; = \; x_{n1}x_{n2}x_{n3}x_{n4}\cdots$$

Suppose that we were to define z to be the number in $(0, 1)$ with the decimal expansion $z_1z_2z_3\cdots$ obtained by assigning $z_n = 1$ if $x_{nn} = 0$, and $z_n = 0$ if $x_{nn} \neq 0$. Then $z \neq d_n$ for each n, so that z would not appear in the listing of members of D. However, there is no assurance that z would be irrational! So we need to be more careful in defining z. Specifically, let z have the decimal expansion $z_1z_2z_3\cdots$ obtained by letting $z_{2^n} = 1$ if $x_{2^n n} = 0$, and $z_{2^n} = 0$ if $x_{2^n n} \neq 0$, for each n. This ensures that $z \neq d_n$ for every n, so z is not in the listing for D. To guarantee that z is irrational, we will insert enough different sized strings of

0's and 1's into the expansion for z to make certain that the expansion is not repeating. We do this by letting $z_k = 0$ for all k strictly between 2^n and 2^{n+1} if n is even, and $z_k = 1$ for all k strictly between 2^n and 2^{n+1} if n is odd. Thus the expansion of z is not repeating, so z is irrational and hence in D. However, z is not in the proposed listing of D, which means that D is uncountable. ■

Since the interval $[0, 1]$ contains the uncountable set of irrationals in $[0, 1]$, the following corollary is immediate.

COROLLARY 2.22. The set of all numbers in $[0, 1]$ is uncountable.

A slight alteration in the proof of Theorem 2.21 yields the following result.

COROLLARY 2.23. The set of all sequences of 0's and 1's is uncountable.

It is possible to show that between any two rational numbers in $(0, 1)$ there are infinitely many irrational numbers, and between any two irrational numbers in $(0, 1)$ there are infinitely many rational numbers. So how can the set of rational numbers in $(0, 1)$ be countable, whereas the set of irrational numbers in $(0, 1)$ is uncountable? There is nothing faulty in our mathematics; the paradox shows that our intuition can sometimes be faulty.

Our next goal is to show that C is uncountable. The uncountable set A of *all* non-finite sequences of 0's and 1's will come in handy. To start with, recall from the proof of Lemma 3 before Theorem 2.17 that if x is in C, then for each n, x is in the subinterval $P_{nk(n)}$ of length $1/3^n$, where $k(n)$ is an appropriate integer such that $1 \le k(n) \le 2^n$. Let $h: C \to A$ be defined by

$$h(x) = x_1 x_2 x_3 \cdots, \quad \text{where } x_n = \begin{cases} 0 \text{ if } k(n) \text{ is an even integer} \\ 1 \text{ if } k(n) \text{ is an odd integer} \end{cases}$$

The function h and the set A are related to the function and set with similar names that appeared in Section 2.2. For convenience we have altered the initial index of the sequence $h(x)$; the set A now includes all finite sequences as well as non-finite sequences. We will prove that h is a **one-to-one correspondence** between C and A, that is, $h: C \to A$ is one-to-one and onto.

THEOREM 2.24. h is a one-to-one correspondence between C and A.

Proof. To show that h is one-to-one, we notice that if

$$x_1 x_2 x_3 \cdots = h(x) = h(y) = y_1 y_2 y_3 \cdots$$

then $x_1 = y_1$, $x_2 = y_2$, $x_3 = y_3$, ..., so that x and y both lie in the same subinterval $P_{nk(n)}$ for all n. Since the length of $P_{nk(n)}$ is $1/3^n$ and approaches 0 as n increases without bound, it follows that h is one-to-one. To show that h is onto, let $z_1 z_2 z_3 \cdots$ be an arbitrary sequence of 0's and 1's. Pick the subintervals $P_{1k(1)}, P_{2k(2)}, \ldots$ satisfying the following conditions:

$$x \text{ in } P_{1k(1)} \text{ implies that } h(x) \text{ begins } z_1$$

$$x \text{ in } P_{2k(2)} \text{ implies that } h(x) \text{ begins } z_1 z_2$$

$$x \text{ in } P_{3k(3)} \text{ implies that } h(x) \text{ begins } z_1 z_2 z_3$$

In general,

$$x \text{ in } P_{nk(n)} \text{ implies that } h(x) \text{ begins } z_1 z_2 z_3 \cdots z_n$$

Since

$$P_{1k(1)} \supseteq P_{2k(2)} \supseteq P_{3k(3)} \supseteq \cdots$$

and each of the sets is closed and bounded, the Heine-Borel Theorem (Theorem 2.7) implies that there is a point x^* that is simultaneously in $P_{nk(n)}$ for all n. By the definition of h, $h(x^*) = z_1 z_2 z_3 \cdots$, so that h is onto. Consequently h is a one-to-one correspondence between C and A. ∎

COROLLARY 2.25. C is uncountable.

Proof. Since A is uncountable by Corollary 2.23, and since h is a one-to-one correspondence between C and A, it follows that C is also uncountable. ∎

Now we have finished presenting the special features of C, including the fact that C is a closed, totally disconnected and perfect set (hence a Cantor set), and is uncountable. We will use the information we have accumulated about C as we discover properties of Q_μ (for $\mu > 4$).

Strong Chaos of Functions in $\{Q_\mu\}$

What lies ahead in this section is to show that the subset C_μ of $[0, 1]$ associated with Q_μ is a Cantor set on which Q_μ is strongly chaotic. First we turn to the set C_μ itself. Recall that both C and C_μ were defined through nested sequences of disjoint, closed subintervals of $[0, 1]$. The proof that C is closed and uncountable carries over to show that C_μ also is a closed and uncountable subset of $[0, 1]$. If we knew that the lengths of the defining subintervals J_{nk} of C_μ shrank to 0 as n increased without bound, then we could also use the corresponding proofs for C to show that C_μ is totally disconnected and is perfect.

Lengths of nested bounded closed intervals need not converge to 0 (see Exercise 6). Although it is possible to prove that whenever $\mu > 4$ the length of J_{nk} converges to 0 as n increases without bound, the proof is much less compli-cated when $\mu > 2 + \sqrt{5} \approx 4.236$. The reason is that

$$\text{if } \mu > 2 + \sqrt{5}, \text{ then } |Q'_\mu(x)| > 1 \text{ for all } x \text{ in } J_{11} \cup J_{12} \tag{4}$$

(Exercise 11). The proof that the length of J_{nk} shrinks to 0 will be based on the observation that if $0 \leq f(x) \leq 1$ and $|f'(x)| > s > 1$ for all x in $[a, b]$, then the length of $[a, b]$ is less than $1/s$. Indeed, since $0 \leq f(x) \leq 1$ for all x in $[a, b]$,

$$|f(b) - f(a)| \leq 1 - 0 = 1 \tag{5}$$

Since $|f'(x)| > s > 1$, the Mean Value Theorem implies that there is a w in (a, b) such that

$$|f(b) - f(a)| = |f'(w)| \, |b - a| > s|b - a| \tag{6}$$

Combining (5) and (6), we deduce that

$$1 \geq |f(b) - f(a)| > s|b - a|$$

so that $|b - a| < 1/s$, as asserted.

THEOREM 2.26. Suppose that $\mu > 2 + \sqrt{5}$. Then there is a constant $r > 1$ such that J_{nk} has length less than $1/r^n$ for each n and k.

Proof. Since $J_{11} \cup J_{12}$ is closed and Q'_μ is continuous, it follows from (4) that there is a constant $r > 1$ such that $|Q'_\mu(x)| > r$ for all x in $J_{11} \cup J_{12}$. Our first goal is to show that $|(Q^{[n]}_\mu)'| > r^n$ on J_{nk} for each k. To prove it, let x be in J_{nk}. By the definition of J_{nk}, $Q^{[j]}_\mu(x)$ is in $J_{11} \cup J_{12}$ for $j = 0, 1, 2, ..., n - 1$. It follows that $|Q'_\mu(Q^{[j]}_\mu(x))| > r$ for $j = 0, 1, 2, ..., n - 1$. By the Chain Rule,

$$|(Q^{[n]}_\mu)'(x)| = \underset{> r}{|Q'_\mu(Q^{[n-1]}_\mu(x))|} \; \underset{> r}{|Q'_\mu(Q^{[n-2]}_\mu(x))|} \cdots \underset{> r}{|Q'_\mu(x)|} > r^n$$

Thus we have proved that $|(Q^{[n]}_\mu)'| > r^n$ on J_{nk}. To complete the proof, we write J_{nk} as $[y, z]$. Since $0 \leq Q^{[n]}_\mu(y) \leq 1$ and $0 \leq Q^{[n]}_\mu(z) \leq 1$, the comment before the lemma implies that $|y - z| < 1/r^n$, so that the length of J_{nk} is less than $1/r^n$. ∎

THEOREM 2.27. Let $\mu > 2 + \sqrt{5}$. Then C_μ is an uncountable Cantor set.

Proof. Theorem 2.26 is the missing link that allows us to deduce that C_μ is totally disconnected and perfect by mimicking the corresponding proofs for C (Lemmas 2 and 3 before Theorem 2.17 and Corollary 2.25). We have already noted that C_μ is closed. Therefore C_μ is a Cantor set. ∎

We know that Q_μ maps C_μ into C_μ, and that C_μ is an uncountable Cantor set. Therefore in theory it is possible for Q_μ to be strongly chaotic on C_μ. In order to address this question we introduce the notion of distance between two sequences.

Let A denote, as before, the collection of sequences of 0's and 1's. To simplify notation we will write X for the sequence $x_1 x_2 x_3 \cdots$ and Y for the sequence $y_1 y_2 y_3 \cdots$. Then the **distance** between X and Y, written $\|X - Y\|$, is defined by the formula

$$\|X - Y\| = \sum_{k=1}^{\infty} \frac{|x_k - y_k|}{2^k}$$

Since $|x_k - y_k| = 0$ or 1, it follows that

$$\sum_{k=1}^{\infty} \frac{|x_k - y_k|}{2^k} \le \sum_{k=1}^{\infty} \frac{1}{2^k} = 1$$

This distance function has several important properties:

i. $0 \le \|X - Y\|$ for all X and Y, and $\|X - Y\| = 0$ only if $X = Y$

ii. $\|X - Y\| = \|Y - X\|$ for all X and Y

iii. $\|X - Z\| \le \|X - Y\| + \|Y - Z\|$ for all X, Y, Z

A distance function with properties (i)–(iii) is called a **metric**. The set on which the metric is defined, along with the metric, is a **metric space**. Thus A with its distance function is a metric space.

Next we notice that

if the initial n terms of X and Y are identical, then $\|X - Y\| \le 1/2^n$ \hfill (7)

The reason is that in this case,

$$\|X - Y\| = \sum_{k=n+1}^{\infty} \frac{|x_k - y_k|}{2^k} \le \sum_{k=n+1}^{\infty} \frac{1}{2^k} = \frac{1}{2^n}$$

A similar argument shows that

if $\|X - Y\| < 1/2^n$, then the initial n terms of X and Y are equal   (8)

(Exercise 14). Consequently X and Y are close together precisely when their initial several terms are identical.

Now we define the **left shift map** $S: A \to A$ by

$$S(x_1 x_2 x_3 \cdots) = x_2 x_3 x_4 \cdots$$

The effect of S is to delete the first term of the given sequence, thereby shifting all terms to the left one place.

We would like to be able to show that S is a continuous function. However, the domain of S is a set of sequences, not a set of real numbers. As a result, we need to define what we mean by continuity for a function from one metric space to another. To that end, let $f: D \to E$, where D has metric $\| \ \|$ and E has metric $\|\| \ \|\|$. We say that f **is continuous at** x in D if for any arbitrary $\varepsilon > 0$, there is a number $\delta > 0$ such that whenever z is in D and $\|x - z\| < \delta$, it follows that $\|\|f(x) - f(z)\|\| < \varepsilon$. Moreover, if f is continuous at each point of its domain, then we say that f is a **continuous function**.

THEOREM 2.28. $S: A \to A$ is continuous.

Proof. Let x be in A, and let $\varepsilon > 0$. Choose a positive integer n such that $1/2^n < \varepsilon$, and let $\delta = 1/2^n$. It follows from (8) that if $\|X - Y\| < \delta = 1/2^n$, then the initial n terms of X and Y are identical, so that

$$\|S(X) - S(Y)\| = \sum_{k=n+1}^{\infty} \frac{|x_{k+1} - y_{k+1}|}{2^k} \leq \sum_{k=n+1}^{\infty} \frac{1}{2^k} = \frac{1}{2^n} < \varepsilon$$

Therefore S is a continuous function. ∎

The notions of sensitive dependence on initial conditions, dense set of periodic points, and dense orbit that were defined for functions whose domains are intervals can be extended to any metric space (Exercise 15). If D is endowed with a metric, then we will say that a function $f: D \to D$ is **strongly chaotic** on D if f has sensitive dependence on initial conditions, a dense set of periodic points, and a dense orbit.

THEOREM 2.29. S is strongly chaotic on the set A.

Proof. To show that S has sensitive dependence on initial conditions, suppose that

X and Y are distinct elements in A. This means that there is some n for which $x_n \neq y_n$. Since S shifts a given sequence left one term, $S^{[n-1]}$ shifts a given sequence left $n - 1$ terms, so that

$$\|S^{[n-1]}(X) - S^{[n-1]}(Y)\| = \|x_n x_{n+1} x_{n+2} \cdots - y_n y_{n+1} y_{n+2} \cdots \| \geq \frac{|x_n - y_n|}{2} = \frac{1}{2}$$

Thus S has sensitive dependence on initial conditions. Next we will show that S admits a dense set of periodic points. Notice that since S shifts terms left, the periodic points for S are the repeating sequences of the form $\overline{x_1 x_2 x_3 \cdots x_n} \cdots$ for $n \geq 1$. Now suppose that $x = x_1 x_2 x_3 \cdots$ is any element of A, and n is any positive integer. Define Y by

$$Y = \text{the repeating sequence } \overline{x_1 x_2 x_3 \cdots x_n} \cdots$$

Then Y is periodic. Moreover, since the initial n terms of X and Y are identical, (7) implies that $\|X - Y\| \leq 1/2^n$. Consequently the periodic points are dense in A. Finally, the sequence

$$s = 0 \ 1 \ \ 00 \ 01 \ 10 \ 11 \ \ 000 \ 001 \ 010 \ 011 \ 100 \ 101 \ 110 \ 101 \cdots$$

(in which all n-tuples of 0's and 1's appear in order, for each positive integer n) has a dense orbit in A, since any partial sequence $x_1 x_2 x_3 \cdots x_n$ appears somewhere inside s. We conclude that S is strongly chaotic. ∎

If D and E are metric spaces, then we say that a function $f\colon D \to E$ is a **homeomorphism** if f is one-to-one, onto, and continuous, and if f^{-1} is also continuous. For example, the function $f\colon [0, 2] \to [0, 4]$ defined by $f(x) = x^2$ is a homeomorphism if the metric on $[0, 2]$ and $[0, 4]$ is the usual distance function. If two metric spaces D and E are homeomorphic, then in many ways D and E are similar. In particular, they contain the same number of elements, and small distances in one space correspond to small distances in the other space. It is with this in mind that we will define a homeomorphism $H_\mu\colon C_\mu \to A$, so that the fact that A is strongly chaotic will yield corresponding information about C_μ.

Recall that x is in C_μ if and only if all the iterates $Q_\mu^{[n]}(x)$ are in $[0, 1]$, which means that for each n, $Q_\mu^{[n]}(x)$ is in either J_{11} or J_{12}. We will associate x in C_μ with the sequence in A whose nth term notes whether $Q_\mu^{[n]}(x)$ is in J_{11} or is in J_{12}. Specifically, let $H_\mu\colon C_\mu \to A$ be defined by

$$H_\mu(x) = x_1 x_2 x_3 \cdots, \text{ where } x_n = \begin{cases} 0 \text{ if } Q_\mu^{[n]}(x) \text{ is in } J_{11} \\ 1 \text{ if } Q_\mu^{[n]}(x) \text{ is in } J_{12} \end{cases}$$

Notice that $x_1 x_2 x_3 \cdots x_n$ identifies the nth stage subinterval J_{nk} in which x is located. The function H_μ is closely related to the function $h : C \to A$ that was proved in Theorem 2.24 to be a one-to-one correspondence.

THEOREM 2.30. Let $\mu > 2 + \sqrt{5}$. Then the function H_μ is a homeomorphism from C_μ onto A.

Proof. To prove that H_μ is one-to-one and onto one can make the needed modifications of the proof of Theorem 2.24 (Exercise 13). Therefore we need only show that H_μ and H_μ^{-1} are continuous. To prove that H_μ is continuous, let $\varepsilon > 0$, and let x be an arbitrary point in C_μ. Furthermore, let n be so large that $1/2^n < \varepsilon$. Next, pick $\delta > 0$ so small that if z is in C_μ and $|x - z| < \delta$, then x and z lie in the same $J_{jk(j)}$ for $j = 0, 1, 2, 3, ..., n - 1$. (Such a δ exists by virtue of Theorem 2.26.) For such a z, the sequences $H_\mu(x)$ and $H_\mu(z)$ have the same initial n terms by the definition of H_μ. Then (7) implies that

$$\|H_\mu(x) - H_\mu(z)\| \le \frac{1}{2^n} < \varepsilon$$

We conclude that H_μ is continuous. That H_μ^{-1} is also continuous can be proved directly, or can be deduced from the fact that H_μ is a one-to-one, onto, continuous function whose domain is a closed, bounded subset of reals. Therefore H_μ is a homeomorphism. ∎

Now we turn to conjugacy of two functions. Let D and E be two metric spaces, and let $f: D \to D$ and $g: E \to E$ be continuous functions. We say that f and g are **conjugate** provided that there is a homeomorphism $h: D \to E$ such that $h \circ f = g \circ h$. Our demonstration that Q_μ is strongly chaotic on C_μ will use this notion of conjugacy.

THEOREM 2.31. Let $\mu > 2 + \sqrt{5}$. Then Q_μ (restricted to C_μ) and S are conjugate.

Proof. We observe that $Q_\mu : C_\mu \to C_\mu$, $S : A \to A$, and $H_\mu : C_\mu \to A$. By Theorem 2.30, H_μ is a homeomorphism from C_μ onto A, so we need only show that $H_\mu \circ Q_\mu = S \circ H_\mu$. To that end, suppose that x is in C_μ, with $H_\mu(x) = x_1 x_2 x_3 \cdots$. Then

$$(S \circ H_\mu)(x) = S(H_\mu(x)) = x_2 x_3 x_4 \cdots$$

so that the nth term of $(S \circ H_\mu)(x)$ is x_{n+1}. By the definition of H_μ, we know that $(H_\mu \circ Q_\mu)(x) = H_\mu(Q_\mu(x)) = z_1 z_2 z_3 \cdots$, where

$$z_n = \begin{cases} 0 \text{ if } Q_\mu^{[n]}(Q_\mu(x)) = Q_\mu^{[n+1]}(x) \text{ is in } J_{11} \\ 1 \text{ if } Q_\mu^{[n]}(Q_\mu(x)) = Q_\mu^{[n+1]}(x) \text{ is in } J_{12} \end{cases}$$

Since $z_n = x_{n+1}$, it follows that for each n, the nth terms of $(S \circ H_\mu)(x)$ and of $(H_\mu \circ Q_\mu)(x)$ are x_{n+1}. Therefore $S \circ H_\mu = H_\mu \circ Q_\mu$, completing the proof. ∎

On closed, bounded subsets of the real line, homeomorphisms automatically transfer all properties of separation (and hence sensitive dependence on initial conditions), periodicity, and density. Therefore we are ready for our final result.

THEOREM 2.32.   If $\mu > 2 + \sqrt{5}$, then Q_μ is strongly chaotic on the uncountable Cantor set C_μ.

Proof. Theorem 2.29 says that S is strongly chaotic on A. Since Q_μ and S are conjugate to one another by Theorem 2.31, it follows that Q_μ is strongly chaotic on C_μ. Finally, C_μ is an uncountable Cantor set by Theorem 2.27. ∎

We can conclude from Theorem 2.32 that C_μ contains nonperiodic points for Q_μ, as we indicated at the outset of the section.

We have now finished our study of the dynamics of functions of one variable. We have discussed many features of our two most illustrious parametrized families, $\{T_\mu\}$ and $\{Q_\mu\}$. Although the members of $\{T_\mu\}$ are polygonal lines and those of $\{Q_\mu\}$ are curved, nevertheless both families display similar dynamic behavior, and members of each family are not only chaotic but even strongly chaotic. In Chapter 3 we will turn to functions of more than one variable, and will use the information we have gained concerning functions of one variable.

EXERCISES 2.4

1. a.   Show that any finite set is closed.
    b.   Show that the union of a finite number of closed intervals is closed.

2. a.   Show that a number in $(0, 1)$ is rational if and only if its decimal expansion is repeating.
    b.   Show that the set of all rational real numbers is countable.

3.   Show that between any two rationals there are infinitely many irrationals, and vice versa.

4. a.   Prove that any subset of a countable set is countable.
    b.   Prove that if a set D is uncountable and if $E \supseteq D$, then E is also uncountable.

5.   Determine whether the set of numbers in $(0, 1)$ whose decimal expansions contain no 1's or 5's is
     a.   closed, open, or neither   b.   totally disconnected

6.   Find an example of a nested set of bounded closed intervals $\{B_n\}_{n=1}^{\infty}$ such that the length of B_n does *not* converge to 0 as n increases without bound.

7.   Let $D_n(C)$ denote the total length of the set L_n defined in (3) and employed in the definition of C. Find $D(C) = \lim_{n \to \infty} D_n(C)$. The number $D(C)$ is sometimes called the length of C. Is that reasonable?

8.   Let $C_{1/5}$ denote the set obtained the same way as is C, but with subintervals of length $1/5^n$ deleted at each stage. Let $D_n(C_{1/5})$ denote the total length of the remaining subintervals at the nth stage.
     a.   Find $D_n(C_{1/5})$.   b.   Find $\lim_{n \to \infty} D_n(C_{1/5})$.

9.   Show that C_μ contains all periodic points of Q_μ whenever $\mu > 4$.

10. Let $\mu = 5$. Show that J_{21} and J_{22} have different lengths.

11. Prove that if $\mu > 2 + \sqrt{5}$, then $|Q_\mu'(x)| > 1$ for all x in $J_{11} \cup J_{12}$.

12. Let $\mu > 2 + \sqrt{5}$. Show that Q_μ has sensitive dependence on initial conditions on the interval $[0, 1]$.

13. By modifying the proof of Theorem 2.24, prove that the function $H_\mu: C_\mu \to A$ is one-to-one and onto.

14. Let $X = x_1 x_2 x_3 \cdots$ and $Y = y_1 y_2 y_3 \cdots$. Show that if $\|X - Y\| < 1/2^n$, then the initial n terms of X and Y are identical.

15. Let D be a metric space, and let $f : D \to D$. Write down a definition, applicable to f and D, of
     a.   sensitive dependence on initial conditions
     b.   a dense set of periodic points
     c.   a dense orbit

CHAPTER
3

TWO-DIMENSIONAL CHAOS

In the first two chapters we discussed functions of one variable, with an eye toward those features of a chaotic nature. Now we turn to functions of two variables. Such functions can exhibit more varied behavior.

In the analysis of higher-dimensional functions, matrices serve as an indispensable tool. As a result, we give a brief review of matrices in Section 3.1, and in particular focus on features of similar matrices. In Section 3.2, with the help of matrices we study the dynamics of the simplest two-dimensional functions, the linear functions. Section 3.3 presents the basic ideas concerning dynamics of more general two-dimensional functions.

The remainder of Chapter 3 is devoted to well-known examples of two-dimensional functions: the baker's function, the Hénon map, and the Smale horse-shoe map. We show that the first and third of these maps are chaotic, and that a Hénon map appears to be chaotic.

3.1 REVIEW OF MATRICES

In the study of nonlinear dynamics in two (and higher) dimensions, the use of matrices cannot easily be avoided. As a result, we devote this section to the basic definitions and properties of matrices, and limit our discussion to 2×2 matrices because the chapter focuses on two-dimensional functions. The section begins with basic definitions relating to 2×2 matrices, and finishes with a longer discussion of similar matrices.

Brief Review of 2×2 Matrices

Let $A = \begin{pmatrix} a & b \\ c & d \end{pmatrix}$ and $B = \begin{pmatrix} e & f \\ g & h \end{pmatrix}$, and let r be a real number. Then

$$A + B = \begin{pmatrix} a+e & b+f \\ c+g & d+h \end{pmatrix}, \quad rA = \begin{pmatrix} ra & rb \\ rc & rd \end{pmatrix} \quad \text{and} \quad AB = \begin{pmatrix} ae+bg & af+bh \\ ce+dg & cf+dh \end{pmatrix}$$

EXAMPLE 1. Let a and d be any real numbers, and let $A = \begin{pmatrix} a & 0 \\ 0 & d \end{pmatrix}$. Find A^n for an arbitrary positive integer n.

Solution. First we have

$$A^2 = A \cdot A = \begin{pmatrix} a & 0 \\ 0 & d \end{pmatrix}\begin{pmatrix} a & 0 \\ 0 & d \end{pmatrix} = \begin{pmatrix} a^2 & 0 \\ 0 & d^2 \end{pmatrix}$$

If $A = \begin{pmatrix} a^{n-1} & 0 \\ 0 & d^{n-1} \end{pmatrix}$, then

$$A^n = A^{n-1} \cdot A = \begin{pmatrix} a^{n-1} & 0 \\ 0 & d^{n-1} \end{pmatrix}\begin{pmatrix} a & 0 \\ 0 & d \end{pmatrix} = \begin{pmatrix} a^n & 0 \\ 0 & d^n \end{pmatrix}$$

By the Law of Induction, $A^n = \begin{pmatrix} a^n & 0 \\ 0 & d^n \end{pmatrix}$ for each positive integer n. ❏

Matrices of the variety given in Example 1 are called **diagonal matrices** because the only nonzero entries appear along the major (upper left to lower right) diagonal. Example 1 shows that any power of a diagonal matrix is also a diagonal matrix. Equally easily we could show that any product of diagonal matrices is also a diagonal matrix.

Normally $AB \neq BA$. However, if A, B, and C are any 2×2 matrices, then $A(BC) = (AB)C$, so we can write ABC for the product of the three matrices.

The matrix $\begin{pmatrix} 1 & 0 \\ 0 & 1 \end{pmatrix}$ is called the **identity matrix**, and is denoted I. The matrix I has the property that $AI = IA = A$ for every 2×2 matrix A. Now suppose that A is an arbitrary 2×2 matrix. If there is a 2×2 matrix B such that $AB = I$ and $BA = I$, then B is the **inverse** of A, and we say that A is **invertible**. (The hypothesis that $BA = I$ is actually superfluous. See Exercise 9.) If such a matrix B exists for a given matrix A, then B is unique, and is written A^{-1}.

Not every matrix has an inverse. In fact, we will show below that $\begin{pmatrix} a & b \\ c & d \end{pmatrix}$

has an inverse precisely when $ad - bc \neq 0$. The expression $ad - bc$ is called the **determinant** of A, and is denoted det A or $|A|$. Thus

$$\det \begin{pmatrix} a & b \\ c & d \end{pmatrix} = ad - bc$$

Therefore if $A = \begin{pmatrix} 3 & 0 \\ 2 & -1 \end{pmatrix}$, then $\det A = (3)(-1) - (0)(2) = -3$. The determinant of the identity matrix I is given by

$$\det I = \det \begin{pmatrix} 1 & 0 \\ 0 & 1 \end{pmatrix} = (1)(1) - (0)(0) = 1$$

Next, assume that A and B are 2×2 matrices. A routine calculation establishes that

$$\det (AB) = (\det A)(\det B) \tag{1}$$

(Exercise 8). One consequence of (1) is that if A has an inverse, then

$$\det (A^{-1}) = \frac{1}{\det A} \tag{2}$$

Formula (1) also helps us to verify the following criterion concerning the existence of inverses.

THEOREM 3.1. A 2×2 matrix A has an inverse if and only if $\det A \neq 0$.

Proof. Let $A = \begin{pmatrix} a & b \\ c & d \end{pmatrix}$. First suppose that $\det A = ad - bc \neq 0$. Then you can check directly that

$$A^{-1} = \frac{1}{ad - bc} \begin{pmatrix} d & -b \\ -c & a \end{pmatrix} \tag{3}$$

Conversely, suppose that $\det A = 0$. Then by (1),

$$(\det AB) = (\det A)(\det B) = 0 \, (\det B) = 0 \text{ for all } B$$

If there were a B such that $AB = I$ then by (1), $1 = \det I = \det (AB) = 0$. This contradiction shows that no such B exists. Therefore A has no inverse, and the proof is complete. ∎

In the sequel we will consider each element of R^2 either as a point, or an ordered pair of real numbers, or a two-dimensional vector. Thus the point whose x and y coordinates are -1 and 2, respectively, represents the ordered pair $(-1, 2)$ as well as the vector $\begin{pmatrix} -1 \\ 2 \end{pmatrix}$. As is customary, bold-face letters such as $\mathbf{v}$, $\mathbf{w}$, $\mathbf{x}$, and $\mathbf{y}$ represent vectors. Three noteworthy, special vectors are

$$\mathbf{0} = \begin{pmatrix} 0 \\ 0 \end{pmatrix}, \quad \mathbf{i} = \begin{pmatrix} 1 \\ 0 \end{pmatrix}, \quad \text{and} \quad \mathbf{j} = \begin{pmatrix} 0 \\ 1 \end{pmatrix}$$

By definition the sum and constant multiple of vectors are given by

$$\begin{pmatrix} x \\ y \end{pmatrix} + \begin{pmatrix} z \\ w \end{pmatrix} = \begin{pmatrix} x+z \\ y+w \end{pmatrix}$$

$$r\begin{pmatrix} x \\ y \end{pmatrix} = \begin{pmatrix} rx \\ ry \end{pmatrix} \quad \text{for every real number } r$$

The product of the matrix $A = \begin{pmatrix} a & b \\ c & d \end{pmatrix}$ and the vector $\mathbf{v} = \begin{pmatrix} x \\ y \end{pmatrix}$ is the 2×1 matrix given by

$$A\mathbf{v} = \begin{pmatrix} a & b \\ c & d \end{pmatrix}\begin{pmatrix} x \\ y \end{pmatrix} = \begin{pmatrix} ax+by \\ cx+dy \end{pmatrix}$$

For example, if $A = \begin{pmatrix} 1 & 2 \\ 3 & 4 \end{pmatrix}$ and $\mathbf{v} = \begin{pmatrix} -2 \\ 1 \end{pmatrix}$, then

$$A\mathbf{v} = \begin{pmatrix} 1 & 2 \\ 3 & 4 \end{pmatrix}\begin{pmatrix} -2 \\ 1 \end{pmatrix} = \begin{pmatrix} 0 \\ -2 \end{pmatrix}$$

In the event that $\mathbf{v} = \mathbf{i}$ or $\mathbf{v} = \mathbf{j}$, we obtain

$$A\mathbf{i} = \begin{pmatrix} a & b \\ c & d \end{pmatrix}\begin{pmatrix} 1 \\ 0 \end{pmatrix} = \begin{pmatrix} a \\ c \end{pmatrix} \quad \text{and} \quad A\mathbf{j} = \begin{pmatrix} a & b \\ c & d \end{pmatrix}\begin{pmatrix} 0 \\ 1 \end{pmatrix} = \begin{pmatrix} b \\ d \end{pmatrix}$$

Consequently

$A\mathbf{i}$ is the first column of A, and $A\mathbf{j}$ is the second column of A (4)

Now we turn to eigenvalues and eigenvectors of 2×2 matrices.

DEFINITION 3.2. Suppose that A is a 2×2 matrix. The real number λ is an **eigenvalue** of A provided that there is a nonzero $\mathbf{v}$ in R^2 such that $A\mathbf{v} = \lambda\mathbf{v}$. In this case $\mathbf{v}$ is an **eigenvector** of A (relative to λ).

If $\mathbf{v}$ satisfies $A\mathbf{v} = \mathbf{v}$, then $\mathbf{v}$ looks very much like a "fixed point" of A. For later use we observe that the eigenvalues of any diagonal matrix $\begin{pmatrix} a & 0 \\ 0 & d \end{pmatrix}$ are the diagonal elements a and d (Exercise 14).

It is easy to demonstrate that the following conditions involving eigenvalues and eigenvectors are equivalent:

   i. λ is an eigenvalue of A.
   ii. $(A - \lambda I)\mathbf{v} = \mathbf{0}$ for some nonzero vector $\mathbf{v}$.
   iii. $\det(A - \lambda I) = 0$.

EXAMPLE 2. Let $A = \begin{pmatrix} 1 & 1 \\ -2 & 4 \end{pmatrix}$. Find the eigenvalues of A, and corresponding eigenvectors.

Solution. By (iii), to find the (real) eigenvalues of A we need only determine the values of λ for which $\det(A - \lambda I) = 0$. To that end, we notice that

$$\det(A - \lambda I) = \det\begin{pmatrix} 1-\lambda & 1 \\ -2 & 4-\lambda \end{pmatrix} = (1-\lambda)(4-\lambda) + 2 = \lambda^2 - 5\lambda + 6 = (\lambda-2)(\lambda-3)$$

Thus 2 and 3 are the eigenvalues of A. To find an eigenvector $\mathbf{v}$ for $\lambda = 2$, we will solve $A\mathbf{v} = \lambda\mathbf{v}$ for $\mathbf{v}$. If $\mathbf{v} = \begin{pmatrix} x \\ y \end{pmatrix}$, this means we must find x and y such that

$$\begin{pmatrix} 1 & 1 \\ -2 & 4 \end{pmatrix}\begin{pmatrix} x \\ y \end{pmatrix} = 2\begin{pmatrix} x \\ y \end{pmatrix}, \text{ which is equivalent to } \begin{pmatrix} x+y \\ -2x+4y \end{pmatrix} = \begin{pmatrix} 2x \\ 2y \end{pmatrix}$$

Thus $x + y = 2x$, so that $y = x$. Therefore any vector of the form $\begin{pmatrix} x \\ x \end{pmatrix}$, like $\begin{pmatrix} 1 \\ 1 \end{pmatrix}$, is an eigenvector for $\lambda = 2$. For $\lambda = 3$, we need x and y to satisfy

$$\begin{pmatrix} 1 & 1 \\ -2 & 4 \end{pmatrix}\begin{pmatrix} x \\ y \end{pmatrix} = 3\begin{pmatrix} x \\ y \end{pmatrix}, \quad \text{which is equivalent to} \quad \begin{pmatrix} x + y \\ -2x + 4y \end{pmatrix} = \begin{pmatrix} 3x \\ 3y \end{pmatrix}$$

Thus $x + y = 3x$, so that $y = 2x$. Consequently any vector of the form $\begin{pmatrix} x \\ 2x \end{pmatrix}$, such as $\begin{pmatrix} 1 \\ 2 \end{pmatrix}$, is an eigenvector for $\lambda = 3$.   ❏

It is easy to find a general formula for the eigenvalues of any 2×2 matrix. To confirm this assertion, let $A = \begin{pmatrix} a & b \\ c & d \end{pmatrix}$. Then

$$\det(A - \lambda I) = \det\begin{pmatrix} a - \lambda & b \\ c & d - \lambda \end{pmatrix} = (a - \lambda)(d - \lambda) - bc = \lambda^2 - (a + d)\lambda + (ad - bc)$$

Thus $\det(A - \lambda I) = 0$ if and only if $\lambda^2 - (a + d)\lambda + (ad - bc) = 0$. The roots of this equation are

$$\lambda = \frac{1}{2}(a + d) \pm \frac{1}{2}\sqrt{(a - d)^2 + 4bc}$$

If the roots are real, they are eigenvalues of A. If the roots are complex, then we call them **complex eigenvalues**. Notice that if λ is a complex eigenvalue of the matrix A, then there are *no* vectors $\mathbf{v}$ in R^2 such that $A\mathbf{v} = \lambda\mathbf{v}$.

Similar Matrices

The next topic we discuss is the notion of similarity, which will play an important role in the study of dynamics of two-dimensional functions.

DEFINITION 3.3.   Two 2×2 matrices A and B are **similar** if there is an invertible 2×2 matrix E such that $EA = BE$. In that case we write $A \approx B$, or we write $A \underset{E}{\approx} B$ if we wish to exhibit E, which is called a **similarity matrix**   (for A and B).

EXAMPLE 3. Let $A = \begin{pmatrix} 1 & 1 \\ -2 & 4 \end{pmatrix}$ and $B = \begin{pmatrix} 2 & 0 \\ 0 & 3 \end{pmatrix}$. Show that $A \approx B$.

Solution. Let $E = \begin{pmatrix} a & b \\ c & d \end{pmatrix}$. We will find numerical values of a, b, c, and d in order that $EA = BE$. Equivalently, we will solve

$$\begin{pmatrix} a & b \\ c & d \end{pmatrix}\begin{pmatrix} 1 & 1 \\ -2 & 4 \end{pmatrix} = EA = BE = \begin{pmatrix} 2 & 0 \\ 0 & 3 \end{pmatrix}\begin{pmatrix} a & b \\ c & d \end{pmatrix}$$

for a, b, c, and d. Since

$$\begin{pmatrix} a & b \\ c & d \end{pmatrix}\begin{pmatrix} 1 & 1 \\ -2 & 4 \end{pmatrix} = \begin{pmatrix} a-2b & a+4b \\ c-2d & c+4d \end{pmatrix} \quad \text{and} \quad \begin{pmatrix} 2 & 0 \\ 0 & 3 \end{pmatrix}\begin{pmatrix} a & b \\ c & d \end{pmatrix} = \begin{pmatrix} 2a & 2b \\ 3c & 3d \end{pmatrix}$$

we must simultaneously solve the equations

$$a - 2b = 2a$$
$$a + 4b = 2b$$
$$c - 2d = 3c$$
$$c + 4d = 3d$$

The first two equations yield $a = -2b$, and the last two equations yield $d = -c$. Therefore we can let $E = \begin{pmatrix} -2 & 1 \\ 1 & -1 \end{pmatrix}$, for example. Then $\det E = 1 \neq 0$, so that by Theorem 3.1, E is invertible. A straightforward computation yields $EA = BE$, so that $A \underset{E}{\approx} B$. ❏

We remark that the invertible matrix E such that $EA = BE$ in Example 3 is not unique. There are infinitely many such matrices, including rE for any $r \neq 0$. In addition we observe that A is similar to B precisely when B is similar to A. After all, $EA = BE$ if and only if $AE^{-1} = E^{-1}B$.

Next, suppose that $A \underset{E}{\approx} B$. Then $EA = BE$, or equivalently, $A = E^{-1}BE$. It follows that for any positive integer n,

$$A^n = (E^{-1}BE)^n = (E^{-1}BE)(E^{-1}BE)\cdots(E^{-1}BE) = E^{-1}BIBI\cdots IBE = E^{-1}B^n E$$
$$n \text{ expressions } E^{-1}BE$$

Consequently $A^n \underset{E}{\approx} B^n$, so that

$$\text{if } A \underset{E}{\approx} B, \text{ then } A^n \underset{E}{\approx} B^n \tag{5}$$

This formula will be useful when we analyze iterates of linear functions on R^2.

Now we will show that any 2×2 matrix A is similar to a matrix of one of three simple forms.

THEOREM 3.4. Let $A = \begin{pmatrix} a & b \\ c & d \end{pmatrix}$. Then A is similar to one of the normal

forms

$$\begin{pmatrix} \lambda & 0 \\ 0 & \mu \end{pmatrix}, \qquad \begin{pmatrix} \lambda & \beta \\ 0 & \lambda \end{pmatrix}, \quad \text{or} \quad \begin{pmatrix} \beta & -\gamma \\ \gamma & \beta \end{pmatrix} \tag{6}$$

| A has two distinct real | A has one real | A has two complex eigen- |
| eigenvalues, λ and μ | eigenvalue, λ | values, $\beta + i\gamma$ and $\beta - i\gamma$ |

for appropriate real numbers λ, μ, β, and γ.

Proof. Suppose that $A = \begin{pmatrix} a & b \\ c & d \end{pmatrix}$ has distinct real eigenvalues, λ and μ. We will

show that $B \approx A$, where $B = \begin{pmatrix} \lambda & 0 \\ 0 & \mu \end{pmatrix}$. To that end, let $\mathbf{v} = \begin{pmatrix} r \\ t \end{pmatrix}$ and $\mathbf{w} = \begin{pmatrix} s \\ u \end{pmatrix}$ be

nonzero eigenvectors of A corresponding to the eigenvalues λ and μ, respec-

tively, and let $E = \begin{pmatrix} r & s \\ t & u \end{pmatrix}$. We will verify that $EB = AE$ and that E is

invertible. First we notice that

$$EB = \begin{pmatrix} r & s \\ t & u \end{pmatrix} \begin{pmatrix} \lambda & 0 \\ 0 & \mu \end{pmatrix} = \begin{pmatrix} r\lambda & s\mu \\ t\lambda & u\mu \end{pmatrix}$$

and

$$AE = \begin{pmatrix} a & b \\ c & d \end{pmatrix} \begin{pmatrix} r & s \\ t & u \end{pmatrix} = \begin{pmatrix} ar + bt & as + bu \\ cr + dt & cs + du \end{pmatrix}$$

Next we observe that $A\mathbf{v} = \lambda\mathbf{v}$ and $A\mathbf{w} = \mu\mathbf{w}$, because $\mathbf{v}$ and $\mathbf{w}$ are eigenvectors of A corresponding to the eigenvalues λ and μ, respectively. It follows
that

$$\begin{pmatrix} ar+bt \\ cr+dt \end{pmatrix} = \begin{pmatrix} a & b \\ c & d \end{pmatrix}\begin{pmatrix} r \\ t \end{pmatrix} = A\mathbf{v} = \lambda\mathbf{v} = \lambda\begin{pmatrix} r \\ t \end{pmatrix} = \begin{pmatrix} \lambda r \\ \lambda t \end{pmatrix}$$

and

$$\begin{pmatrix} as+bu \\ cs+du \end{pmatrix} = \begin{pmatrix} a & b \\ c & d \end{pmatrix}\begin{pmatrix} s \\ u \end{pmatrix} = A\mathbf{w} = \mu\mathbf{w} = \mu\begin{pmatrix} s \\ u \end{pmatrix} = \begin{pmatrix} \mu s \\ \mu u \end{pmatrix}$$

Therefore we can substitute in the columns of AE to obtain

$$\begin{pmatrix} ar+bt & as+bu \\ cr+dt & cs+du \end{pmatrix} = \begin{pmatrix} \lambda r & \mu s \\ \lambda t & \mu u \end{pmatrix}$$

Consequently

$$EB = \begin{pmatrix} r\lambda & s\mu \\ t\lambda & u\mu \end{pmatrix} = \begin{pmatrix} ar+bt & as+bu \\ cr+dt & cs+du \end{pmatrix} = AE$$

To finish the proof we need to certify that E is invertible, which by Theorem 3.1 is equivalent to showing that $\det E \neq 0$. However, if $\det E = 0$, then $ru - st = 0$. Since $\mathbf{v} \neq \mathbf{0}$ and $\mathbf{w} \neq \mathbf{0}$, this means that either

$$\frac{r}{t} = \frac{s}{u} \quad \text{or} \quad \frac{t}{r} = \frac{u}{s}$$

Either way, $\mathbf{v}$ would be a constant multiple of $\mathbf{w}$, say, $\mathbf{v} = q\mathbf{w}$. But then

$$\lambda\mathbf{v} = A\mathbf{v} = A(q\mathbf{w}) = qA\mathbf{w} = q\mu\mathbf{w} = \mu(q\mathbf{w}) = \mu\mathbf{v}$$

so that $\lambda = \mu$. But this contradicts the hypothesis that λ and μ are distinct eigenvalues of A. Consequently $\det E \neq 0$, so that E is invertible.

We defer to the exercises the proofs involving one real eigenvalue and complex eigenvalues (see Exercises 21 and 22). ∎

What makes the normal forms of matrices appearing in (6) important for the study in dynamics is the following result.

THEOREM 3.5. If $A \approx B$, then A and B have identical eigenvalues.

Proof. Suppose that $A \underset{E}{\approx} B$, so that $EA = BE$, or equivalently, $A = E^{-1}BE$. Then by (1) and (2),

$$\det (A - \lambda I) = \det (E^{-1}BE - \lambda I) = \det [E^{-1}BE - E^{-1}(\lambda I)E] = \det [E^{-1}(B - \lambda I)E]$$

$$\underset{(1)}{=} [\det (E^{-1})] [\det (B - \lambda I)] [\det E] \underset{(2)}{=} \frac{1}{\det E} [\det (B - \lambda I)][\det E]$$

$$= \det (B - \lambda I)$$

Therefore λ is an eigenvalue of A if and only if λ is an eigenvalue of B. ∎

EXERCISES 3.1

In Exercises 1–3, find the eigenvalues and the normal form, of the given matrix. Also find an eigenvector for each eigenvalue that is a real number.

1. $\begin{pmatrix} 0 & 1 \\ 1 & 0 \end{pmatrix}$ 2. $\begin{pmatrix} 3 & 1 \\ -1 & 1 \end{pmatrix}$ 3. $\begin{pmatrix} 1 & 1 \\ -1 & 1 \end{pmatrix}$

4. Let $A = \begin{pmatrix} 1 & -2 \\ 0 & 3 \end{pmatrix}$. Find $\mathbf{v} = \begin{pmatrix} x \\ y \end{pmatrix}$ such that $A\mathbf{v} = \begin{pmatrix} 0 \\ 1 \end{pmatrix}$.

5. Let a and b be real numbers, and $A = \begin{pmatrix} -2a & 1 \\ b & 0 \end{pmatrix}$. Determine the relationship that must hold between a and b in order for A to have a real eigenvalue.

6. Let θ denote an angle in radians, and let $A = \begin{pmatrix} \cos \theta & \sin \theta \\ -\sin \theta & \cos \theta \end{pmatrix}$.

   a. Show that A is invertible, and find A^{-1}.
   b. Show that A has a real eigenvalue if and only if θ is a multiple of π.

7. Let $A = \begin{pmatrix} a & b \\ c & d \end{pmatrix}$. Describe the relationships that must hold between a, b, c, and d in order for $AB = BA$ for every 2×2 matrix B.

8. Let A and B be 2×2 matrices. Prove that $\det (AB) = (\det A)(\det B)$.

9. Let A be a 2×2 matrix, and suppose that there is a 2×2 matrix B such

that $AB = I$. Show that A has an inverse. (Thus A has an inverse if and only if there is a B such that $AB = I$.)

10. Let A be a 2×2 matrix. Show that if B and C are both inverses of A, then $B = C$. (This means that the inverse of A is unique.)

11. Let A and B be invertible 2×2 matrices. Show that AB is also invertible.

12. Suppose that A, B, and C are 2×2 matrices.
    a. Assume that A is invertible, and that $AB = AC$. Show that $B = C$.
    b. Show that if A is *not* invertible, then there are numbers a, b, and r such that $A = \begin{pmatrix} a & b \\ ra & rb \end{pmatrix}$.
    c. Suppose that A is *not* invertible. Show that there are two distinct matrices B and C such that $AB = AC$.

13. Let A be a 2×2 matrix. Show that the following three conditions are equivalent.

    a. λ is an eigenvalue of A.
    b. There is a nonzero vector $\mathbf{v}$ such that $(A - \lambda I)\mathbf{v} = \mathbf{0}$.
    c. $\det (A - \lambda I) = 0$.

14. Let $A = \begin{pmatrix} a & 0 \\ 0 & d \end{pmatrix}$. Show that the eigenvalues of A are the diagonal elements a and d.

15. Show that 0 is an eigenvalue of the matrix A if and only if A is not invertible.

16. Suppose that λ is a real eigenvalue of A.
    a. Show that λ^n is an eigenvalue of A^n, for each integer $n \geq 1$.
    b. Show that A and A^n have the same eigenvectors.

17. Let A and B be 2×2 matrices. Show that AB and BA have the same eigenvalues.

18. Let A be a 2×2 matrix. Show that if $A \approx I$ then $A = I$.

19. Let $A = \begin{pmatrix} \lambda & a \\ 0 & \lambda \end{pmatrix}$ and $B = \begin{pmatrix} \lambda & b \\ 0 & \lambda \end{pmatrix}$, where a, b, and λ are real numbers,

and a and b nonzero. Show that $A \underset{E}{\approx} B$, where E is a diagonal matrix.

20.   Let b and c be nonzero real numbers. Show that $\begin{pmatrix} 0 & b \\ 0 & 0 \end{pmatrix} = \begin{pmatrix} 0 & 0 \\ c & 0 \end{pmatrix}$.

21.   Suppose that the 2×2 matrix A has the single real eigenvalue λ. Show that $A \approx B$ for some matrix $B \approx \begin{pmatrix} \lambda & \beta \\ 0 & \lambda \end{pmatrix}$.

22.   Suppose that the 2×2 matrix A has complex eigenvalues. Show that $A \approx B$ for some matrix $B = \begin{pmatrix} \beta & -\gamma \\ \gamma & \beta \end{pmatrix}$, where β and γ are real numbers.

3.2 DYNAMICS OF LINEAR FUNCTIONS

Functions of the form $f(x) = ax$, where a is a constant and x is a real number, are linear functions because their graphs are lines. The dynamics of such functions are simple to analyze (see Exercise 13 of Section 1.2). For linear functions in two dimensions, the analysis is also reasonable.   It is the dynamics of such functions that we study in this section.

Linear Functions

We begin with the definition of a linear function defined on R^2.

DEFINITION 3.6. The function $L: R^2 \to R^2$ is **linear** if

$$L(b\mathbf{v} + c\mathbf{w}) = bL(\mathbf{v}) + cL(\mathbf{w}) \tag{1}$$

for all $\mathbf{v}$ and $\mathbf{w}$ in R^2, and all real numbers b and c. A linear function is also called a **linear map**.

The reason that such functions are called linear is that the image of any line is a line or, in special cases, a point. Before we show this in Example 1, we recall that a line in the plane can be represented parametrically as the collection of all vectors of the form $t\mathbf{v} + \mathbf{w}$, where $\mathbf{v}$ and $\mathbf{w}$ are fixed vectors with $\mathbf{v} \neq \mathbf{0}$, and where t is any real number. The line passes through $\mathbf{w}$, and if $\mathbf{w} = \mathbf{0}$, then the line passes through the origin. The vector $\mathbf{v}$ tells the slope of the line. Specifically, let $\mathbf{v} = \begin{pmatrix} r \\ s \end{pmatrix}$. If $r \neq 0$, then the slope of the line is s/r; if $r = 0$, then the line is vertical.

EXAMPLE 1. Let $L: R^2 \to R^2$ be linear. Show that the image of any line γ is a line or a point.

Solution. Let γ be the collection of all vectors of the form $t\mathbf{v} + \mathbf{w}$, for all real t. By (1),

$$L(t\mathbf{v} + \mathbf{w}) = tL(\mathbf{v}) + L(\mathbf{w})$$

On the one hand, if $L(\mathbf{v}) \neq \mathbf{0}$, then the image of γ is the line passing through $L(\mathbf{w})$ with slope determined by the coordinates of $L(\mathbf{v})$. On the other hand, if $L(\mathbf{v}) = \mathbf{0}$, then the image of γ is the point $L(\mathbf{w})$. ❏

Any vector $\mathbf{v} = \begin{pmatrix} x \\ y \end{pmatrix}$ in R^2 can be written in terms of the coordinate vectors $\mathbf{i} = \begin{pmatrix} 1 \\ 0 \end{pmatrix}$ and $\mathbf{j} = \begin{pmatrix} 0 \\ 1 \end{pmatrix}$. Indeed,

$$\mathbf{v} = \begin{pmatrix} x \\ y \end{pmatrix} = x \begin{pmatrix} 1 \\ 0 \end{pmatrix} + y \begin{pmatrix} 0 \\ 1 \end{pmatrix} = x\mathbf{i} + y\mathbf{j}$$

As a result, if $L: R^2 \to R^2$ is any linear function, and if $\mathbf{v} = \begin{pmatrix} x \\ y \end{pmatrix}$, then by (1),

$$L(\mathbf{v}) = L(x\mathbf{i} + y\mathbf{j}) = xL(\mathbf{i}) + yL(\mathbf{j}) \tag{2}$$

Therefore $L(\mathbf{i})$ and $L(\mathbf{j})$ uniquely define the linear function L.

Assume once again that $L: R^2 \to R^2$ is a linear function, and suppose that $L(\mathbf{i}) = \begin{pmatrix} a \\ c \end{pmatrix}$ and $L(\mathbf{j}) = \begin{pmatrix} b \\ d \end{pmatrix}$. Then we define the 2×2 matrix A_L by

$$A_L = \begin{pmatrix} a & b \\ c & d \end{pmatrix} \tag{3}$$

EXAMPLE 2.   Let $L\begin{pmatrix} x \\ y \end{pmatrix} = \begin{pmatrix} x + y \\ -2x + 4y \end{pmatrix}$ for all x and y in R. Find A_L.

Solution.   Since

$$L(\mathbf{i}) = L\begin{pmatrix} 1 \\ 0 \end{pmatrix} = \begin{pmatrix} 1 \\ -2 \end{pmatrix} \quad \text{and} \quad L(\mathbf{j}) = L\begin{pmatrix} 0 \\ 1 \end{pmatrix} = \begin{pmatrix} 1 \\ 4 \end{pmatrix}$$

it follows from (3) that $A_L = \begin{pmatrix} 1 & 1 \\ -2 & 4 \end{pmatrix}$.   ❏

A natural relationship between a linear function L on R^2 and its associated matrix A_L is given in the next theorem.

THEOREM 3.7.   Let $L:R^2 \to R^2$ be an arbitrary linear function, and A_L the corresponding matrix defined in (3). Then

$$L(\mathbf{v}) = A_L\mathbf{v} \quad \text{for all } \mathbf{v} \text{ in } R^2$$

Proof. If $L(\mathbf{i}) = \begin{pmatrix} a \\ c \end{pmatrix}$ and $L(\mathbf{j}) = \begin{pmatrix} b \\ d \end{pmatrix}$, then by (3), $A_L = \begin{pmatrix} a & b \\ c & d \end{pmatrix}$. Thus for any

$\mathbf{v} = \begin{pmatrix} x \\ y \end{pmatrix}$ we have

$$A_L\mathbf{v} = \begin{pmatrix} a & b \\ c & d \end{pmatrix}\begin{pmatrix} x \\ y \end{pmatrix} = \begin{pmatrix} ax + by \\ cx + dy \end{pmatrix}$$

Next, the formula in (2) tells us that

$$L(\mathbf{v}) = L(x\mathbf{i} + y\mathbf{j}) \underset{(2)}{=} xL(\mathbf{i}) + yL(\mathbf{j}) = x\begin{pmatrix} a \\ c \end{pmatrix} + y\begin{pmatrix} b \\ d \end{pmatrix} = \begin{pmatrix} xa + yb \\ xc + yd \end{pmatrix}$$

Consequently $L(\mathbf{v}) = A_L\mathbf{v}$, which completes the proof.   ∎

The following result is a nearly immediate consequence of the theorem.

COROLLARY 3.8.   If L is any linear function, then the matrix A_L is unique. Conversely, if A is any matrix, then there is a unique linear function L such that $A_L = A$.

Proof. Let L be an arbitrary linear function. Since

$$A_L \mathbf{i} = \text{first column of } A_L \quad \text{and} \quad A_L \mathbf{j} = \text{second column of } A_L$$

it follows that the matrix A_L is unique. Conversely, if A is any given matrix, then the function L defined by $L(\mathbf{v}) = A\mathbf{v}$ for all $\mathbf{v}$ in R^2 is linear, and has the property that $A_L = A$ (Exercise 12). ∎

We know from Example 1 that if L is a linear function, then the image of a line is a line or a point. One of our goals is to determine conditions under which the image of a line is contained in the same line. We say that a line γ is **invariant** under L if $L(\gamma) \subseteq \gamma$. Thus we seek lines that are invariant for a given linear function.

Suppose that L is a linear function. If λ is a real eigenvalue of A_L and corresponds to the eigenvector $\mathbf{v}$, then the line $\gamma_\mathbf{v}$ through the origin and parallel to $\mathbf{v}$ is invariant under L. The reason is that by Theorem 3.7,

$$L(t\mathbf{v}) = A_L(t\mathbf{v}) = \lambda t\mathbf{v} \text{ for any real number } t \tag{4}$$

Therefore the image of $\gamma_\mathbf{v}$ is contained in the line $\gamma_\mathbf{v}$.

EXAMPLE 3. Let $L\begin{pmatrix} x \\ y \end{pmatrix} = \begin{pmatrix} x+y \\ -2x+4y \end{pmatrix}$. Find lines through the origin that are invariant under L.

Solution. By Example 2, $A_L = \begin{pmatrix} 1 & 1 \\ -2 & 4 \end{pmatrix}$. In Example 2 of Section 3.1 we found that $\mathbf{v} = \begin{pmatrix} 1 \\ 1 \end{pmatrix}$ and $\mathbf{w} = \begin{pmatrix} 1 \\ 2 \end{pmatrix}$ are eigenvectors for the eigenvalues (respectively 2 and 3) of A_L. Consequently by (4) the lines $\gamma_\mathbf{v}$ and $\gamma_\mathbf{w}$ determined by $\mathbf{v}$ and $\mathbf{w}$ and passing through the origin are invariant under L. ❑

The linear function L in Example 3 doubles distances between points in $\gamma_\mathbf{v}$ because $\mathbf{v}$ corresponds to the eigenvalue 2 of A_L. Similarly, L triples distances between points on $\gamma_\mathbf{w}$, since the corresponding eigenvalue of A_L is 3. Despite these results, we have not completed our analysis of the dynamics of L, because we do not as yet know the behavior of iterates of points not on $\gamma_\mathbf{v}$ or $\gamma_\mathbf{w}$. Later we will be able to resolve this issue.

Eigenvalues of A_L play an integral part in the analysis of L. For convenience we will refer to λ as an **eigenvalue** of L if λ is an eigenvalue of the corresponding matrix A_L.

The next theorem shows that composition of linear functions corresponds to multiplication of matrices.

THEOREM 3.9. Let $L: R^2 \rightarrow R^2$ and $M: R^2 \rightarrow R^2$ be linear functions. Then $A_{L \circ M} = A_L A_M$.

Proof. If $\mathbf{v}$ is arbitrary, then by repeated applications of Theorem 3.7 we have

$$A_{L \circ M} \mathbf{v} = (L \circ M)\mathbf{v} = L(M(\mathbf{v})) = L(A_M \mathbf{v}) = A_L (A_M \mathbf{v}) = A_L A_M \mathbf{v}$$

Since

$$\text{1st column of } A_{L \circ M} = A_{L \circ M} \mathbf{i} = A_L A_M \mathbf{i} = \text{1st column of } A_L A_M$$

and similarly,

$$\text{2nd column of } A_{L \circ M} = \text{2nd column of } A_L A_M$$

we conclude that $A_{L \circ M} = A_L A_M$. ∎

In the unlikely event that $A_L A_M = I$, Theorems 3.7 and 3.9 tell us that

$$(L \circ M)\mathbf{v} = A_{L \circ M} \mathbf{v} = (A_L A_M)\mathbf{v} = I\mathbf{v} = \mathbf{v} \text{ for all } \mathbf{v}$$

Consequently $(L \circ M) = I$. By analogy with functions of one variable, we say that L is **invertible**, and that M is the **inverse** of L. We write L^{-1} for the inverse of L. Notice that the linear functions L and M are inverses of one another if their associated matrices A_L and A_M are inverses of each other (see Exercise 13).

EXAMPLE 4. Let $A_L = \begin{pmatrix} 3 & 0 \\ 2 & -1 \end{pmatrix}$ and $A_M = \begin{pmatrix} 1/3 & 0 \\ 2/3 & -1 \end{pmatrix}$. Show that L and M are inverses of one another.

Solution. By Theorem 3.9,

$$A_{L \circ M} = A_L A_M = \begin{pmatrix} 3 & 0 \\ 2 & -1 \end{pmatrix} \begin{pmatrix} 1/3 & 0 \\ 2/3 & -1 \end{pmatrix} = \begin{pmatrix} 1 & 0 \\ 0 & 1 \end{pmatrix}$$

By the remarks preceding the example, L and M are inverses of one another. ❏

If $L = M$, then Theorem 3.9 implies that $A_{L^{[2]}} = (A_L)^2$. More generally,

$$A_{L^{[n]}} = (A_L)^n \quad \text{for any } n \geq 2 \tag{5}$$

To illustrate (5), let $L\begin{pmatrix} x \\ y \end{pmatrix} = \begin{pmatrix} x + y \\ -2x + 4y \end{pmatrix}$, so that $A_L = \begin{pmatrix} 1 & 1 \\ -2 & 4 \end{pmatrix}$. Therefore

$$A_{L^{[2]}} = \begin{pmatrix} 1 & 1 \\ -2 & 4 \end{pmatrix}^2 = \begin{pmatrix} -1 & 5 \\ -10 & 14 \end{pmatrix} \quad \text{and} \quad A_{L^{[3]}} = \begin{pmatrix} 1 & 1 \\ -2 & 4 \end{pmatrix}^3 = \begin{pmatrix} -11 & 19 \\ -38 & -46 \end{pmatrix}$$

Calculating the entries of $A_{L^{[n]}}$ by using (5) becomes cumbersome as n increases. The notion of conjugacy will allow us to evaluate $A_{L^{[n]}}$ with a minimum of effort.

DEFINITION 3.10. Let $L: R^2 \rightarrow R^2$ and $M: R^2 \rightarrow R^2$ be linear functions. Then L and M are **linearly conjugate** if there exists an invertible linear function $P: R^2 \rightarrow R^2$ such that $P \circ L = M \circ P$. In that case we write $L \approx M$, or if we wish to emphasize the role of P, we write $L \underset{P}{\approx} M$.

Linear conjugacy of functions corresponds to similarity of the associated matrices, as Theorem 3.11 suggests.

THEOREM 3.11. Let $L: R^2 \rightarrow R^2$ and $M: R^2 \rightarrow R^2$ be linear functions. Then $L \underset{P}{\approx} M$ if and only if $A_L \underset{A_P}{\approx} A_M$.

Proof. By definition, $L \underset{P}{\approx} M$ if and only if $P \circ L = M \circ P$, which is equivalent to $A_{P \circ L} = A_{M \circ P}$ by Corollary 3.8. By Theorem 3.9, this is tantamount to $A_P A_L = A_M A_P$, which means that $A_L \underset{A_P}{\approx} A_M$. ∎

Theorem 3.11 and (5) help us to solve the following example expeditiously.

EXAMPLE 5. Let $A_L = \begin{pmatrix} 1 & 1 \\ -2 & 4 \end{pmatrix}$. Find the entries of $A_{L^{[n]}}$.

Solution. By Example 3 of Section 3.1, $A_L \underset{E}{\approx} A_M$, where $A_M = \begin{pmatrix} 2 & 0 \\ 0 & 3 \end{pmatrix}$ and $E = \begin{pmatrix} -2 & 1 \\ 1 & -1 \end{pmatrix}$. You can check that $E^{-1} = \begin{pmatrix} -1 & -1 \\ -1 & -2 \end{pmatrix}$. Then (5) in the present section and (5) in Section 3.1 together yield

$$A_{L^{[n]}} = (A_L)^n = E^{-1}(A_M)^n E = \begin{pmatrix} -1 & -1 \\ -1 & -2 \end{pmatrix} \begin{pmatrix} 2 & 0 \\ 0 & 3 \end{pmatrix}^n \begin{pmatrix} -2 & 1 \\ 1 & -1 \end{pmatrix}$$

$$= \begin{pmatrix} -1 & -1 \\ -1 & -2 \end{pmatrix} \begin{pmatrix} 2^n & 0 \\ 0 & 3^n \end{pmatrix} \begin{pmatrix} -2 & 1 \\ 1 & -1 \end{pmatrix}$$

$$= \begin{pmatrix} 2^{n+1} - 3^n & -2^n + 3^n \\ 2^{n+1} - 2(3^n) & -2^n + 2(3^n) \end{pmatrix} \qquad \square$$

Now we are ready to discuss the dynamics of linear functions defined on R^2.

Dynamics of Linear Functions

Every linear function L has the fixed point $\mathbf{0}$, since $L(\mathbf{0}) = \mathbf{0}$. Before we can say whether $\mathbf{0}$ is attracting or repelling (or neither), we need to indicate a distance on R^2. To that end, let $\mathbf{v} = \begin{pmatrix} x \\ y \end{pmatrix}$ and $\mathbf{w} = \begin{pmatrix} r \\ s \end{pmatrix}$. As is usual, we let the **distance** $\|\mathbf{v} - \mathbf{w}\|$ between $\mathbf{v}$ and $\mathbf{w}$ be the distance between the corresponding points in R^2, that is,

$$\|\mathbf{v} - \mathbf{w}\| = \sqrt{(r-x)^2 + (s-y)^2}$$

If $\mathbf{w} = \mathbf{0}$, then we find that $\|\mathbf{v}\| = \sqrt{x^2 + y^2}$, which is the distance between the point corresponding to $\mathbf{v}$ and the origin. We observe that $\| \ \|$ is a metric on R^2, because for any $\mathbf{v}$ and $\mathbf{w}$,

$$\|\mathbf{v}\| \geq 0, \text{ and } \|\mathbf{v}\| = 0 \text{ if and only if } \mathbf{v} = \mathbf{0}$$

$$\|r\mathbf{v}\| = |r| \|\mathbf{v}\| \text{ for any real number } r$$

$$\|\mathbf{v} + \mathbf{w}\| \leq \|\mathbf{v}\| + \|\mathbf{w}\| \qquad \text{(triangle inequality)} \qquad (6)$$

In addition, if $(\mathbf{v}_n)_{n=0}^{\infty}$ is a sequence of elements of R^2 and $\mathbf{w}$ is in R^2, then $(\mathbf{v}_n)_{n=0}^{\infty}$ **converges** to $\mathbf{w}$ if $\|\mathbf{v}_n - \mathbf{w}\| \to 0$ as n increases without bound. In this case we write $\mathbf{v}_n \to \mathbf{w}$.

We say that $L: R^2 \to R^2$ is **continuous** at $\mathbf{0}$ if

for all $\varepsilon > 0$ there is a $\delta > 0$ such that
if $\|\mathbf{v}\| < \delta$ then $\|L(\mathbf{v})\| < \varepsilon$

This corresponds to the notion of continuity for functions of one variable, and is equivalent to the following:

$$\text{if } \mathbf{v}_n \to \mathbf{0}, \quad \text{then } L(\mathbf{v}_n) \to \mathbf{0} \tag{7}$$

In order to prove that every linear function L on R^2 is continuous at $\mathbf{0}$, we need the following preliminary result.

THEOREM 3.12. Let $A = \begin{pmatrix} a & b \\ c & d \end{pmatrix}$ be any 2×2 matrix. Suppose that $\mathbf{v}_n = \begin{pmatrix} x_n \\ y_n \end{pmatrix}$ for $n = 1, 2, \ldots,$ and that $\mathbf{v}_n \to \mathbf{0}$. Then $A\mathbf{v}_n \to \mathbf{0}$.

Proof. Let $\varepsilon > 0$, and let $r = |a| + |b| + |c| + |d|$. If $r = 0$, then $a = b = c = d = 0$, so that A is the 0 matrix and hence $A\mathbf{v}_n = \mathbf{0}$ for all n. Therefore we assume from now on that $r > 0$. Since $\mathbf{v}_n \to \mathbf{0}$ by hypothesis, there is an N such that if $n \geq N$, then

$$\|\mathbf{v}_n\| = \sqrt{x_n^2 + y_n^2} \; < \; \frac{\varepsilon}{r}$$

This means that $|x_n| < \varepsilon/r$ and $|y_n| < \varepsilon/r$ for all $n \geq N$. It follows that

$$\|A\mathbf{v}_n\| = \left\| \begin{pmatrix} a & b \\ c & d \end{pmatrix} \begin{pmatrix} x_n \\ y_n \end{pmatrix} \right\| = \sqrt{(ax_n + by_n)^2 + (cx_n + dy_n)^2}$$

$$< \sqrt{(\frac{\varepsilon}{r})^2 \, [(|a| + |b|)^2 + (|c| + |d|)^2]} \; \leq \; \frac{\varepsilon}{r} \sqrt{(|a| + |b| + |c| + |d|)^2}$$

$$= \frac{\varepsilon}{r} r = \varepsilon \quad \blacksquare$$

COROLLARY 3.13. Let $L: R^2 \to R^2$ be a linear function. Then L is continuous at $\mathbf{0}$.

Proof. Let $\mathbf{v}_n \to \mathbf{0}$. Then $L(\mathbf{v}_n) = A_L \mathbf{v}_n \to \mathbf{0}$ by Theorem 3.12. Therefore (7) implies that L is continuous at $\mathbf{0}$. $\quad \blacksquare$

Our next goal is to show that if L has two real eigenvalues λ and μ with $|\lambda| < 1$ and $|\mu| < 1$, then for each $\mathbf{v}$ in R^2, $L^{[n]}(\mathbf{v}) \to \mathbf{0}$. In this case we call $\mathbf{0}$ an **attracting fixed point** of L.

THEOREM 3.14. Let $L: R^2 \to R^2$ have the property that A_L has distinct real eigenvalues λ and μ, with $|\lambda| < 1$ and $|\mu| < 1$. Then for every $\mathbf{v}$ in R^2, $L^{[n]}(\mathbf{v}) \to \mathbf{0}$. Therefore $\mathbf{0}$ is an attracting fixed point of L, and R^2 is the basin of attraction of $\mathbf{0}$.

Proof. Let $\mathbf{v}$ be arbitrary in R^2. Since $L^{[n]}(\mathbf{v}) = (A_L)^n \mathbf{v}$ by (5), we need only

show that $(A_L)^n \mathbf{v} \to \mathbf{0}$. Now let $B = \begin{pmatrix} \lambda & 0 \\ 0 & \mu \end{pmatrix}$. By Theorem 3.4, $A_L \approx B$, so that

there is an invertible matrix E such that $A_L \underset{E}{\approx} B$. Then (5) in Section 3.1 implies

that $(A_L)^n \underset{E}{\approx} B^n$, so that $(A_L)^n = E^{-1} B^n E$, and hence

$$(A_L)^n \mathbf{v} \;=\; E^{-1} B^n \, (E \, \mathbf{v}) \tag{8}$$

The proof will be complete if we show that $E^{-1} B^n (E\mathbf{v}) \to \mathbf{0}$.

From Example 1 in Section 3.1 we know that $B^n = \begin{pmatrix} \lambda^n & 0 \\ 0 & \mu^n \end{pmatrix}$.   Next, let

$E \, \mathbf{v} = \begin{pmatrix} x \\ y \end{pmatrix}$. Then

$$B^n \, E \, \mathbf{v} \;=\; \begin{pmatrix} \lambda^n & 0 \\ 0 & \mu^n \end{pmatrix} \begin{pmatrix} x \\ y \end{pmatrix} \;=\; \begin{pmatrix} \lambda^n x \\ \mu^n y \end{pmatrix}$$

Since $|\lambda| < 1$ and $|\mu| < 1$ by hypothesis, we find that

$$\|B^n \, E \, \mathbf{v}\| \;=\; \left\| \begin{pmatrix} \lambda^n x \\ \mu^n y \end{pmatrix} \right\| \;=\; \sqrt{\lambda^{2n} x^2 + \mu^{2n} y^2} \;\to\; 0$$

as n increases without bound.   Therefore (8) and Theorem 3.12, with $\mathbf{v}_n$ replaced by $B^n E \mathbf{v}$, tell us that

$$L^{[n]}(\mathbf{v}) \;=\; (A_L)^n \, \mathbf{v} \;=\; E^{-1}(B^n \, E \, \mathbf{v}) \to \mathbf{0}$$

Thus not only is $\mathbf{0}$ an attracting fixed point of L, but also the basin of attraction of $\mathbf{0}$ is R^2 itself. This completes the proof.   ∎

The conclusions of Theorem 3.14 remain valid if the eigenvalues are not necessarily real, as the next result states.

COROLLARY 3.15. Let $L: R^2 \to R^2$ be a linear function with eigenvalues λ and μ such that $|\lambda| < 1$ and $|\mu| < 1$. Then $L^{[n]}(\mathbf{v}) \to \mathbf{0}$ for each $\mathbf{v}$ in R^2. Therefore $\mathbf{0}$ is an attracting fixed point whose basin of attraction is R^2.

Proof. Theorem 3.14 proves the result if λ and μ are real and distinct. The cases in which L has one real eigenvalue, or complex eigenvalues, are left as exercises (Exercises 17 and 19). ∎

We observe that if $|\lambda| > 1$ and $|\mu| \geq 1$, then the proof of Theorem 3.14 shows that $\|L^{[n]}(\mathbf{v})\| \to \infty$ as n increases without bound. In this case we say that $\mathbf{0}$ is a **repelling fixed point** of L. For example, if $L\begin{pmatrix} x \\ y \end{pmatrix} = \begin{pmatrix} x+y \\ -2x+4y \end{pmatrix}$, then the eigenvalues are 2 and 3, by Example 2 of Section 3.1. It follows that $\mathbf{0}$ is a repelling fixed point, and iterates of all points except $\mathbf{0}$ recede from $\mathbf{0}$.

Lines determined by eigenvectors are invariant under a linear function L, as we have already seen. We can extend the notion of invariance to other sets. A set C in R^2 is **invariant** under a linear function $L: R^2 \to R^2$ if $L(C) \subseteq C$. Thus C is invariant under L if C is mapped into itself by L. Using invariant parabolas, we can determine the dynamics of the linear function in the following example.

EXAMPLE 6. Let $L\begin{pmatrix} x \\ y \end{pmatrix} = \begin{pmatrix} x/2 \\ y/4 \end{pmatrix}$. Show that $\mathbf{0}$ is an attracting fixed point whose basin of attraction is R^2. Also show that for any number $r \neq 0$, the parabola $y = rx^2$ is invariant under L.

Solution. We find that

$$L\begin{pmatrix} x \\ y \end{pmatrix} = \begin{pmatrix} 1/2 & 0 \\ 0 & 1/4 \end{pmatrix}\begin{pmatrix} x \\ y \end{pmatrix}$$

so that the eigenvalues of L are $1/2$ and $1/4$. Theorem 3.14 implies that $\mathbf{0}$ is an attracting fixed point whose basin of attraction is R^2. If $y = rx^2$ with $r \neq 0$, then

$$L\begin{pmatrix} x \\ y \end{pmatrix} = L\begin{pmatrix} x \\ rx^2 \end{pmatrix} = \begin{pmatrix} x/2 \\ rx^2/4 \end{pmatrix} = \begin{pmatrix} x/2 \\ r(x/2)^2 \end{pmatrix}$$

The latter point lies on the parabola $y = rx^2$, so it is invariant under L. ❏

In addition to the invariant parabolas, the axes are also invariant under L in Example 6. Since every point in R^2 lies on some parabola $y = rx^2$ or on a coordinate axis, it follows that the iterates of every vector $\mathbf{v}$ converge to $\mathbf{0}$ along a suitable parabola or along a coordinate axis. With these observations, we are ready

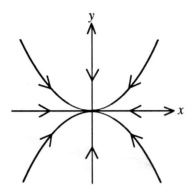

Figure 3.1

to draw a figure describing the dynamics of L. Figure 3.1 is called a **portrait** of L.

More generally, when a linear function L has two real eigenvalues whose absolute values are less than 1, the portrait of L can have the same general form as that in Figure 3.1, the invariant lines and curves adjusted appropriately. To support this claim, let $L\begin{pmatrix} x \\ y \end{pmatrix} = \begin{pmatrix} ax \\ dy \end{pmatrix}$, so that $A_L = \begin{pmatrix} a & 0 \\ 0 & d \end{pmatrix}$. Suppose $a > 0$ and $d > 0$. We will determine a value of α for which the graph C of $y = rx^\alpha$ is invariant under L for every real number r and all positive x. If $\begin{pmatrix} x \\ y \end{pmatrix}$ lies on C, then $L\begin{pmatrix} x \\ y \end{pmatrix} = \begin{pmatrix} ax \\ drx^\alpha \end{pmatrix}$. This latter point lies on C only if

$$drx^\alpha = r(ax)^\alpha, \text{ so that } \alpha = \frac{\ln d}{\ln a} \tag{9}$$

As a result, curves of the form $y = rx^\alpha$ are invariant under L.

Now suppose that the two eigenvalues of L satisfy $|\lambda| < 1$ and $|\mu| < 1$, but the associated matrix is not necessarily diagonal. This is where similar matrices play a decisive role. The reason is that if M is a diagonal matrix such that $L \underset{P}{\approx} M$, and if C is a set that is invariant under M, then

$$(PLP^{-1})(C) = M(C) \subseteq C, \text{ so that } L(P^{-1}(C)) \subseteq P^{-1}(C)$$

This means that $P^{-1}(C)$ is the adjusted curve, and is invariant under L.

EXAMPLE 7. Let $A_L = \begin{pmatrix} 1/2 & 1/8 \\ 1/2 & 1/2 \end{pmatrix}$. Analyze the dynamics of L.

Solution. The eigenvalues of A_L are 1/4 and 3/4, and corresponding eigenvectors

are $\begin{pmatrix} 1 \\ -2 \end{pmatrix}$ and $\begin{pmatrix} 1 \\ 2 \end{pmatrix}$. By Theorem 3.4, $A_L \approx A_M$, where $A_M = \begin{pmatrix} 1/4 & 0 \\ 0 & 3/4 \end{pmatrix}$.

Next, notice that A_L and A_M have the same eigenvalues. Using (9), we calculate that with respect to M, iterates of points converge to **0** along curves C of the form $y = rx^\alpha$, where $\alpha = 1 - (\ln 3)/(\ln 4) \approx .2075$. If P is a linear function such that $L \underset{P}{\approx} M$, then from the remarks preceding the example, $P^{-1}(C)$ is invariant under L. As a result, we can draw a portrait of L in Figure 3.2.   ❑

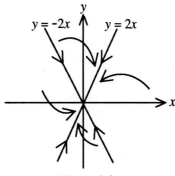

Figure 3.2

Next we will discover what happens when L has eigenvalues λ and μ with $|\lambda| > 1$ and $|\mu| < 1$.

EXAMPLE 8. Let $L\begin{pmatrix} x \\ y \end{pmatrix} = \begin{pmatrix} 2x \\ y/2 \end{pmatrix}$. Show that for each $r \neq 0$, the hyperbola $y = r/x$ is invariant under L. Then analyze the dynamics of L and draw a portrait of L.

Solution. Let $r \neq 0$. If $y = r/x$, then

$$L\begin{pmatrix} x \\ y \end{pmatrix} = L\begin{pmatrix} x \\ r/x \end{pmatrix} = \begin{pmatrix} 2x \\ r/(2x) \end{pmatrix}$$

Since the latter point lies on the hyperbola $y = r/x$, it follows that the hyperbola is invariant under L. Next, we notice that if $y = r/x$, then

$$L^{[n]}\begin{pmatrix} x \\ y \end{pmatrix} = \begin{pmatrix} 2^n x \\ r/(2^n x) \end{pmatrix}$$

As n increases without bound, $L^{[n]}\begin{pmatrix} x \\ y \end{pmatrix}$ approaches the x axis (since $r/(2^n x) \to 0$) and recedes from the y axis (because $|2^n x| \to \infty$). Every point not on the x or y axis is on a hyperbola $y = r/x$ for an appropriate value of r. We conclude that the iterates of all points not on either the x or y axis eventually recede from the origin (Figure 3.3). Since

$$L^{-[n]}\begin{pmatrix} x \\ 0 \end{pmatrix} = \begin{pmatrix} 2^{-n}x \\ 0 \end{pmatrix} \quad \text{and} \quad L^{[n]}\begin{pmatrix} 0 \\ y \end{pmatrix} \to \mathbf{0}$$

iterates of all points on the x axis except for $\mathbf{0}$ also recede from $\mathbf{0}$, whereas iterates of all points on the y axis converge to $\mathbf{0}$. These conclusions are registered in the portrait of L given in Figure 3.3.   ❏

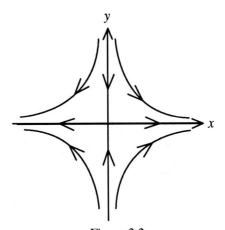

Figure 3.3

If L has real eigenvalues λ and μ with $|\lambda| > 1$ and $|\mu| < 1$, then $\mathbf{0}$ is a **saddle point**. Moreover, one can show that there is an $\alpha < 0$ such that the curves $y = rx^\alpha$ are invariant, for each $r \neq 0$ (Exercise 20). The portrait of such an L has the same general appearance as that in Figure 3.3, with appropriate distortions.

We turn to linear functions whose associated eigenvalues are not real.

EXAMPLE 9. Let $L\begin{pmatrix} x \\ y \end{pmatrix} = \begin{pmatrix} -y \\ x \end{pmatrix}$. Find the eigenvalues of L, and discuss the dynamics of L.

Solution. Notice that

$$L\begin{pmatrix} x \\ y \end{pmatrix} = \begin{pmatrix} 0 & -1 \\ 1 & 0 \end{pmatrix}\begin{pmatrix} x \\ y \end{pmatrix}$$

Since

$$\det\begin{pmatrix} 0-\lambda & -1 \\ 1 & 0-\lambda \end{pmatrix} = \lambda^2 + 1 = 0 \text{ if and only if } \lambda = \pm i$$

it follows that the eigenvalues of L are the complex numbers i and $-i$. Next, we use polar coordinates for x and y:

$$x = r \cos \theta \quad \text{and} \quad y = r \sin \theta$$

From trigonometry we know that

$$-\sin \theta = \cos (\theta + \pi/2) \quad \text{and} \quad \cos \theta = \sin (\theta + \pi/2)$$

so that

$$L\begin{pmatrix} r \cos \theta \\ r \sin \theta \end{pmatrix} = \begin{pmatrix} -r \sin \theta \\ r \cos \theta \end{pmatrix} = \begin{pmatrix} r \cos (\theta + \pi/2) \\ r \sin (\theta + \pi/2) \end{pmatrix} \qquad (10)$$

The formula in (10) tells us that if $\mathbf{v}$ has polar coordinates (r, θ), then $L(\mathbf{v})$ has polar coordinates $(r, \theta + \pi/2)$. In other words, $L(\mathbf{v})$ lies on the same circle of radius r centered at $\mathbf{0}$, but is rotated counterclockwise $\pi/2$ radians (Figure 3.4). Evidently each such circle is invariant under L.   ☐

In general, if $L: R^2 \rightarrow R^2$ is linear, and if

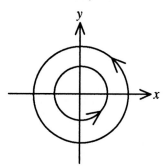

Figure 3.4

$$A_L = \begin{pmatrix} a & -b \\ b & a \end{pmatrix} \text{ with } a^2 + b^2 = 1 \text{ and } a \neq 0$$

then it can be shown that L acts as a rotation through an angle of arctan (b/a). Equivalently, if

$$A_L = \begin{pmatrix} \cos \theta & -\sin \theta \\ \sin \theta & \cos \theta \end{pmatrix}$$

then L acts as a rotation about the origin through an angle of θ in the counterclockwise direction (Exercise 7). For such L, there are *no* real-valued eigenvectors.

In our final example we discuss what can happen if L has but one real eigenvalue.

EXAMPLE 10. Let $L\begin{pmatrix} x \\ y \end{pmatrix} = \begin{pmatrix} 2x + y \\ 2y \end{pmatrix}$. Discuss the dynamics of L.

Solution. Here we have

$$L\begin{pmatrix} x \\ y \end{pmatrix} = \begin{pmatrix} 2 & 1 \\ 0 & 2 \end{pmatrix}\begin{pmatrix} x \\ y \end{pmatrix} = 2\begin{pmatrix} x \\ y \end{pmatrix} + \begin{pmatrix} y \\ 0 \end{pmatrix}$$

Letting

$$L_1\begin{pmatrix} x \\ y \end{pmatrix} = 2\begin{pmatrix} x \\ y \end{pmatrix} \quad \text{and} \quad L_2\begin{pmatrix} x \\ y \end{pmatrix} = \begin{pmatrix} y \\ 0 \end{pmatrix}$$

we notice that L_1 multiplies the distance between $\begin{pmatrix} x \\ y \end{pmatrix}$ and $\mathbf{0}$ by a factor of 2,

and L_2 projects $\begin{pmatrix} x \\ y \end{pmatrix}$ onto the y axis and then rotates $90°$ counterclockwise. ❑

As you can see, the values of the eigenvalues λ and μ of L determine the behavior of the iterates of L. We summarize some of them below:

i. Suppose that λ and μ are real. If $|\lambda| < 1$ and $|\mu| < 1$, we have $L^{[n]}(\mathbf{v}) \to \mathbf{0}$ for all $\mathbf{v}$ in R^2, so that $\mathbf{0}$ is an attracting fixed point of L. If $|\lambda| > 1$ and $|\mu| \geq 1$, then $\|L^{[n]}(\mathbf{v})\| \to \infty$ for all nonzero $\mathbf{v}$ in R^2, so that $\mathbf{0}$ is a repelling fixed point of L.

ii. Suppose that λ and μ are real. If $|\lambda| > 1$ and $|\mu| < 1$, then $\mathbf{0}$ is a saddle point.

iii.   If the eigenvalues λ and μ are complex, then L has a rotation component.

EXERCISES 3.2

1. Let $L\begin{pmatrix} x \\ y \end{pmatrix} = \begin{pmatrix} x/2 + y/8 \\ x/2 + y/2 \end{pmatrix}$, so that L is the linear function in Example 7.
    a.   Confirm the eigenvalues and corresponding eigenvectors of L.
    b.   Confirm the value of α such that the graph of $y = rx^{\alpha}$ is invariant under L for each $r \neq 0$.
    c.   Confirm that Figure 3.2 is a portrait of L.

In Exercises 2–6, let $L: R^2 \rightarrow R^2$. Find the eigenvalues of L, and the eigenvectors where they exist. Draw a portrait of L.

2. $L\begin{pmatrix} x \\ y \end{pmatrix} = \begin{pmatrix} x/2 \\ y/3 \end{pmatrix}$ 3. $L\begin{pmatrix} x \\ y \end{pmatrix} = \begin{pmatrix} x/3 \\ 3y \end{pmatrix}$ 4. $L\begin{pmatrix} x \\ y \end{pmatrix} = \begin{pmatrix} 4y \\ -x \end{pmatrix}$

5. $L\begin{pmatrix} x \\ y \end{pmatrix} = \begin{pmatrix} x + y \\ -2x + 4y \end{pmatrix}$ 6. $L\begin{pmatrix} x \\ y \end{pmatrix} = \begin{pmatrix} -x/2 \\ 9x/2 + 4y \end{pmatrix}$

7. Let $L: R^2 \rightarrow R^2$ be linear, with

$$A_L = \begin{pmatrix} \cos \theta & -\sin \theta \\ \sin \theta & \cos \theta \end{pmatrix}$$

    Show that L represents a rotation of angle θ in the counterclockwise direction about the origin.

8. Find a linear function $L: R^2 \rightarrow R^2$ such that $L^{[3]} = I$ and $L \neq I$. (*Hint*: Use the result of Exercise 7.)

9. Let $L\begin{pmatrix} x \\ y \end{pmatrix} = \begin{pmatrix} ax + by \\ dy \end{pmatrix}$. Show that if $L^{[3]} = I$, then $L = I$.

10. Let $L\begin{pmatrix} x \\ y \end{pmatrix} = \begin{pmatrix} ax + by \\ dy \end{pmatrix}$. Find nonzero values of a, b, and d such that $L^{[4]} = I$.

11.   Let $a \neq 0$ and $b \neq 0$, and suppose that $L\begin{pmatrix} x \\ y \end{pmatrix} = \begin{pmatrix} ax \\ by \end{pmatrix}$ for all x and y in R. Show that the image of any circle centered at the origin is an ellipse centered at the origin.

12.   Let A be an arbitrary 2×2 matrix. Define $L: R^2 \to R^2$ by $L(\mathbf{v}) = A\mathbf{v}$, for all $\mathbf{v}$ in R^2. Show that L is a linear function and that $A_L = A$.

13.   Let $L: R^2 \to R^2$ and $M: R^2 \to R^2$. Show that $M = L^{-1}$ if and only if $A_M = (A_L)^{-1}$.

14.   Let $L: R^2 \to R^2$. Show that L is linear if and only if there are real numbers a, b, c, and d such that $L\begin{pmatrix} x \\ y \end{pmatrix} = \begin{pmatrix} ax + by \\ cx + dy \end{pmatrix}$ for all real x and y.

15.   Let $L: R^2 \to R^2$ be a linear function.
   a.   Suppose that there is a fixed point $\mathbf{v}_0$ that is not the origin. Show that L need not be I.
   b.   Suppose that there are nonzero fixed points $\mathbf{v}_0$ and $\mathbf{w}_0$ such that $\mathbf{w}_0$ is not a multiple of $\mathbf{v}_0$. Determine whether or not L must be I.

16.   Use (6) to show that $\|\mathbf{v} - \mathbf{w}\| \geq \big| \|\mathbf{v}\| - \|\mathbf{w}\| \big|$ for all $\mathbf{v}$ and $\mathbf{w}$ in R^2.

17.   Let $L: R^2 \to R^2$, and suppose that A_L has complex eigenvalues $\beta + i\gamma$ and $\beta - i\gamma$, with $\beta^2 + \gamma^2 < 1$. Show that $\lim_{n \to \infty} L^{[n]}(\mathbf{v}) = \mathbf{0}$ for all $\mathbf{v}$ in R^2.

18.   Suppose that $|\lambda| < 1$, and let $\varepsilon > 0$ be so small that $|\lambda| + \varepsilon < 1$. Also, let $A = \begin{pmatrix} \lambda & \varepsilon \\ 0 & \lambda \end{pmatrix}$. Show that if $c = |\lambda| + \varepsilon$, then $\|A^n \mathbf{v}\| \leq c^n \|\mathbf{v}\|$ for every integer $n \geq 1$, and hence that $\lim_{n \to \infty} L^{[n]}(\mathbf{v}) = \mathbf{0}$ for all $\mathbf{v}$ in R^2.

19.   Let $L: R^2 \to R^2$, and suppose that A_L has but one real eigenvalue λ, with $|\lambda| < 1$. Show that $\lim_{n \to \infty} L^{[n]}(\mathbf{v}) = \mathbf{0}$ for all $\mathbf{v}$ in R^2.

20.   Let the linear function $L: R^2 \to R^2$ have real eigenvalues λ and μ such that $|\lambda| > 1$ and $|\mu| < 1$. Show that there is an $\alpha < 0$ such that the curves $y = rx^\alpha$ are invariant under L, for all $r \neq 0$.

3.3 NONLINEAR MAPS

Section 3.2 was devoted to linear functions defined on R^2. In the remainder of the chapter we will discuss functions defined in R^2 that are not necessarily linear. In contrast to linear functions, whose dynamics are relatively tame, nonlinear functions can have very rich dynamics. The present section is preparatory, formulating concepts that will play a role in the ensuing discussions of nonlinear functions.

Let V be a subset of R^2, and let $F: V \rightarrow R^2$. Frequently such a function is called a **map**. The function F can always be represented in the form

$$F(\mathbf{v}) = \begin{pmatrix} f(\mathbf{v}) \\ g(\mathbf{v}) \end{pmatrix} \quad \text{for all } \mathbf{v} \text{ in } V$$

where f and g are real-valued **coordinate functions** of F. For example, if

$$F\begin{pmatrix} x \\ y \end{pmatrix} = \begin{pmatrix} y \\ a \sin x + by \end{pmatrix} \tag{1}$$

then $V = R^2$, $f\begin{pmatrix} x \\ y \end{pmatrix} = y$ and $g\begin{pmatrix} x \\ y \end{pmatrix} = a \sin x + by$. If the constants a and b in (1) are negative real numbers, then F could represent the motion of a damped, unforced pendulum, which we will study in more detail in Chapter 5.

Just as the derivative is important in the analysis of a function of one variable, the differential plays a key role in the study of functions of several variables.

DEFINITION 3.16. Let V be a subset of R^2, and consider $F: V \rightarrow R^2$. Assume that the first partials of the coordinate functions f and g of F exist at $\mathbf{v}_0$. The **differential** of F at $\mathbf{v}_0$ is the linear function $DF(\mathbf{v}_0)$ defined on R^2 by

$$[DF(\mathbf{v}_0)](\mathbf{v}) = \begin{pmatrix} \dfrac{\partial f}{\partial x}(\mathbf{v}_0) & \dfrac{\partial f}{\partial y}(\mathbf{v}_0) \\ \dfrac{\partial g}{\partial x}(\mathbf{v}_0) & \dfrac{\partial g}{\partial y}(\mathbf{v}_0) \end{pmatrix} \mathbf{v} \quad \text{for all } \mathbf{v} \text{ in } R^2$$

Notice that $DF(\mathbf{v}_0)$ is a linear function, by Corollary 3.8. We will normally identify $DF(\mathbf{v}_0)$ with the associated **Jacobian matrix**

$$
\begin{pmatrix}
\dfrac{\partial f}{\partial x}(\mathbf{v}_0) & \dfrac{\partial f}{\partial y}(\mathbf{v}_0) \\[3mm]
\dfrac{\partial g}{\partial x}(\mathbf{v}_0) & \dfrac{\partial g}{\partial y}(\mathbf{v}_0)
\end{pmatrix}
$$

The one-dimensional version of $DF(\mathbf{v}_0)$ is the derivative $f'(x_0)$ of the linear function L defined by $L(x) = f'(x_0)x$ for all real numbers x. Analogously, $DF(\mathbf{v}_0)$ can be considered as a two-dimensional derivative.

EXAMPLE 1. Let $F\begin{pmatrix} x \\ y \end{pmatrix} = \begin{pmatrix} y \\ a \sin x + by \end{pmatrix}$. Find $DF\begin{pmatrix} x_0 \\ y_0 \end{pmatrix}$.

Solution. We have $f\begin{pmatrix} x \\ y \end{pmatrix} = y$ and $g\begin{pmatrix} x \\ y \end{pmatrix} = a \sin x + by,$ so that

$$
\frac{\partial f}{\partial x} = 0, \quad \frac{\partial f}{\partial y} = 1, \quad \frac{\partial g}{\partial x} = a \cos x, \quad \frac{\partial g}{\partial y} = b
$$

Therefore at $\begin{pmatrix} x_0 \\ y_0 \end{pmatrix}$ the partials are

$$
\frac{\partial f}{\partial x} = 0, \quad \frac{\partial f}{\partial y} = 1, \quad \frac{\partial g}{\partial x} = a \cos x_0, \quad \text{and} \quad \frac{\partial g}{\partial y} = b
$$

Consequently

$$
DF\begin{pmatrix} x_0 \\ y_0 \end{pmatrix} \mathbf{v} = \begin{pmatrix} 0 & 1 \\ a \cos x_0 & b \end{pmatrix} \mathbf{v} \quad \text{for all } \mathbf{v} \text{ in } R^2
$$

or in Jacobian matrix form,

$$
DF\begin{pmatrix} x_0 \\ y_0 \end{pmatrix} = \begin{pmatrix} 0 & 1 \\ a \cos x_0 & b \end{pmatrix} \qquad \square
$$

EXAMPLE 2. Let $L\begin{pmatrix} x \\ y \end{pmatrix} = \begin{pmatrix} ax + by \\ cx + dy \end{pmatrix} = \begin{pmatrix} a & b \\ c & d \end{pmatrix}\begin{pmatrix} x \\ y \end{pmatrix}$. Find $DL\begin{pmatrix} 0 \\ 0 \end{pmatrix}$.

Solution. In this case we have $f\begin{pmatrix} x \\ y \end{pmatrix} = ax + by$ and $g\begin{pmatrix} x \\ y \end{pmatrix} = cx + dy$. Therefore

$$\frac{\partial f}{\partial x} = a, \qquad \frac{\partial f}{\partial y} = b, \qquad \frac{\partial g}{\partial x} = c, \qquad \frac{\partial g}{\partial y} = d$$

It follows that

$$DL\begin{pmatrix} 0 \\ 0 \end{pmatrix} = \begin{pmatrix} a & b \\ c & d \end{pmatrix}$$

so that $DL\begin{pmatrix} 0 \\ 0 \end{pmatrix} = A_L$. ❑

Example 2 tells us that for a linear function L on R^2, the differential at the origin $\mathbf{0}$ is the same linear function L, because the associated matrix is A_L. Is the same true for the differential of L at any $\mathbf{v}_0$?

If the partials of the coordinate functions f and g are continuous in a neighborhood of $\mathbf{v}_0$, then it is possible to prove that

$$\frac{F(\mathbf{v}) - F(\mathbf{v}_0) - [DF(\mathbf{v}_0)](\mathbf{v} - \mathbf{v}_0)}{\|\mathbf{v} - \mathbf{v}_0\|} \qquad \text{approaches } \mathbf{0} \text{ as } \mathbf{v} \text{ approaches } \mathbf{v}_0 \qquad (2)$$

Therefore $F(\mathbf{v}_0) + [DF(\mathbf{v}_0)](\mathbf{v} - \mathbf{v}_0)$ is a good approximation to $F(\mathbf{v})$ if $\mathbf{v}$ is near to $\mathbf{v}_0$. Thus the differential $DF(\mathbf{v}_0)$ can tell us something about the behavior of $F(\mathbf{v})$ when $\mathbf{v}$ is near to $\mathbf{v}_0$.

Moreover, $DF(\mathbf{v}_0)$ indicates whether F significantly contracts or expands areas of regions near to $\mathbf{v}_0$. To be more specific, we consider a function $F: V \to R^2$ with coordinate functions f and g whose partials exist at $\mathbf{v}_0$. The determinant of $DF(\mathbf{v}_0)$ is called the **Jacobian** of F at $\mathbf{v}_0$, and is given by

$$\det DF(\mathbf{v}_0) = \det \begin{pmatrix} \dfrac{\partial f}{\partial x}(\mathbf{v}_0) & \dfrac{\partial f}{\partial y}(\mathbf{v}_0) \\[2ex] \dfrac{\partial g}{\partial x}(\mathbf{v}_0) & \dfrac{\partial g}{\partial y}(\mathbf{v}_0) \end{pmatrix} \qquad (3)$$

If $|\det DF(\mathbf{v}_0)| < 1$, then F is **area-contracting** at $\mathbf{v}_0$, in the sense that F shrinks small regions containing $\mathbf{v}_0$. Similarly, if $|\det DF(\mathbf{v}_0)| > 1$, then F is **area-expanding** at $\mathbf{v}_0$.

EXAMPLE 3.   Let $F\begin{pmatrix} x \\ y \end{pmatrix} = \begin{pmatrix} y \\ a \sin x + by \end{pmatrix}$. Find the Jacobian of the function F

at $\begin{pmatrix} \pi/3 \\ 4 \end{pmatrix}$, and determine conditions on a and b that imply that F is area-

contracting at $\begin{pmatrix} \pi/3 \\ 4 \end{pmatrix}$.

Solution. From Example 1 with $x_0 = \pi/3$ and $y_0 = 4$, we have

$$DF\begin{pmatrix} \pi/3 \\ 4 \end{pmatrix} = \begin{pmatrix} 0 & 1 \\ \dfrac{a}{2} & b \end{pmatrix}$$

This means that $\det DF\begin{pmatrix} \pi/3 \\ 4 \end{pmatrix} = -\dfrac{a}{2}$, so that F is area-contracting at $\begin{pmatrix} \pi/3 \\ 4 \end{pmatrix}$
provided that $|a| < 2$.   ❏

Now we turn to fixed points of F.

DEFINITION 3.17. Let **p** be a fixed point of F. Then **p** is **attracting** if
and only if there is a disk centered at **p** such that $F^{[n]}(\mathbf{v}) \to \mathbf{p}$, for every **v** in the
disk. By contrast, **p** is **repelling** if and only if there is a disk centered at **p** such
that $\|F(\mathbf{v}) - F(\mathbf{p})\| > \|\mathbf{v} - \mathbf{p}\|$ for every **v** in the disk for which $\mathbf{v} \neq \mathbf{p}$.

These definitions extend to two dimensions the notions of attracting and repelling fixed points presented in Section 1.2. Often attracting fixed points of multi-
variable functions are called **sinks**, and repelling fixed points are called **sources**.
Recall that a fixed point p of a function f defined in R is attracting
provided that $|f'(p)| < 1$. Below we will prove a two-dimensional version of this
criterion. Before we do it, we need to recall the Chain Rule from (3) in Section 1.3:

$$(f^{[n]})'(x) = [f'(f^{[n-1]}(x))]\, [f'(f^{[n-2]}(x))] \cdots [f'(f(x))]\, [f'(x)]$$

If x is a fixed point, say $x = p$, then the formula reduces to

$$(f^{[n]})'(p) = [f'(p)]^n \tag{4}$$

The two-dimensional analogue of (4), applied to the function F defined in R^2, is

$$(DF^{[n]})(\mathbf{p}) = [(DF(\mathbf{p})]^{[n]} \tag{5}$$

(You can find a proof of (5) in multi-variable analysis texts such as Buck, 1978.) Now we are ready for the promised theorem.

THEOREM 3.18. Let $\mathbf{p}$ be a fixed point of F. Assume that $DF(\mathbf{p})$ exists, with eigenvalues λ and μ such that $|\lambda| < 1$ and $|\mu| < 1$. Then $\mathbf{p}$ is attracting.

Proof. Let $\varepsilon > 0$. By (5), and by (2) with F replaced by $F^{[n]}$ and $\mathbf{v}_0$ replaced by $\mathbf{p}$, we find that

$$\left\| \frac{F^{[n]}(\mathbf{v}) - \mathbf{p} - [(DF)(\mathbf{p})]^{[n]}(\mathbf{v} - \mathbf{p})}{\|\mathbf{v} - \mathbf{p}\|} \right\| = \left\| \frac{F^{[n]}(\mathbf{v}) - F^{[n]}(\mathbf{p}) - [(DF^{[n]})(\mathbf{p})](\mathbf{v} - \mathbf{p})}{\|\mathbf{v} - \mathbf{p}\|} \right\| < \varepsilon$$

if $\mathbf{v}$ is in a sufficiently small neighborhood U of $\mathbf{p}$, and $\mathbf{v} \neq \mathbf{p}$. However, the eigenvalues of $[DF(\mathbf{p})]^{[n]}$ are λ^n and μ^n. Since $|\lambda| < 1$ and $|\mu| < 1$ by hypothesis, Corollary 3.15 then implies that $[DF(\mathbf{p})]^{[n]} \dfrac{\mathbf{v} - \mathbf{p}}{\|\mathbf{v} - \mathbf{p}\|} \to \mathbf{0}$. Therefore if n is large enough and $\mathbf{v}$ is in U with $\mathbf{v} \neq \mathbf{p}$, then

$$\left\| \frac{F^{[n]}(\mathbf{v}) - \mathbf{p}}{\|\mathbf{v} - \mathbf{p}\|} \right\| < 2\varepsilon$$

It follows that $F^{[n]}(\mathbf{v}) \to \mathbf{p}$ for all $\mathbf{v}$ in U. Consequently $\mathbf{p}$ is attracting. ■

Using the same kind of argument as appears in the proof of Theorem 3.18, one can show that if the eigenvalues of $DF(\mathbf{p})$ satisfy $|\lambda| > 1$ and $|\mu| \geq 1$, then the fixed point $\mathbf{p}$ is repelling.

If $\mathbf{p}$ is a fixed point of F with eigenvalues λ and μ such that $|\lambda| < 1$ and $|\mu| > 1$, then $\mathbf{p}$ is called a **saddle point**. This corresponds to the definition of saddle point given in Section 3.2 for linear functions. In order to understand the behavior of F near a saddle point, let $\mathbf{v}_\lambda$ and $\mathbf{v}_\mu$ be eigenvectors for λ and μ, respectively. If $\mathbf{v}$ is near to $\mathbf{p}$, then

$$\|F(\mathbf{v}) - \mathbf{p}\| < \|\mathbf{v} - \mathbf{p}\| \text{ if } \mathbf{v} \text{ lies in the direction of } \mathbf{v}_\lambda$$
$$\|F(\mathbf{v}) - \mathbf{p}\| > \|\mathbf{v} - \mathbf{p}\| \text{ if } \mathbf{v} \text{ lies in the direction of } \mathbf{v}_\mu$$

If F is a linear function, then the iterates of points on the line through $\mathbf{p}$ in the direction of $\mathbf{v}_\lambda$ converge to $\mathbf{p}$, and the iterates of points on the line through $\mathbf{p}$ in the direction of $\mathbf{v}_\mu$ separate from $\mathbf{p}$. However, if F is not linear, then there is no assurance that these conclusions hold. Nevertheless there is a famous theorem, with the imposing name **Stable and Unstable Manifold Theorem**, which states that there are differentiable curves C_λ and C_μ through $\mathbf{p}$ such that

C_λ is tangent to $\mathbf{v}_\lambda$ at $\mathbf{p}$, and $F^{[n]}(\mathbf{v}) \to \mathbf{p}$ for all $\mathbf{v}$ on C_λ

C_μ is tangent to $\mathbf{v}_\mu$ at $\mathbf{p}$, and $F^{-[n]}(\mathbf{v}) \to \mathbf{p}$ for all $\mathbf{v}$ on C_μ

The expression $F^{-[n]}(\mathbf{v}) \to \mathbf{p}$ signifies that the pre-images of $\mathbf{v}$ on C_μ approach $\mathbf{p}$, and suggests that the iterates of points on C_μ recede from $\mathbf{p}$. The curve C_λ is often called a **local stable manifold** for $\mathbf{p}$ and is denoted $W^s_{\text{loc}}(\mathbf{p})$. Similarly, the curve C_μ is often called a **local unstable manifold** for $\mathbf{p}$ and is denoted $W^u_{\text{loc}}(\mathbf{p})$. (For more details, see the books by Devaney, 1989 or Guckenheimer and Holmes, 1983.) The geometrical interpretation of these ideas is given in Figure 3.5.

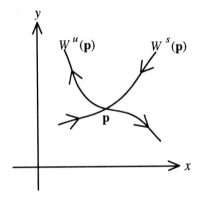

Figure 3.5

We return once again to the function F defined in Example 1, with b assigned the value -1.

EXAMPLE 4. Let $F\begin{pmatrix} x \\ y \end{pmatrix} = \begin{pmatrix} y \\ a\sin x - y \end{pmatrix}$. Determine the values of a for which $DF(\mathbf{0})$ has real eigenvalues and the fixed point $\mathbf{0}$ is

a. attracting   b. repelling   c. a saddle point

Solution. By the result of Example 1 we have

$$DF(\mathbf{0}) = \begin{pmatrix} 0 & 1 \\ a\cos 0 & -1 \end{pmatrix} = \begin{pmatrix} 0 & 1 \\ a & -1 \end{pmatrix}$$

so that

$$\det[DF(\mathbf{0}) - \lambda I] = \det \begin{pmatrix} -\lambda & 1 \\ a & -1-\lambda \end{pmatrix} = \lambda^2 + \lambda - a$$

Now $\lambda^2 + \lambda - a = 0$ if and only if

$$\lambda = \frac{-1 \pm \sqrt{1 + 4a}}{2}$$

Next let

$$\lambda = \frac{-1 - \sqrt{1 + 4a}}{2} \quad \text{and} \quad \mu = \frac{-1 + \sqrt{1 + 4a}}{2}$$

A little calculation yields

if $a < -1/4$, then $1 + 4a < 0$, so that λ and μ are not real
if $-1/4 < a < 0$, then $-1 < \lambda < -1/2$ and $-1/2 < \mu < 0$
if $0 < a < 2$, then $-2 < \lambda < -1$ and $0 < \mu < 1$
if $a > 2$, then $\lambda < -2$ and $\mu > 1$

Therefore when λ and μ are real, we have the following results:

0 is an attracting fixed point if $-1/4 < a < 0$
0 is a saddle point if $0 < a < 2$
0 is a repelling fixed point if $a > 2$

This completes the solution. ❏

You may have observed that our discussion has revolved around those fixed points **p** whose eigenvalues satisfy $|\lambda| \neq 1$ and $|\mu| \neq 1$. Such fixed points are called **hyperbolic**; they are either attracting or repelling fixed points or saddle points, and the behavior of the function near such a point is amenable to analysis. If the eigenvalues satisfy $|\lambda| = 1$ or $|\mu| = 1$, then **p** is **non-hyperbolic**, and the analysis of the function near such a point is much more difficult.

Baker's Functions

We complete this section with a discussion of a two-dimensional version of the baker's function described in Section 1.3. We define the function $B_0: R^2 \to R^2$ by the two formulas

$$B_0 \begin{pmatrix} x \\ y \end{pmatrix} = \begin{pmatrix} \dfrac{1}{4}x \\ 3y \end{pmatrix} \quad \text{for } 0 \leq x \leq 1 \text{ and } 0 \leq y < \frac{1}{3}$$

and

$$B_0 \begin{pmatrix} x \\ y \end{pmatrix} = \begin{pmatrix} \dfrac{1}{2} + \dfrac{1}{3}\, x \\[2mm] \dfrac{3}{2}\, (y - \dfrac{1}{3}) \end{pmatrix} \quad \text{for } 0 \le x \le 1 \text{ and } \dfrac{1}{3} \le y \le 1$$

Then B_0 is called a **baker's function** because it is linear on two portions of the domain the way the one-dimensional baker's function is. The domain of B_0 consists of the unit square $S = [0, 1] \times [0, 1]$. The effect of B_0 on S can be seen graphically in Figure 3.6, where the rectangles in the right graph are the images of the rectangles with similar shading in the left graph:

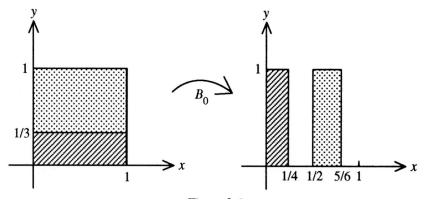

Figure 3.6

The image $B_0^{[2]}(S)$ is depicted on the right in Figure 3.7, and suggests that

$$S \supseteq B_0(S) \supseteq B_0^{[2]}(S) \supseteq \cdots$$

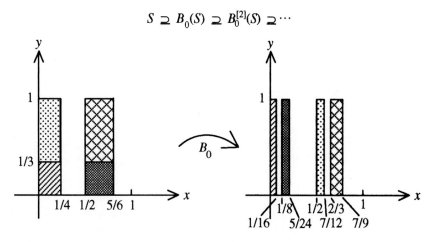

Figure 3.7

One can prove that $\{B_0^{[n]}(S)\}_{n=0}^{\infty}$ is in fact a nested sequence of strips in S. Next, let

A_{B_0} = the collection of $\mathbf{v}$ that are in $B_0^{[n]}(S)$ for all $n \geq 0$, including boundary

Then A_{B_0} is composed of a set of vertical line segments of length 1. The intersection of A_{B_0} and the x axis is a one-dimensional Cantor set (though *not* the Cantor ternary set). The set A_{B_0} is called an **attractor** because the iterates of every point in S approach A_{B_0}. Notice that $\mathbf{0}$ is a fixed point of B_0.

EXAMPLE 5. Determine whether $\mathbf{0}$ is an attracting or repelling fixed point, or a saddle point.

Solution. By the first formula that defines B_0,

$$B_0(\mathbf{v}) = \begin{pmatrix} f(\mathbf{v}) \\ g(\mathbf{v}) \end{pmatrix}, \text{ where } f\begin{pmatrix} x \\ y \end{pmatrix} = \frac{1}{4}x \text{ and } g\begin{pmatrix} x \\ y \end{pmatrix} = 3y$$

Because $\mathbf{0}$ is on the border of the domain, we can only take one-sided partial derivatives of B_0 at $\mathbf{0}$. We obtain

$$DB_0(\mathbf{0}) = \begin{pmatrix} 1/4 & 0 \\ 0 & 3 \end{pmatrix}$$

Therefore $\mathbf{0}$ is a saddle point, contracting by a factor of $1/4$ along the x axis and expanding by a factor of 3 along the y axis. ❑

It follows from the solution to Example 5 that the interval $[0, 1]$ along the x axis is a local stable manifold $W_{\text{loc}}^s(\mathbf{0})$. By contrast, the interval $[0, 1/3]$ along the y axis is a local unstable manifold $W_{\text{loc}}^u(\mathbf{0})$.

The point $\mathbf{0}$ is not the only periodic point of B_0. There is another fixed point, as well as periodic points that are not fixed points (Exercises 13–14). In addition, B_0 has sensitive dependence on initial conditions because B_0 separates nearby points in the domain by at least a factor of $3/2$ (Exercise 15).

The function B_0 described above is but one of a whole family of functions. The full family is defined in the following way. Let a, b, and c be constants, with $0 < a \leq 1/2$, and $0 < b < c < 1/2$. Then a (generalized) **baker's function**, which we designate by B rather than the more accurate but cumbersome B_{abc}, is defined on the unit square S by the two formulas

$$B\begin{pmatrix} x \\ y \end{pmatrix} = \begin{pmatrix} bx \\ \dfrac{1}{a} y \end{pmatrix} \quad \text{for } 0 \le x \le 1 \text{ and } 0 \le y < a$$

and

$$B\begin{pmatrix} x \\ y \end{pmatrix} = \begin{pmatrix} \dfrac{1}{2} + cx \\ \dfrac{1}{1-a}(y-a) \end{pmatrix} \quad \text{for } 0 \le x \le 1 \text{ and } a \le y \le 1$$

The effect of B on the domain S is shown in Figure 3.8. Like B_0, B has an attractor A_B whose intersection with the x axis is a one-dimensional Cantor set. If $\mathbf{v}_0 = \begin{pmatrix} x_0 \\ y_0 \end{pmatrix}$ is any vector in S such that $x_0 \ne 0$ or 1 and $y_0 \ne 0$ or a or 1, then $DB(\mathbf{v}_0)$ exists, and moreover,

$$DB(\mathbf{v}_0) = \begin{pmatrix} b & 0 \\ 0 & \dfrac{1}{a} \end{pmatrix} \text{ if } 0 < y_0 < a \quad \text{and} \quad DB(\mathbf{v}_0) = \begin{pmatrix} c & 0 \\ 0 & \dfrac{1}{1-a} \end{pmatrix} \text{ if } a < y_0 < 1$$

Exercises 17–19 are devoted to features of B.

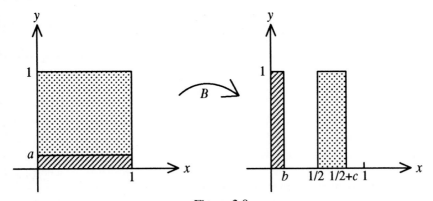

Figure 3.8

EXERCISES 3.3

In Exercises 1–4, find the fixed points of F, and determine whether F is area-expanding, area-contracting, or neither at these points.

1. $F\begin{pmatrix} x \\ y \end{pmatrix} = \begin{pmatrix} x^2 + \dfrac{1}{4} \\ 4x - y^2 \end{pmatrix}$

2. $F\begin{pmatrix} x \\ y \end{pmatrix} = \begin{pmatrix} 1 - x^2 + y \\ x \end{pmatrix}$

3. $F\begin{pmatrix} x \\ y \end{pmatrix} = \begin{pmatrix} y^2 - \dfrac{1}{2}\,x \\ \dfrac{1}{4}\,x + \dfrac{1}{2}\,y \end{pmatrix}$

4. $F\begin{pmatrix} x \\ y \end{pmatrix} = \begin{pmatrix} 2xy + y \\ 3y - x \end{pmatrix}$

In Exercises 5–8, determine whether the fixed points of F are attracting, repelling, or saddle points.

5. F in Exercise 1

6. F in Exercise 2

7. F in Exercise 3

8. F in Exercise 4

9. Let $L\begin{pmatrix} x \\ y \end{pmatrix} = \begin{pmatrix} 2y \\ 0 \end{pmatrix}$.

 a. Show that 0 is the only eigenvalue of L, but $\|L(\mathbf{v})\| > \|\mathbf{v}\|$ for all nonzero $\mathbf{v}$ on the y axis.
 b. Does the result of part (a) contradict Corollary 3.15? Explain your answer.

10. Let $L\begin{pmatrix} x \\ y \end{pmatrix} = \begin{pmatrix} \dfrac{1}{2}\,x \\ 10x + \dfrac{1}{4}\,y \end{pmatrix}$.

 a. Show that the eigenvalues of L are less than 1 in absolute value.
 b. Find a vector $\mathbf{v}$ such that $\|L(\mathbf{v})\| > \|\mathbf{v}\|$.
 c. Do (a) and (b) contradict Theorem 3.18? Explain your answer.

11. Let $F\begin{pmatrix} x \\ y \end{pmatrix} = \begin{pmatrix} \sin x \\ y^2 \end{pmatrix}$. Show that 0 is an attracting fixed point of F, although $DF(0)$ has an eigenvalue that is not less than 1 in absolute value.

12. Find an example of a function $F: R^2 \to R^2$ that has a saddle point $\mathbf{p}$, such that F is
 a. area-expanding at $\mathbf{p}$ b. area-contracting at $\mathbf{p}$

13.   Determine a fixed point of B_0 that is not $\mathbf{0}$.

14.   Find a 2-cycle for B_0.

15.   Show that B_0 has sensitive dependence on initial conditions.

16.   Use the formulas that deine B_0 to show explicitly that $S \supseteq B_0(S)$.

Exercises 17–19 are devoted to the (generalized) baker's function B.

17.   Determine the fixed points of B.

18.   a.   Use the formulas that define B to show explicitly that $S \supseteq B(S)$.
       b.   Find the areas of $B(S)$ and $B^{[2]}(S)$.

19.   Show that $\mathbf{0}$ is a saddle point for B, for any choice of a, b, and c.

20.   Let $L: R^2 \rightarrow R^2$ be a linear function, and let $\mathbf{v}_0$ be an arbitrary element of R^2. Is $DL(\mathbf{v}_0) = L$? Explain your answer.

The **cat map** C, introduced by the Russian mathematician V. I. Arnold, is defined on the half-open unit square $[0, 1) \times [0, 1)$, using the "mod" notation. By "z mod 1" we mean $z - n$, where n is the integer such that $0 \le z - n < 1$. Then

$$C\begin{pmatrix} x \\ y \end{pmatrix} = \begin{pmatrix} (x + y) \bmod 1 \\ (x + 2y) \bmod 1 \end{pmatrix}$$

Exercises 21–22 explore features of C.

21.   Show that $\mathbf{0}$ is the only fixed point of C, and find two 2-cycles for C.

22.   Note that $DC\begin{pmatrix} x \\ y \end{pmatrix} = \begin{pmatrix} 1 & 1 \\ 1 & 2 \end{pmatrix}$, for any $\mathbf{v}$ at which C is continuous.

       a.   Find the eigenvalues of $DC(\mathbf{0})$, and show that $\mathbf{0}$ is a saddle point.
       b.   Let $c_1 = 1 = c_2$, and recursively let $c_n = c_{n-1} + c_{n-2}$ for $n \ge 3$. Then $c_3 = 2$, $c_4 = 3$, $c_5 = 5$, etc., and the sequence of c_n's is the **Fibonacci sequence**. Use induction to show that

$$DC^{[n]}(\mathbf{v}) = \begin{pmatrix} c_{2n-1} & c_{2n} \\ c_{2n} & c_{2n+1} \end{pmatrix} \quad \text{for } n \ge 2$$

3.4 THE HÉNON MAP

About 15 years ago the French astronomer-mathematician Michel Hénon was searching for a simple two-dimensional function possessing special properties of more complicated systems. The result was a family of functions denoted by H_{ab} and given by

$$H_{ab}\begin{pmatrix} x \\ y \end{pmatrix} = \begin{pmatrix} 1 - ax^2 + y \\ bx \end{pmatrix}, \text{ where } a \text{ and } b \text{ are real numbers} \qquad (1)$$

The maps defined in (1) are called **Hénon maps**. Many authors refer to the Hénon maps as H, and just call them the Hénon map.

Notice that if $b = 1$, $x = t$ and $y = 0$, then (1) becomes

$$H_{ab}\begin{pmatrix} t \\ 0 \end{pmatrix} = \begin{pmatrix} 1 - at^2 \\ t \end{pmatrix}$$

Thus the image of the real line is the parabola given parametrically by $x = 1 - at^2$ and $y = t$ (Figure 3.9). As a result, the Hénon maps constitute a 2-dimensional generalization of the family $\{F_C\}$ mentioned in Section 2.3, where $F_C(x) = 1 - Cx^2$. Next we will find the Jacobian of H_{ab}.

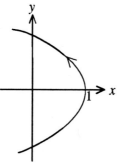

Figure 3.9

THEOREM 3.19. Let a and b be any fixed real numbers. Then $\det DH_{ab}\begin{pmatrix} x \\ y \end{pmatrix}$

$= -b$ for all x, y in R^2. If $a^2x^2 + b \geq 0$, then the eigenvalues of $DH_{ab}\begin{pmatrix} x \\ y \end{pmatrix}$ are

the real numbers $-ax \pm \sqrt{a^2x^2 + b}$.

Proof. Since the coordinate functions of H_{ab} are given by

$$f\begin{pmatrix} x \\ y \end{pmatrix} = 1 - ax^2 + y \quad \text{and} \quad g\begin{pmatrix} x \\ y \end{pmatrix} = bx$$

we find that

$$DH_{ab}\begin{pmatrix} x \\ y \end{pmatrix} = \begin{pmatrix} -2ax & 1 \\ b & 0 \end{pmatrix}$$

so that

$$\det DH_{ab}\begin{pmatrix} x \\ y \end{pmatrix} = \det\begin{pmatrix} -2ax & 1 \\ b & 0 \end{pmatrix} = -b$$

To determine the eigenvalues of $DH_{ab}\begin{pmatrix} x \\ y \end{pmatrix}$ we observe that

$$\det\left(DH_{ab}\begin{pmatrix} x \\ y \end{pmatrix} - \lambda I\right) = \det\begin{pmatrix} -2ax - \lambda & 1 \\ b & -\lambda \end{pmatrix} = \lambda^2 + 2ax\lambda - b$$

Therefore λ is an eigenvalue of $DH_{ab}\begin{pmatrix} x \\ y \end{pmatrix}$ if $\lambda^2 + 2ax\lambda - b = 0$. This means that

$$\lambda = \frac{-2ax \pm \sqrt{4a^2x^2 + 4b}}{2} = -ax \pm \sqrt{a^2x^2 + b}$$

Thus the eigenvalues are real if $a^2x^2 + b \geq 0$. ∎

The map H_{ab} has a constant Jacobian $\det DH_{ab}$. Hénon noted in his original paper (1976) that H_{ab} is the "most general quadratic mapping [on R^2] with constant Jacobian." Next, recall that the Jacobian of H_{ab} determines whether H_{ab} is area-expanding or area-contracting (or neither). By Theorem 3.19, the map H_{ab} is area-contracting if $0 \leq b < 1$; it is genuinely a two-dimensional map if $b \neq 0$. Thus we will henceforth assume $0 < b < 1$. For such values of b, $H_{ab}(\mathbf{v})$ has distinct real eigenvalues for every value of the parameter a, and all $\mathbf{v}$.

It is straightforward to show that H_{ab} is one-to-one.

THEOREM 3.20. H_{ab} is one-to-one.

Proof. Let x, y, z, and w be real numbers. Then

$$H_{ab}\begin{pmatrix} x \\ y \end{pmatrix} = H_{ab}\begin{pmatrix} z \\ w \end{pmatrix} \text{ if and only if } \begin{pmatrix} 1 - ax^2 + y \\ bx \end{pmatrix} = \begin{pmatrix} 1 - az^2 + w \\ bz \end{pmatrix}$$

that is,

$$1 - ax^2 + y = 1 - az^2 + w \quad \text{and} \quad bx = bz$$

Since $b \neq 0$, it follows that $x = z$. Therefore $y = w$ as well, so that $\begin{pmatrix} x \\ y \end{pmatrix} = \begin{pmatrix} z \\ w \end{pmatrix}$

Consequently H_{ab} is one-to-one. ∎

The map H_{ab} is composed of the three functions H_1, H_2, and H_3, where

$$H_1\begin{pmatrix} x \\ y \end{pmatrix} = \begin{pmatrix} x \\ 1 - ax^2 + y \end{pmatrix}, \quad H_2\begin{pmatrix} x \\ y \end{pmatrix} = \begin{pmatrix} bx \\ y \end{pmatrix}, \quad \text{and} \quad H_3\begin{pmatrix} x \\ y \end{pmatrix} = \begin{pmatrix} y \\ x \end{pmatrix}$$

More precisely, $H_{ab} = H_3 \circ H_2 \circ H_1$ (where for convenience we suppress the subscript ab on H_1, H_2, and H_3). To interpret H_1, H_2, and H_3 geometrically, suppose that $a > 1$. Then H_1 begins the folding process. The effect of H_1 on the ellipse in Figure 3.10(a) is shown in Figure 3.10(b). Next, H_2 contracts curves in the x direction, since $0 < b < 1$ by hypothesis (Figure 3.10(c)). The folding started by H_1 is enhanced by H_2. Finally, H_3 flips shapes across the line $y = x$. The total effect of H_1, H_2, and H_3 (that is, of H_{ab}) on an ellipse is shown in Figure 3.10(d).

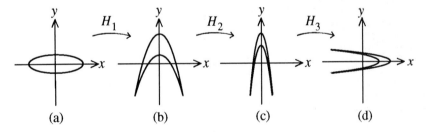

Figure 3.10

Now we will prove that H_{ab} is invertible.

THEOREM 3.21. H_{ab} is invertible, and $H_{ab}^{-1}\begin{pmatrix} x \\ y \end{pmatrix} = \begin{pmatrix} \dfrac{1}{b} y \\ -1 + \dfrac{a}{b^2} y^2 + x \end{pmatrix}$.

Proof. We could show that H_1, H_2, and H_3 are invertible, and then that

$$H_{ab}^{-1} = (H_3 \circ H_2 \circ H_1)^{-1} = H_1^{-1} \circ H_2^{-1} \circ H_3^{-1}$$

It is easier to show that by the formula for H_{ab}^{-1}, we have $H_{ab} \circ H_{ab}^{-1} = I$:

$$(H_{ab} \circ H_{ab}^{-1})\begin{pmatrix} x \\ y \end{pmatrix} = H_{ab}\begin{pmatrix} \dfrac{1}{b}\, y \\ -1 + \dfrac{a}{b^2}\, y^2 + x \end{pmatrix} = \begin{pmatrix} x \\ y \end{pmatrix} \quad \text{for all } x \text{ and } y \quad \blacksquare$$

Next we determine the values of a and b for which H_{ab} has fixed points.

THEOREM 3.22. Let $a \neq 0$. Then H_{ab} has a fixed point if $a \geq -\dfrac{1}{4}(1-b)^2$.

Proof. The point $\begin{pmatrix} x \\ y \end{pmatrix}$ is a fixed point of H_{ab} provided that

$$\begin{pmatrix} x \\ y \end{pmatrix} = H_{ab}\begin{pmatrix} x \\ y \end{pmatrix} = \begin{pmatrix} 1 - ax^2 + y \\ bx \end{pmatrix}$$

The left-hand and right-hand vectors are equal if $y = bx$ and $x = 1 - ax^2 + y$, which implies that $x = 1 - ax^2 + bx$. This is equivalent to $ax^2 + (1-b)x - 1 = 0$, which by the quadratic formula yields

$$x = \frac{1}{2a}\left(b - 1 \pm \sqrt{(1-b)^2 + 4a}\right)$$

Such an x exists if $(1-b)^2 + 4a \geq 0$, that is, if $a \geq -\dfrac{1}{4}(1-b)^2$. $\blacksquare$

In the event that H_{ab} has two fixed points **p** and **q**, they are given by

$$\mathbf{p} = \begin{pmatrix} \dfrac{1}{2a}(b - 1 + \sqrt{(1-b)^2 + 4a}) \\ \dfrac{b}{2a}(b - 1 + \sqrt{(1-b)^2 + 4a}) \end{pmatrix}, \quad \mathbf{q} = \begin{pmatrix} \dfrac{1}{2a}(b - 1 - \sqrt{(1-b)^2 + 4a}) \\ \dfrac{b}{2a}(b - 1 - \sqrt{(1-b)^2 + 4a}) \end{pmatrix} \quad (2)$$

Since we know the fixed points of H_{ab} and the eigenvalues of $DH_{ab}\begin{pmatrix} x \\ y \end{pmatrix}$ for all x and y, we can determine conditions under which the fixed point $\mathbf{p}$ is attracting.

THEOREM 3.23. The fixed point $\mathbf{p}$ is attracting provided that a is a nonzero number lying in the interval

$$J = \left(-\frac{1}{4}(1-b)^2,\ \frac{3}{4}(1-b)^2\right)$$

Proof. Theorem 3.18 tells us that $\mathbf{p}$ is attracting if the eigenvalues of $DH_{ab}(\mathbf{p})$ are less than 1 in absolute value. Letting $\mathbf{p} = \begin{pmatrix} p_1 \\ p_2 \end{pmatrix}$, we know from (2) that

$$p_1 = \frac{1}{2a}\left(b - 1 + \sqrt{(1-b)^2 + 4a}\right) \tag{3}$$

so that $2ap_1 = b - 1 + \sqrt{(1-b)^2 + 4a}$. Therefore

$$2ap_1 > b - 1, \text{ or equivalently, } 2ap_1 + 1 > b \tag{4}$$

By Theorem 3.19 the eigenvalues of $DH_{ab}(\mathbf{p})$ are less than 1 in absolute value if $|-ap_1 \pm \sqrt{a^2p_1^2 + b}| < 1$. We will show that if a is in the interval J, then

$$0 \le -ap_1 + \sqrt{a^2p_1^2 + b} < 1 \tag{5}$$

On the one hand, because $b > 0$ we have

$$-ap_1 + \sqrt{a^2p_1^2 + b} \ge -ap_1 + \sqrt{a^2p_1^2} = -ap_1 + |ap_1| \ge 0$$

On the other hand, $a > -(1-b)^2/4$ by hypothesis, so that $(1-b)^2 + 4a > 0$. Consequently p_1 is a real number by (3), and by (4),

$$(ap_1 + 1)^2 = a^2p_1^2 + 2ap_1 + 1 > a^2p_1^2 + b > 0$$

It follows that $ap_1 + 1 > \sqrt{a^2p_1^2 + b}$, so that $-ap_1 + \sqrt{a^2p_1^2 + b} < 1$. Therefore (5) is proved. An analogous argument proves that

$$-1 < -ap_1 - \sqrt{a^2 p_1^2 + b} < 0$$

(see Exercise 9). Consequently the eigenvalues of $DH_{ab}(\mathbf{p})$ are less than 1 in absolute value, so that $\mathbf{p}$ is an attracting fixed point.   ■

From Theorem 3.23, the fixed point $\mathbf{p}$ is attracting for certain values of a. By contrast the fixed point $\mathbf{q}$ given in (2) is a saddle point (Exercise 8). Thus we have the following situation for a given value of b in $(0, 1)$:

If $a < -\dfrac{1}{4} (1 - b)^2$, then H_{ab} has no fixed points.

If $-\dfrac{1}{4} (1 - b)^2 < a < \dfrac{3}{4} (1 - b)^2$ and $a \neq 0$, then H_{ab} has two fixed

points, $\mathbf{p}$ and $\mathbf{q}$, of which $\mathbf{p}$ is attracting and $\mathbf{q}$ is a saddle point.

For the present, let b be fixed in the interval $(0, 1)$, and let the parameter a increase. In addition to the bifurcation at $-(1 - b)^2/4$, H_{ab} has a bifurcation at $a = 3(1 - b)^2/4$, because one of the two eigenvalues of $H_{ab}(\mathbf{p})$ descends through -1 (Exercise 10), so that $\mathbf{p}$ is transformed from an attracting fixed point to a saddle point. Recollect that an attracting 2-cycle for the quadratic family $\{Q_\mu\}$ emerges as μ increases and passes through 3. Thus we might suspect that as a passes through $3(1 - b)^2/4$, an attracting 2-cycle for H_{ab} would be born. This is the case. In order to prove it, one would have to solve the equation

$$\begin{pmatrix} x \\ y \end{pmatrix} = H_{ab}^{[2]} \begin{pmatrix} x \\ y \end{pmatrix} = \begin{pmatrix} 1 - a(1 - ax^2 + y)^2 + bx \\ b(1 - ax^2 + y) \end{pmatrix}$$

for x and y. Of course this entails solving a fourth-degree equation in x, which is possible because two roots are known from the two fixed points of H_{ab}. The result is that H_{ab} has a period-doubling bifurcation at $a = 3(1 - b)^2/4$.

As a increases further, H_{ab} undergoes a period-doubling cascade. For certain special values of b the bifurcation values of a, as well as the Feigenbaum constant, are known. In particular, Derrida, Gervois, and Pomeau (1979) have calculated the following bifurcation values of a for $b = 0.3$:

bifurcation point	period-n cycle appears
− 0.1225	1
0.3675	2
0.9125	4
1.0260···	8
1.0510···	16
1.0565···	32

The cascade terminates at approximately 1.0580459, which is the "Feigenbaum constant" for the Hénon map.

How does $H_{a(.3)}$ behave when $a > 1.06$? One might imagine that for an arbitrary $a > 1.06$, the iterates of virtually any initial point would be sprinkled unpredictably throughout a region in the plane. However, that does not happen. For example, let $a = 1.4$, and designate $H_{(1.4)(.3)}$ by H. If we neglect the first few iterates of **0** and plot the next 10,000 iterates, then we obtain the shape A_H appearing in Figure 3.11, on the front cover, and in Color Plate 1. The set A_H is called the **Hénon attractor** of the map, because the iterates of every point in a certain quadrilateral Q surrounding A_H approach the attractor.

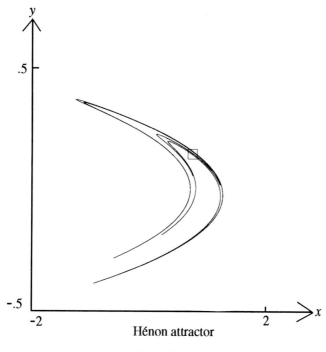

Hénon attractor

Figure 3.11

Although A_H may appear to consist of a few fairly simple curves, when we zoom in on a small rectangle containing the fixed point **p**, we see that there are several strands (Figure 3.12(a)). No matter how much we magnify the region, nearly identical new sets of strands appear (Figures 3.12(b) and (c)). It turns out that there are in reality an infinite number of such strands that make the region near to **p** look like a product of a line and a Cantor set.

The iterates of nearly all points in a rectangular region Q containing A_H not only converge to A_H but seem to trace out a dense subset of A_H. Thus whether the initial point is **0** or another point, after a few initial iterates the next several thousand iterates yield a virtually identical shape.

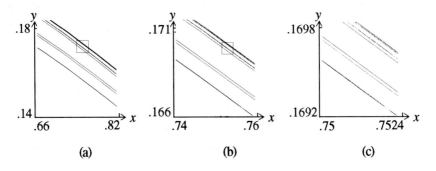

Figure 3.12

Two significant features of the Hénon attractor are evident from Figure 3.13, which displays iterates 34 through 40 of the origin (denoted by an isolated asterisk). First, the figure shows that iterates bounce around the attractor erratically. Second, the figure suggests that the Hénon attractor has sensitive dependence on initial conditions. Iterates of the origin identified by asterisks were computed by a CRAY supercomputer with single precision accuracy, whereas iterates identified by little squares were computed by the same CRAY supercomputer, but with double precision. The difference in accuracy between single and double precision is approximately 10^{-14} units. However, after 40 iterates the results are completely unrelated. Because of its sensitive dependence, A_H is called a **chaotic attractor**.

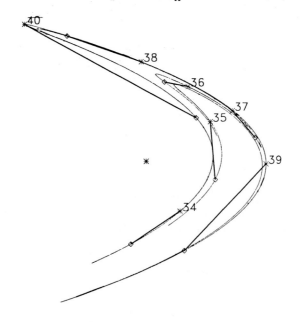

Figure 3.13

In his famous article of 1976, Hénon featured the attractor A_H with $a = 1.4$ and $b = 0.3$. Why did he select these parameters? On the one hand, he noticed that if a lies somewhere near 1.4, then as b increases from 0 to 0.3 the attractor A_H grows from nothing to a robust set in R^2; moreover, if b is much larger than 0.3, then the iterates of various points near **0** are unbounded. On the other hand, he found that if $b = 0.3$, and if $a <$ the Feigenbaum constant ≈ 1.06, then H_{ab} has an attracting cycle, so the attractor is a finite point set; however, if $a > 1.55$ then iterates of all points are unbounded. These observations led Hénon to select $a = 1.4$ and $b = 0.3$ as parameters that provide an attractor that is full and has interesting characteristics. Nevertheless, other parameters near 1.4 and 0.3 yield interesting attractors. In fact, some colleagues use $a = 1.42$ instead of $a = 1.4$. You might be interested to see what differences there are between the attractors corresponding to these two values of a.

EXERCISES 3.4

1. Using the program HENON, determine whether it takes 100, 500, 1000, 2000, or 5000 iterates of **0** to fill out A_H so that it appears as a collection of (relatively) complete curves.

2. Using the program HENON, show that after the first few iterates, the iterates of the points **0** and $\begin{pmatrix} 0.1 \\ 0.2 \end{pmatrix}$ yield virtually the same shape for A_H.

3. Recall that H has parameters $a = 1.4$ and $b = 0.3$.
   a. Find the (approximate) coordinates of the fixed point **p** for H.
   b. Find the eigenvalues λ and μ of $DH(\mathbf{p})$, with $\lambda < \mu$.
   c. Find eigenvectors $\mathbf{v}_\lambda$ and $\mathbf{v}_\mu$ of $DH(\mathbf{p})$. Confirm that H stretches distances along (that is, in the direction of) the attractor at **p**, and contracts distances in a direction oblique to the direction of the attractor at **p**.

4. Let $a = 1.4$. Use the program HENON to discuss what happens to the attractor of H_{ab} when
   a. b increases from 0.3 to 0.5
   b. b decreases from 0.3 to 0.1

5. Let $b = 0.3$. Use the program HENON to discuss what happens to the attractor of H_{ab} when
   a. a increases from 1.4
   b. a decreases from 1.4

6.   Let $b = 0.3$. Use the program HENON to find a value of a such that H_{ab} has an attracting n-cycle, and determine an n-cycle.

    a. $n = 4$ b. $n = 8$ c. $n = 7$ d. $n = 3$

7.   Let $a = 0$.

    a.   Show that for each b in the interval $(0, 1)$, H_{ab} has a unique fixed point $\mathbf{p}$, and find $\mathbf{p}$.

    b.   Determine whether $\mathbf{p}$ is attracting, repelling, or a saddle point.

    c.   Find an eigenvector corresponding to each eigenvalue of H_{ab}.

8.   a.   Let $a = -(1-b)^2/4$ and $0 < b < 1$. Find the eigenvalues of $H_{ab}(\mathbf{q})$.

    b.   Let $a > -(1-b)^2/4$ and $a \neq 0$. Show that the fixed point $\mathbf{q}$ in (2) is a saddle point of H_{ab} for each b in the interval $(0, 1)$.

    c.   Use the results of (a) and (b) to discuss the type of bifurcation of H_{ab} that occurs at $a = -0.1225$.

9.   Assume that $0 < b < 1$ and $-(1-b)^2/4 < a < 3(1-b)^2/4$. Let

$$p_1 = \frac{1}{2a}\left(b - 1 + \sqrt{(1-b)^2 + 4a}\right)$$

Show that $-1 < -ap_1 - \sqrt{a^2 p_1^2 + b} < 0$.

10.   Let b be in $(0, 1)$ and $a = 3(1-b)^2/4$. Find the eigenvalues of H_{ab}, and convince yourself that H_{ab} has a bifurcation at this value of a.

11.   Let $H^*\begin{pmatrix} x \\ y \end{pmatrix} = \begin{pmatrix} a - x^2 + by \\ x \end{pmatrix}$.

    a.   Show that $H_{ab} \underset{E}{\approx} H^*$, where E is an appropriate linear function.

    b.   What does the result of (a) tell you about the attractor of H^*? (H^* is often used as an alternative to H_{ab}.)

12.   Use the bifurcation table for $H_{a(.3)}$ that appears in this section to compute an approximate Hénon version of the Feigenbaum number (see Section 1.5). Compare your answer with the Feigenbaum number for the quadratic family.

3.5   THE HORSESHOE MAP

One of the earliest examples of a function defined on R^2 that exhibits interesting dynamics is the **horseshoe map** described in the 1960's by the American mathematician Stephen Smale (1967).

The horseshoe map will be denoted by M. Its domain is the set S in R^2 composed of the unit square $T = [0, 1] \times [0, 1]$, bounded on the left and right by semicircles B and E (Figure 3.14(a)). We assume that S contains its boundary. The function M shrinks S vertically by a factor of $a < 1/3$, and expands S horizontally by a factor of $b = 3$, with the semicircles B and E altered so as to continue to be semicircular. The resulting figure is folded by M so that it fits again inside S, with only the semicircles protruding to the left of T (Figure 3.14(b)). Thus the range of M looks like a horseshoe. When S is partitioned as in Figure 3.14(c), we can see the effect of M on each member of the partition (Figure 3.14(d)). Specifically, M sends semicircles B and E into B, and sends the square T into two strips inside T plus a curved strip inside E.

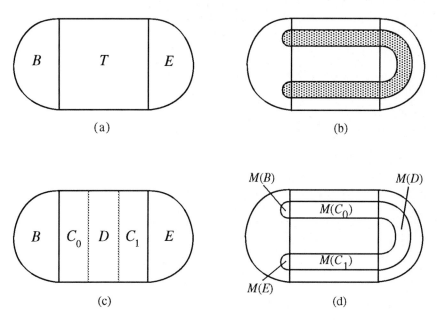

Figure 3.14

Even though we have not defined M by a formula or a series of formulas, we are able to systematically analyze it. To start with, we note that M is well-defined and the range is contained in S. Next we will show that M is a homeomorphism.

THEOREM 3.24. $M: S \to M(S)$ is a homeomorphism.

Proof. By definition, M maps onto $M(S)$. That M is one-to-one follows from the fact that stretching and contracting are one-to-one operations, and M folds in a non-overlapping manner. To prove that M is continuous, we compare subregions in Figure 3.14(c) with their corresponding images in Figure 3.14(d). The map M expands distances in C_0 and C_1 by a factor of 3, and shrinks distances in B and E. The largest expansion of distances for points in D occurs at the top boundary, which maps onto the exterior boundary γ of $M(D)$. Since the length of the top boundary of D is 1/3 and the length of γ is less than $\pi/2$, it follows that M expands distances in D by no more than a factor of 6. Consequently if $\mathbf{v}$ and $\mathbf{w}$ are in S and $\|\mathbf{v} - \mathbf{w}\| < \varepsilon$, then $\|M(\mathbf{v}) - M(\mathbf{w})\| < 6\varepsilon$. Therefore M is continuous. The inverse map M^{-1} is continuous by the same kind of argument. Therefore M is a homeomorphism. ∎

Assuming that B in Figure 3.14(c) contains its boundary, we show next that M has a fixed point that lies in B and on that boundary.

THEOREM 3.25. M has a unique fixed point in B, to which the iterates of all points in B and E converge.

Proof. Since M shrinks the domain vertically by a factor of $a < 1/3$, this means that for each n, $M^{[n]}(B)$ is closed and semicircular with diameter a^n. In addition,

$$B \supseteq M(B) \supseteq M^{[2]}(B) \supseteq \cdots \supseteq M^{[n]}(B)$$

The two-dimensional version of the Heine-Borel Theorem (Theorem 2.7) implies that a nested sequence of closed, bounded sets in R^2 has a common point. Thus the intersection B_∞ of the sets $M^{[n]}(B)$ for $n \geq 1$ has at least one element. Notice that the diameter of $M^{[n]}(B)$ is a^n, and $\lim_{n\to\infty} a^n = 0$. Thus B_∞ contains exactly one element, which we denote by $\mathbf{p}$. Now $\mathbf{p}$ lies on the boundary of B and C_0. Since $\mathbf{p}$ and $M(\mathbf{p})$ are in $M^{[n]}(B)$ for all n, it follows that $\|\mathbf{p} - M(\mathbf{p})\| \leq a^n \to 0$ as n grows without bound. Consequently $\mathbf{p} = M(\mathbf{p})$, which means that $\mathbf{p}$ is a fixed point. Finally, because $B \supseteq M(E)$ and all elements of B converge to $\mathbf{p}$, we conclude that all elements of E also converge to $\mathbf{p}$. ∎

Although $\mathbf{p}$ attracts all points in B and E, $\mathbf{p}$ is not an attracting fixed point because the iterates of points in the interior of C_0 are drawn away from $\mathbf{p}$.

Figure 3.15 shows the image of $M^{[2]}$, which is composed of two connected horseshoes. Similarly, for each positive integer n, the image of $M^{[n]}$ contains 2^{n-1} connected horseshoes whose width is approximately a^n.

From the definition of M, if $\mathbf{v}$ is in T, then $M(\mathbf{v})$ is also in T only if $\mathbf{v}$ is in $C_0 \cup C_1$. Let C_+ denote the collection of points in $C_0 \cup C_1$ *all* of whose iterates lie in $C_0 \cup C_1$. Thus

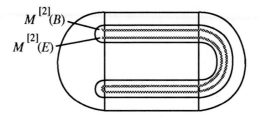

$M^{[2]}(B)$

$M^{[2]}(E)$

Figure 3.15

$$C_+ = \{\mathbf{v} \text{ in } C_0 \cup C_1 : M^{[n]}(\mathbf{v}) \text{ is in } C_0 \cup C_1 \text{ for } n = 0, 1, 2, 3, \ldots \}$$

All points in the domain of M either migrate toward $\mathbf{p}$, which is the most illustrious member of C_+, or start out in C_+ and stay there. Consequently C_+ serves as the attractor A_M of M. What does C_+ look like geometrically?

The set C_+ is an intersection of a nested sequence of ever thinner vertical strips. To show this we notice first that the collection of all points $\mathbf{v}$ such that $M(\mathbf{v})$ is in $C_0 \cup C_1$ consists of the four strips in Figure 3.16(a), where C_{ik} is the set of all $\mathbf{v}$ such that $\mathbf{v}$ is in C_i and $M(\mathbf{v})$ is in C_k, for $i, k = 0, 1$. (For example, $\mathbf{v}$ is in C_{01} if and only if $\mathbf{v}$ is in C_0 and $M(\mathbf{v})$ is in C_1.) Similarly, the collection of all $\mathbf{v}$ such that $M^{[2]}(\mathbf{v})$ is in $C_0 \cup C_1$ consists of the eight strips in Figure 3.16(b), and so forth. Therefore C_+, and hence the attractor A_M, is a collection of vertical lines in the square $T = [0, 1] \times [0, 1]$ whose intersection with the x axis is a Cantor-like set.

If $\mathbf{v}$ is in C_+, then each iterate of $\mathbf{v}$ lies either in C_0 or in C_1, so we can associate with $\mathbf{v}$ the **forward sequence** $z = z_0 z_1 z_2 \cdots$, where

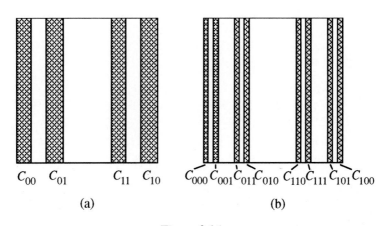

C_{00} C_{01} C_{11} C_{10} C_{000} C_{001} C_{011} C_{010} C_{110} C_{111} C_{101} C_{100}

(a)   (b)

Figure 3.16

$$z_n = \begin{cases} 0 & \text{if } M^{[n]}(\mathbf{v}) \text{ is in } C_0 \\ 1 & \text{if } M^{[n]}(\mathbf{v}) \text{ is in } C_1 \end{cases}$$

Notice that the sequence $z_0 z_1 z_2 \cdots$ identifies the forward iterates of $\mathbf{v}$. This is reminiscent of the sequence identified with the numbers in $[0, 1]$ for the tent function described in Section 2.2. However, since C_0 and C_1 are separated by a rectangle of width $1/3$, *every* sequence of 0's and 1's is the image of an element of C_+. (The corresponding result is false for T!)

In contrast to the tent function, each sequence of 0's and 1's corresponds to a whole vertical line, not just an individual point. To identify sequences of 0's and 1's with unique points in S, we need to examine the pre-images of points in S.

Figure 3.14(d) indicates that $M(C_0)$ and $M(C_1)$ are horizontal strips in T that we denote by V_0 and V_1, respectively (Figure 3.17(a)). Similarly, $M^{[2]}(C_{00})$, $M^{[2]}(C_{01})$, $M^{[2]}(C_{11})$, and $M^{[2]}(C_{10})$ are horizontal strips in Figure 3.15, and we designate these strips as V_{00}, V_{01}, V_{11}, and V_{10} respectively (Figure 3.17(b)). Letting $M^{-[m]}(P)$ denote the set Q such that $M^{[m]}(Q) = P$, we find that

$$M^{-[1]}(V_i) = C_i \quad \text{and} \quad M^{-[2]}(V_{ik}) = C_{ik} \quad \text{for } i = 0, 1 \text{ and } k = 0, 1$$

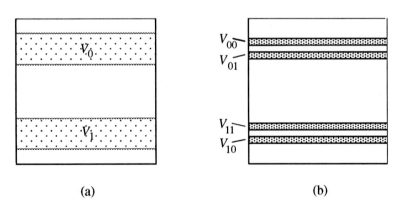

(a)   (b)

Figure 3.17

Continuing in this fashion, we obtain a nested sequence of ever thinner horizontal strips whose intersection we denote by C_-. Thus

$$C_- = \{\mathbf{v} \text{ in } S: M^{-[n]}(\mathbf{v}) \text{ is in } C_0 \cup C_1 \text{ for } n = 1, 2, 3, \dots \}$$

In the same way as we identified a sequence of 0's and 1's for each element of C_+, we now assign the **backward sequence** $\cdots z_{-3} z_{-2} z_{-1}$ to each $\mathbf{v}$ in C_-, where

$$z_{-n} = \begin{cases} 0 \text{ if } M^{-[n]}(\mathbf{v}) \text{ is in } C_0 \\ 1 \text{ if } M^{-[n]}(\mathbf{v}) \text{ is in } C_1 \end{cases}$$

Each such sequence corresponds to a horizontal line in T, and C_- is a collection of horizontal lines in T whose intersection with the y axis is a Cantor-like set.

Combining the forward sequence $z_0 z_1 z_2 \cdots$ and the backward sequence $\cdots z_{-3} z_{-2} z_{-1}$, we obtain the **two-sided sequence** (or **bi-infinite sequence**) $\cdots z_{-3} z_{-2} z_{-1} . z_0 z_1 z_2 \cdots$, where the decimal point separates the backward part from the forward part of the sequence. Finally, we define the set C^* by

$$C^* = C_+ \cap C_-$$

The set C^* consists of the points in T all of whose forward and backward iterates lie in $C_0 \cup C_1$.

Let Z denote the collection of all two-sided sequences of 0's and 1's, and define $h: C^* \to Z$ by

$$h(\mathbf{v}) = \text{ the two-sided sequence corresponding to } \mathbf{v}, \text{ for } \mathbf{v} \text{ in } C^* \qquad (1)$$

Then h is well-defined, and identifies C^* with Z, as Lemma 1 asserts.

LEMMA 1. $h: C^* \to Z$ is one-to-one and onto.

Proof. If $\mathbf{v}$ and $\mathbf{w}$ are in C^* and $h(\mathbf{v}) = h(\mathbf{w})$, then because $h(\mathbf{v})$ and $h(\mathbf{w})$ have the same forward (backward) sequence, they lie on the same vertical (horizontal) line in T. Therefore $\mathbf{v} = \mathbf{w}$, so that h is one-to-one. To show that h is onto, assume that $x = \cdots x_{-3} x_{-2} x_{-1} . x_0 x_1 x_2 \cdots$ is in Z. For $n \geq 0$, let

$$J_n = \{\mathbf{v} \text{ in } C_0 \cup C_1 : h(\mathbf{v}) = \cdots z_{-3} z_{-2} z_{-1} . z_0 z_1 z_2 \cdots \text{ and } z_0 z_1 z_2 \cdots z_n = x_0 x_1 x_2 \cdots x_n\}$$

and

$$J_{-n} = \{\mathbf{v} \text{ in } C_0 \cup C_1 : h(\mathbf{v}) = \cdots z_{-3} z_{-2} z_{-1} . z_0 z_1 z_2 \cdots \text{ and } z_{-n} \cdots z_{-3} z_{-2} z_{-1} = x_{-n} \cdots x_{-3} x_{-2} x_{-1}\}$$

Then J_n and J_{-n} are closed for all n. Because $\bigcap_{n \geq 0} J_n$ is a single vertical line and $\bigcap_{n < 0} J_n$ is a single horizontal line in T, it follows that $\bigcap_{\text{all } n} J_n$ is a unique point $\mathbf{v}^*$. By construction, $h(\mathbf{v}^*) = x$, so that h is onto. ∎

The function h has the property that

if $h(\mathbf{v}) = \cdots z_{-3} z_{-2} z_{-1} . z_0 z_1 z_2 \cdots$, then $h(M(\mathbf{v})) = \cdots z_{-2} z_{-1} z_0 . z_1 z_2 z_3 \cdots$

This means that the sequence associated with $M(\mathbf{v})$ is the sequence associated with $\mathbf{v}$, shifted to the left one place with respect to the decimal point. This fact, along with the association between points of C^* and two-sided sequences of 0's and 1's, bears immediate fruit. In particular, the two doubly-repeated sequences

$$\cdots \overline{0}.\overline{0}\cdots \quad \text{and} \quad \cdots \overline{1}.\overline{1}\cdots$$

correspond to fixed points of M. You can check that the fixed point $\mathbf{p}$ on the border between B and C_0 corresponds to $\cdots \overline{0}.\overline{0}\cdots$. Can you locate the other fixed point in T? Next, the sequences $\cdots \overline{10}.\overline{10}\cdots$ and $\cdots \overline{01}.\overline{01}\cdots$ comprise a 2-cycle for M. Using these sequences as models, one can exhibit two-sided sequences corresponding to n-cycles for any positive integer n. From the definition of $\cdots z_{-3}z_{-2}z_{-1}.z_0z_1z_2\cdots$, one can even indicate where in T members of such a cycle lie.

Next we will introduce a distance on the set Z of two-sided sequences. This distance will make it possible for us to show that C^* and Z are homeomorphic. Let $x = \cdots x_{-3}x_{-2}x_{-1}.x_0x_1x_2\cdots$ and $z = \cdots z_{-3}z_{-2}z_{-1}.z_0z_1z_2\cdots$. Then we define the **distance** $\|x - z\|$ between x and z by the formula

$$\|x - z\| = \sum_{k=-\infty}^{\infty} \frac{|x_k - z_k|}{2^{|k|}} \tag{2}$$

The distance is a metric on the space of two-sided sequences (Exercise 6). If $x_k = z_k$ for $|k| \le n$, then $\|x - z\| \le 1/2^{n-1}$. Moreover, if $\|x - z\| \le 1/2^n$, then $x_k = z_k$ for $|k| \le n + 1$. Thus the distance between x and z is small provided that the central blocks of x and z are identical.

THEOREM 3.26. $h: C^* \to Z$ is a homeomorphism.

Proof. By Lemma 1, h is one-to-one and onto. Therefore we need only show that h and h^{-1} are continuous. Let $\varepsilon > 0$, and choose n so large that $1/2^{n-1} < \varepsilon$. Next, let $\mathbf{v}$ and $\mathbf{w}$ be in C^*, with

$$h(\mathbf{v}) = x = \cdots x_{-3}x_{-2}x_{-1}.x_0x_1x_2\cdots \quad \text{and} \quad h(\mathbf{w}) = z = \cdots z_{-3}z_{-2}z_{-1}.z_0z_1z_2\cdots$$

If $\|\mathbf{v} - \mathbf{w}\| < 1/3^{n+1}$, then $\mathbf{v}$ and $\mathbf{w}$ lie in the same vertical strip of width $1/3^{n+1}$, so that $x_k = z_k$ for $k = 0, 1, 2, ..., n$. Similarly, there is a $\delta_1 > 0$ such that if $\|\mathbf{v} - \mathbf{w}\| < \delta_1$, then $\mathbf{v}$ and $\mathbf{w}$ lie in the same horizontal strip at the nth stage, which means that $x_k = z_k$ for $k = -1, -2, ..., -n$. Now choose $\delta > 0$ such that $\delta < 1/3^{n+1}$ and $\delta \le \delta_1$. It follows that if $\|\mathbf{v} - \mathbf{w}\| < \delta$, then $x_k = z_k$ for $|k| \le n$, and thus

$$\|x - z\| = \sum_{|k|=n+1}^{\infty} \frac{|x_k - z_k|}{2^{|k|}} \le \frac{1}{2^{n-1}} < \varepsilon$$

Consequently h is continuous. The proof that h^{-1} is continuous follows by a similar argument (but with the roles of vertical contraction and horizontal expansion interchanged). ■

The **left shift map** $\sigma: Z \to Z$ is defined as the name suggests:

if $z = \cdots z_{-3}z_{-2}z_{-1}.z_0 z_1 z_2 \cdots$, then $\sigma(z) = \cdots z_{-3}z_{-2}z_{-1}z_0.z_1 z_2 \cdots$

Thus σ shifts the entries to the left one place with respect to the decimal point, or equivalently, shifts the decimal point one place to the right. That σ is a homeomorphism on Z can be proved in a straightforward manner (Exercise 7). Moreover, we can show that σ is **strongly chaotic**, which by definition means that

i. its domain has a dense set of periodic points
ii. it has sensitive dependence on initial conditions
iii. it is transitive (that is, it has an element with dense orbit)

THEOREM 3.27. σ is strongly chaotic on Z.

Proof. First we will show that the set of periodic points of σ is dense in Z. To that end, let $z = \cdots z_{-3}z_{-2}z_{-1}.z_0 z_1 z_2 \cdots$ be an arbitrary element of Z, and let n be an arbitrary positive integer. If x is the doubly-repeating two-sided sequence $\overline{z_{-n}\cdots z_{-3}z_{-2}z_{-1}.z_0 z_1 z_2 \cdots z_n}$, then it follows that x is periodic (with period $2n+1$). Moreover, $z_k = x_k$ for $|k| \le n$, so that $\|x - z\| \le 1/2^{n-1}$. Thus the periodic points are dense in Z. To show that σ has sensitive dependence on initial conditions, let z be in Z, $\varepsilon = 1/2$, $\delta > 0$, and n so large that $1/2^{n-1} < \delta$. If x is chosen with $x_k = z_k$ for all k and such that $|k| \le n$ but $x_{n+1} \ne z_{n+1}$, then $\|x - z\| \le 1/2^{n-1} < \delta$. However,

$$\sigma^{[n+1]}(x) = \cdots x_n . x_{n+1} \cdots \quad \text{and} \quad \sigma^{[n+1]}(z) = \cdots z_n . z_{n+1} \cdots$$

so that $\|\sigma^{[n+1]}(x) - \sigma^{[n+1]}(x)\| \ge 1 > \varepsilon$. Therefore σ has sensitive dependence on initial conditions. Finally, we will show that σ is transitive. To see this, let the forward portion of the two-sided sequence z^* have the form

0 1 000 001 010 011 100 101 110 111 00000 00001 $\cdots$

(where for each positive odd integer n, all possible n-tuples appear in order), and let the backward portion of z^* have the form

$\cdots$ 0100 0011 0010 0001 0000 11 10 01 00

(where for each positive even integer n, all possible n-tuples appear in backward order). Then it is possible to show that the orbit of z^* is dense in Z (Exercise 4). With this we have completed the proof that σ is strongly chaotic on Z. ■

The functions $M: C^* \rightarrow C^*$ and $\sigma: Z \rightarrow Z$ are **conjugate**, in the sense that there is a homeomorphism $h: C^* \rightarrow Z$ such that $h \circ M = \sigma \circ h$. The homeomorphism we have in mind, of course, is the one defined in (1). You should check that indeed $h(M(\mathbf{v})) = \sigma(h(\mathbf{v}))$ for all $\mathbf{v}$ in C^* (Exercise 8). The strong chaotic nature of one function is inherited by another function conjugate to it (as we saw in Section 2.3). Thus we obtain the following consequence of Theorem 3.27.

COROLLARY 3.28. M is strongly chaotic on C^*.

The Smale horseshoe map is justifiably famous, for several reasons. First, it has a very simple geometric definition and is easy to visualize. Second, it has all the characteristics of a strongly chaotic map: sensitive dependence, plenty of periodic points, and a dense orbit. It also has, in the most obvious way, the telltale signs of a chaotic map: stretching and folding. The stretching yields sensitive dependence, and the folding allows the map to be bounded. It is conjectured that in some sense every map that is strongly chaotic has a subset of its domain on which the map acts like the horseshoe map M.

Homoclinic Points

Let f be a function, and p a fixed point of f. It can happen that there is a point q in the domain of f whose forward iterates converge to p and such that a sequence of backward iterates of q also converges to p. The following example illustrates such behavior.

EXAMPLE 1. Let T be the tent function, defined by

$$T(x) = \begin{cases} 2x & \text{for } 0 \leq x \leq 1/2 \\ 2(1-x) & \text{for } 1/2 < x \leq 1 \end{cases}$$

Show that the forward and backward iterates of $1/4$ converge to the fixed point 0.

Solution. First we notice that

$$T(\frac{1}{4}) = \frac{1}{2}, \quad T^{[2]}(\frac{1}{4}) = 1, \text{ and } T^{[3]}(\frac{1}{4}) = 0$$

so that the forward iterates of $1/4$ are eventually 0, and hence converge to 0. By

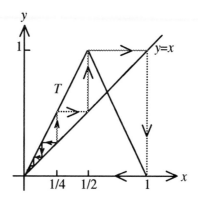

Forward and backward iterates of 1/4

Figure 3.18

contrast, $T^{-1}(1/4)$ contains 1/8, $T^{-[2]}(1/4)$ contains 1/16, and in general, $T^{-[n]}(1/4)$ contains $1/2^{n+2}$. Thus there are backward iterates of 1/4 that converge to 0. Figure 3.18 shows both forward and backward iterates of 1/4. ❏

Since the forward iterates of 1/4 converge to 0, 1/4 is in the local stable manifold $W_{loc}^s(0)$ of 0. In the same manner, 1/4 is in the local unstable manifold $W_{loc}^u(0)$ of 0, because there are backward iterates of 1/4 that converge to 0. This means that 1/4 is in both $W_{loc}^s(0)$ and $W_{loc}^u(0)$. Points with this property are called **homoclinic points**.

DEFINITION 3.29. Let F be defined on a subset of R^n, where $n = 1$ or 2, and let **p** be a fixed point of F. A point **q** is **homoclinic to p,** or is a **homoclinic point,** if $\mathbf{q} \neq \mathbf{p}$ but **q** is in both $W_{loc}^s(\mathbf{p})$ and $W_{loc}^u(\mathbf{p})$.

From Example 1 we know that 1/4 is homoclinic to 0 for T. In fact, there are infinitely many numbers in $(0, 1)$ that are homoclinic to 0 for T (Exercise 11). Analogously, there are numbers in $(0, 1)$ that are homoclinic to 0 for the quadratic function Q_4 (Exercise 12). By contrast, there are no numbers in $(0, 1)$ that are homoclinic to 0 for the quadratic function Q_μ if $0 < \mu < 4$ (also Exercise 12).

The horseshoe map M has points homoclinic to the fixed point **p** in B. Since M is conjugate to the left shift σ, we can verify this by showing, equivalently, that the left shift σ has points homoclinic to the fixed point $z^{**} = \cdots \overline{0}.\overline{0} \cdots$ of σ. We will indicate how one can determine such homoclinic points of σ, but will leave the details to be completed in Exercise 10. On the one hand, $W_{loc}^s(z^{**})$ consists of all two-sided sequences all of whose entries to the right of some entry are identical to those of z^{**}. On the other hand, $W_{loc}^u(z^{**})$ consists of those two-sided sequences all of whose entries to the left of some entry are identical to those of z^{**}. It follows from these facts that σ has points homoclinic to z^{**}. What is more, there are infinitely many points homo-clinic to z^{**}.

Homoclinic points play a significant role in the study of the dynamics of higher-dimensional functions.

Conclusion

Chapter 3 has introduced dynamics of functions defined in R^2. Besides the linear functions, which exhibit regularity, we have focused on three prominent functions: the baker's function, the Hénon map, and the Smale horseshoe map. For each of these three functions there is an attractor to which the iterates of all points in the domain converge, and which contains in a natural way a two-dimensional version of a Cantor set. In each there is a stretching and some form of folding, which are characteristic of chaotic behavior. Finally, each represents a two-dimensional analogue of a one-dimensional chaotic function studied in Chapters 1–2.

EXERCISES 3.5

1. Find the approximate location of the point in S corresponding to the sequence $\cdots \overline{1}.\overline{1} \cdots$ in Z.

2. Determine the number of n-cycles in C^*.

3. Let $x = \cdots x_{-3} x_{-2} x_{-1}.x_0 x_1 x_2 \cdots$ and $z = \cdots z_{-3} z_{-2} z_{-1}.z_0 z_1 z_2 \cdots$.

    a. Show that if $x_k = z_k$ for $|k| \le n$, then $\|x - z\| \le 1/2^{n-1}$.

    b. Find x and z such that $x_k = z_k$ for $|k| \le n$ and $\|x - z\| = 1/2^{n-1}$.
    c. Find x and z in Z such that $\|x - z\| = 3$.

4. Show that the sequence z^* in the proof of Theorem 3.27 has a dense orbit.

5. Show that there is an element of C^* that neither is periodic nor has a dense orbit under M.

6. Show that the distance defined in (2) has the properties of a metric as defined in Section 2.4.

7. Show that the left shift $\sigma : Z \to Z$ is a homeomorphism.

8. Show that $h(M(\mathbf{v})) = \sigma(h(\mathbf{v}))$ for all $\mathbf{v}$ in C^*.

9. Show that there is a one-to-one function whose domain is C^* and whose range is the Cantor ternary set C.

10. Let $z^{**} = \cdots \overline{0}.\overline{0}\cdots$, one of the two fixed points of σ.

   a. Show that $W^s_{loc}(z^{**})$ consists of all two-sided sequences all of whose entries to the right of some entry are identical to those of z^{**}.

   b. Show that $W^u_{loc}(z^{**})$ consists of those two-sided sequences all of whose entries to the left of some entry are identical to those of z^{**}.

   c. Show that there are infinitely many points homoclinic to z^{**}.

11. Show that there are infinitely many numbers in the interval $(0, 1)$ that are homoclinic to 0 for the tent function T.

12. Let $Q_\mu(x) = \mu x(1 - x)$ for $0 \le x \le 1$.

   a. Show that there are no points homoclinic to the fixed point 0 whenever $0 < \mu < 4$.

   b. Show that there are infinitely many points homoclinic to 0 if $\mu = 4$.

   c. Show that there are infinitely many points homoclinic to the fixed point $p_\mu = 1 - 1/\mu$ if $\mu = 4$.

CHAPTER
4

FRACTALS

In the study of chaotic dynamics one inevitably encounters sets in the plane or in space that have very complicated and interesting structures. Many such sets are so complicated that they can be realized only with the help of a computer, and thus have come into prominence during the past decade. An attribute common to these sets is that one cannot easily ascribe an integer dimension to them. Rather, such a set can be assigned a fractional dimension, which gave rise to the term "fractal."

The present chapter is devoted to sets that are fractal. In the first section we define one kind of dimension, the capacity dimension. It is especially valuable in the analysis of sets that have the highly geometric property that as one focuses on smaller and smaller areas within the set, one continues to see the structure of the whole. Such sets are called self-similar and play an important role in the study of chaotic dynamics. The second section focuses on another notion of dimension called the Lyapunov dimension, which is tailored to the study of higher-dimensional maps. The third section is devoted to the famous twin topics: Julia sets and the Mandelbrot set. These are perhaps the most famous of all fractals. The final section involves a discussion of sets that are derived by iterating several functions simultaneously. As a group these functions are called iterated function systems.

4.1 CAPACITY DIMENSION

Dimension is a word common to everyday life. We say that a wire has one dimension, paper has two dimensions, and a box has three dimensions. However, it is not clear what dimension the shape in Figure 4.1, which is called the **Sierpinski carpet**, should be assigned. It is at least one-dimensional because the boundary contains lines. However, it appears to be nearly a filled-in square, so perhaps it should be two-dimensional. We will resolve this question in the present section, as we define a notion of dimension that is especially applicable to geometric figures. A discussion of dimension for sets defined by maps will occur in Section 4.2.

Let S be a subset of R^n, where $n = 1$, 2, or 3. By an **n-dimensional box** in R^n we mean a closed interval if $n = 1$, square if $n = 2$, and cube if $n = 3$.

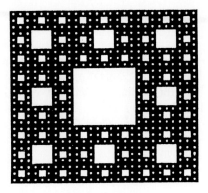

Sierpinski carpet

Figure 4.1

Normally we will refer to such a set as a **box** if n is understood. For each $\varepsilon > 0$, we denote by $N(\varepsilon)$ the smallest number of n-dimensional boxes of side length ε required in order to completely cover S. Thus it takes eight boxes of side length $\varepsilon = 1/3$ to cover the cut-out square S in Figure 4.2(a), whereas it takes seventy-two boxes of side length $\varepsilon = 1/9$ to cover S (Figure 4.2(b)). Therefore $N(1/3) = 8$ and $N(1/9) = 72$. As ε approaches 0, the number $N(\varepsilon)$ increases without bound.

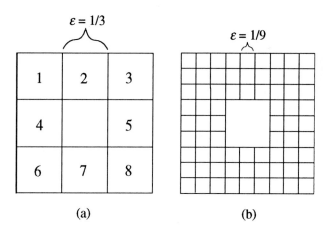

(a)   (b)

Figure 4.2

If S is the interval $[0, 2]$, then for small $\varepsilon > 0$ it takes approximately $1/\varepsilon$ boxes to cover each unit interval, so it takes approximately $2/\varepsilon$ boxes to cover the line segment in Figure 4.3(a). Thus $N(\varepsilon) \approx 2/\varepsilon$. Of course, if $2/\varepsilon$ is not an integer, then $N(\varepsilon) \neq 2/\varepsilon$. For example, if $\varepsilon = 3/32$, then the number of boxes necessary to cover the interval $[0, 2]$ is $N(3/32) = 22$, whereas $2/\varepsilon = 64/3 \neq 22$. Nevertheless, $N(3/32) \approx 64/3$. Moreover, as ε approaches 0, $N(\varepsilon)$ is proportional to $1/\varepsilon$. We say that $N(\varepsilon)$ **scales** as $1/\varepsilon$. Similarly, if S is the rectangle $[0, 2] \times [0, 3]$, then for small $\varepsilon > 0$ it takes approximately $6/\varepsilon^2$ boxes to cover S (Figure 4.3(b)).

For a given subset S of R^n, it may happen that there are numbers $C > 0$ and $D \geq 0$ such that $N(\varepsilon) \approx C(1/\varepsilon)^D$ as ε approaches 0, that is, $N(\varepsilon)$ scales as $(1/\varepsilon)^D$. Since $N(\varepsilon) \approx 2/\varepsilon$ for the line segment $[0, 2]$ in Figure 4.3(a), it follows that $N(\varepsilon) \approx C(1/\varepsilon)^D$ with $C = 2$ and $D = 1$. Similarly, for the rectangle in Figure 4.3(b) we have $N(\varepsilon) \approx C/\varepsilon^2 = 6(1/\varepsilon)^D$ with $C = 6$ and $D = 2$.

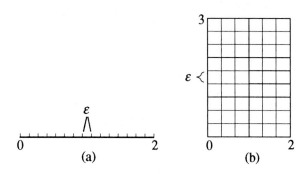

Figure 4.3

Assume that $N(\varepsilon) = C(1/\varepsilon)^D$ for a given set S. Taking logarithms of both sides, we find that

$$\ln N(\varepsilon) \approx D \ln \frac{1}{\varepsilon} + \ln C$$

Solving for D, we obtain

$$D \approx \frac{\ln N(\varepsilon)}{\ln (1/\varepsilon)} - \frac{\ln C}{\ln (1/\varepsilon)} \approx \frac{\ln N(\varepsilon)}{\ln (1/\varepsilon)} \tag{1}$$

provided that ε is very small (which implies that $\ln (1/\varepsilon)$ is very large). If the limit of the final fraction in (1) exists as ε approaches 0, we define D to be the dimension of S.

DEFINITION 4.1. Let S be a subset of R^n, where $n = 1, 2$, or 3. The **capacity dimension** (or **box dimension** or **box-counting dimension**) of S is given by

$$\dim_c S = \lim_{\varepsilon \to 0} \frac{\ln N(\varepsilon)}{\ln (1/\varepsilon)}$$

if the limit exists. If the capacity dimension of S exists and is *not* an integer, then S is said to have **fractal dimension**.

The term "fractal" was introduced by the mathematician Benoit Mandelbrot in the late 1970's to refer to a set with fractal dimension. If $D = \dim_c S$ exists, then whether or not it is an integer, we say that the number $N(\varepsilon)$ of boxes necessary to cover S **scales** as $(1/\varepsilon)^D$.

From our discussion, we see that the dimensions of the shapes in Figure 4.3 are given by

$$\dim_c \text{ (line segment)} = 1 \quad \text{and} \quad \dim_c \text{ (filled-in rectangle)} = 2$$

However, in general it is no easier to evaluate the capacity dimension by Definition 4.1 than it is to evaluate limits in calculus by the $\varepsilon\text{-}\delta$ method. The next result will simplify calculation of capacity dimensions.

THEOREM 4.2. Let $0 < r < 1$. Consider a subset S of R^n, with $n = 1$, 2, or 3. Then $\lim_{k \to \infty} [\ln N(r^k)/\ln (1/r^k)]$ exists if and only if $\dim_c S$ exists; in that case,

$$\dim_c S \;=\; \lim_{k \to \infty} \frac{\ln N(r^k)}{\ln (1/r^k)} \tag{2}$$

Proof. Let $0 < \varepsilon < r$, and let k be a positive integer so large that $r^{k+1} < \varepsilon \le r^k$. Then $N(r^k) \le N(\varepsilon) \le N(r^{k+1})$. The fact that the natural logarithm is an increasing function and $0 < r < 1$ implies that

$$\ln (1/r^k) \;\le\; \ln (1/\varepsilon) \;<\; \ln (1/r^{k+1}) \quad \text{and} \quad \ln N(r^k) \;\le\; \ln N(\varepsilon) \;\le\; \ln N(r^{k+1})$$

Therefore

$$\frac{\ln N(r^k)}{\ln (1/r^k) + \ln (1/r)} = \frac{\ln N(r^k)}{\ln (1/r^{k+1})} \le \frac{\ln N(\varepsilon)}{\ln (1/\varepsilon)} \le \frac{\ln N(r^{k+1})}{\ln (1/r^k)} = \frac{\ln N(r^{k+1})}{\ln (1/r^{k+1}) - \ln (1/r)}$$

If one of the following limits exists, then they all do, and

$$\lim_{k \to \infty} \frac{\ln N(r^k)}{\ln (1/r^k) + \ln (1/r)} = \lim_{k \to \infty} \frac{\ln N(r^k)}{\ln (1/r^k)} = \lim_{k \to \infty} \frac{\ln N(r^{k+1})}{\ln (1/r^{k+1}) - \ln (1/r)}$$

Therefore

$$\lim_{k \to \infty} \frac{\ln N(r^k)}{\ln (1/r^k)} \quad \text{exists if and only if} \quad \lim_{\varepsilon \to 0} \frac{\ln N(\varepsilon)}{\ln (1/\varepsilon)} \quad \text{exists}$$

and when they do, they are equal. Thus the formula in (2) is verified. ∎

We will put Theorem 4.2 to immediate use.

EXAMPLE 1. Let C = the Cantor ternary set, discussed in Section 2.4. Show that $\dim_c C = (\ln 2)/(\ln 3)$.

Solution. Recall that C is obtained from the interval $[0, 1]$ by deleting the middle third, then the middle third of each remaining interval, and so on. As Figure 4.4 indicates, at each stage there are twice as many segments as in the preceding stage, and each segment is one-third as long. Since $N(1/3) = 2$, it follows by induction that $N(1/3^k) = 2^k$ for each $k \geq 1$. If we let $r = 1/3$ in Theorem 4.2, we find that

$$\dim_c C = \lim_{k \to \infty} \frac{\ln N(1/3^k)}{\ln (1/(1/3^k))} = \lim_{k \to \infty} \frac{\ln 2^k}{\ln 3^k} = \lim_{k \to \infty} \frac{k \ln 2}{k \ln 3} = \frac{\ln 2}{\ln 3} \approx 0.63 \quad \square$$

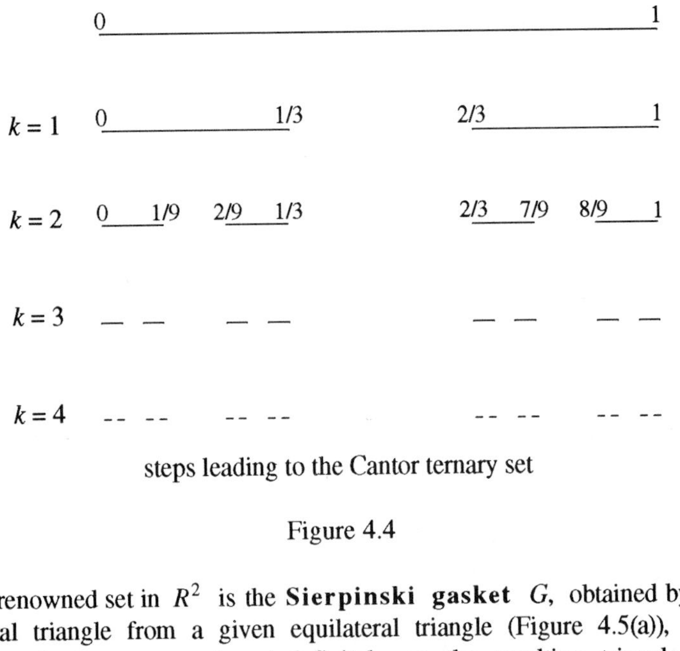

steps leading to the Cantor ternary set

Figure 4.4

A renowned set in R^2 is the **Sierpinski gasket** G, obtained by deleting the central triangle from a given equilateral triangle (Figure 4.5(a)), and then performing the same operation indefinitely on the resulting triangles (Figure 4.5(b)–(f)). The set is named after the Polish mathematician Waclaw Sierpinski, who studied it over half a century ago.

EXAMPLE 2. Show that $\dim_c G = (\ln 3)/(\ln 2)$.

Solution. Assume that the legs of the triangle have length 1. At each stage in the creation of the Sierpinski gasket, there are three times as many triangles as in the preceding stage, the legs of the triangles having half the length of those in the preceding stage. Thus $N(1/2) = 3$, and by induction, $N(1/2^k) = 3^k$. Consequently by Theorem 4.2 with $r = 1/2$,

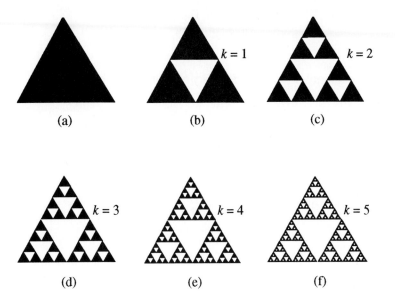

steps leading to the Sierpinski gasket

Figure 4.5

$$\dim_c G = \lim_{k \to \infty} \frac{\ln N(1/2^k)}{\ln (1/(1/2^k))} = \lim_{k \to \infty} \frac{\ln 3^k}{\ln 2^k} = \lim_{k \to \infty} \frac{k \ln 3}{k \ln 2} = \frac{\ln 3}{\ln 2} \approx 1.58 \quad \square$$

In a similar way one can show that the Sierpinski carpet shown in Figure 4.1 has capacity dimension $(\ln 8)/(\ln 3) \approx 1.89$ (Exercise 5).

We also observe that if S is a subset of R^2 or R^3, then in order to calculate $\dim_c S$, one can use disks of radius ε (in R^2) or balls of radius ε (in R^3), and the limit defining $\dim_c S$ is unaltered (Exercises 12(a) and 13). In R^2, one can also use grids of congruent rectangles, provided that as the grid becomes finer the rectangles are similar in shape (Exercise 12(b)).

For geometrical shapes that are very regular, such as those appearing so far in the section, there is another way of calculating the capacity dimension. To describe it, we first let S be a set in R^n, where $n = 1$, 2, or 3. As before, the distance between two points a and b in S is denoted by $\|a - b\|$. Suppose that $f: S \to S$ has the property that for some constant r with $0 < r < 1$,

$$\|f(x) - f(y)\| = r\|x - y\|, \quad \text{for all } x \text{ and } y \text{ in } S$$

Then f is called a **similarity** of S, since $f(S)$ has a shape similar to that of S. The constant r is the **similarity constant** of f. Because distances between points in S are contracted (by a factor of r), the function is also called a **contraction.** For example, if S is the interval $[0, 1]$, and if

$$f(x) = \frac{1}{3} x + \frac{1}{2}$$

then f is a similarity of S with similarity constant $1/3$ because

$$|f(x) - f(y)| = \left|\left(\frac{1}{3} x + \frac{1}{2}\right) - \left(\frac{1}{3} y + \frac{1}{2}\right)\right| = \frac{1}{3} |x - y| \text{ for all } x \text{ and } y \text{ in } S$$

In this case f first contracts the interval by a factor of $1/3$ and then translates it to the right by $1/2$.

If there are similarities $f_1, f_2, ..., f_m$ of S such that

$$S = f_1(S) \cup f_2(S) \cup \cdots \cup f_m(S)$$

and the images $f_1(S), f_2(S), ..., f_m(S)$ are non-overlapping (except possibly for simple boundaries), then S is a **self-similar set**. In this case, S is composed of m (shrunk) copies of itself.

EXAMPLE 3. Show that the Cantor ternary set C is self-similar.

Solution. Let

$$f_1(x) = \frac{1}{3} x \quad \text{and} \quad f_2(x) = \frac{2}{3} + \frac{1}{3} x$$

Then $f_1(C) = C \cap [0, 1/3]$ and $f_2(C) = C \cap [2/3, 1]$. It follows that f_1 and f_2 are similarities, with similarity constant $1/3$. Moreover, $f_1(C) \cup f_2(C) = C$. Therefore C is self-similar. ❏

Suppose that S is a self-similar subset of R^n, where $n = 1, 2,$ or 3. Assume also that $f_1, f_2, ..., f_m$ are similarity functions, each with similarity constant r. If $f_1(S), f_2(S), ..., f_m(S)$ are non-overlapping and

$$S = f_1(S) \cup f_2(S) \cup \cdots \cup f_m(S)$$

then determination of the capacity dimension by means of the similarities proceeds as follows. Since $N(r^k)$ is proportional to m^k, we conclude that

$$\dim_c S = \lim_{k \to \infty} \frac{\ln N(r^k)}{\ln (1/r^k)} = \lim_{k \to \infty} \frac{\ln m^k}{\ln (1/r)^k} = \frac{\ln m}{\ln (1/r)} \tag{3}$$

To apply (3) to the Sierpinski gasket G, we must first show that it is self-similar. That it is self-similar can be deduced by letting $f_1, f_2,$ and f_3 map G onto the three subtriangles in Figure 4.5(b). Then f_1, f_2 and f_3 are similarities with similarity constant 1/2; it follows that G is self-similar. Since $r = 1/2$ and $m = 3$, it follows from (3) that $\dim_c G = (\ln 3)/(\ln 2)$. This is the same value determined in Example 2.

A self-similar set constructed by accretion is the **von Koch curve** K, described in 1904 by the Swedish mathematician Helge von Koch. To create K, we begin with the unit line segment in Figure 4.6(a), and replace it by the polygonal line in Figure 4.6(b). We repeat the procedure on each segment that results (Figure 4.6(c)). Continuing the process indefinitely (Figures 4.6(d)–(e)), we obtain a curve K that at each point is non-differentiable.

To show that K is self-similar, we notice that the polygonal curve in Figure 4.6(c) is composed of four copies of the polygonal curve in Figure 4.6(b). If we let the graphs of f_1, f_2, f_3 and f_4 be these four copies, respectively, then each of the functions so defined is a similarity, with $r = 1/3$, so that K is self-similar. Since there are four functions, $m = 4$, so (3) implies that $\dim_c K = (\ln 4)/(\ln 3)$. The von Koch curve is employed in making the von Koch snowflake (Exercise 3).

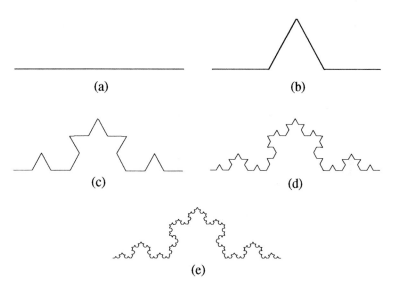

(a)   (b)

(c)   (d)

(e)

steps leading to the von Koch curve

Figure 4.6

An example of a higher-dimensional self-similar set with fractal dimension is provided by the **Menger sponge** M. The sponge is obtained from a unit cube by boring out the middle ninth of the cube in each direction (Figure 4.7(a)), and then continuing the process indefinitely on the remaining subcubes. The result is M (Figure 4.7(b)). In Exercise 4 you are asked to find the capacity dimension of M.

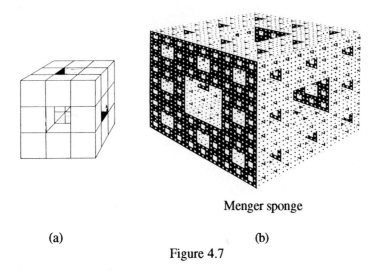

Menger sponge

(a)   (b)

Figure 4.7

Shapes related to the Sierpinski gasket arise in the "chaos game," which is played as follows. Let B, C, and D be three non-collinear points arbitrarily placed in the plane, and let x_0 be any point whatsoever in the plane (Figure 4.8(a)). Assign each side of a fair die one of the letters B, C, and D, so that each letter appears twice on the die. Then roll the die. Let the face that shows up be denoted by A_1 (so that A_1 is B, C, or D). Define x_1 to be the midpoint of the line segment joining x_0 and the point A_1. Proceed inductively: if the nth roll of the die

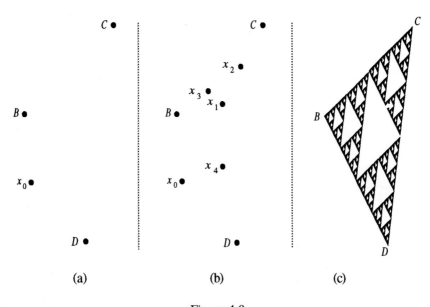

(a)   (b)   (c)

Figure 4.8

produces the face denoted A_n, then x_n is the midpoint of the line segment joining x_{n-1} and A_n. Figure 4.8(b) shows the first few steps of the procedure for a given selection of x_0 and faces of a rolled die. If we ignore the first few points that arise from the procedure, but identify the next several thousand, then amazingly a deformed Sierpinski gasket arises (Figure 4.8(c)). In fact, the actual Sierpinski gasket results if B, C, and D form an equilateral triangle. The procedure outlined above has been dubbed the **chaos game** by the mathematician Michael Barnsley.

EXERCISES 4.1

1. Suppose that S consists of k line segments in R^2, each with positive length. Show that $\dim_c S = 1$.

2. a. On a line of length b erect an isosceles triangle of height $b/2$, and then delete the base. On each remaining line segment, erect an isosceles triangle whose height is half the base, and delete the base. Repeat the procedure indefinitely to obtain a set S. Find $\dim_c S$.
    b. Suppose we began with an equilateral triangle. Could we define the capacity dimension for the set that results? Explain why or why not.

3. Start with an equilateral triangle of side length L. Suppose that we replace each side with the von Koch curve described in the text. The figure that results is the **von Koch snowflake** S_K (Figure 4.9).
    a. Show that S_K has infinite length.
    b. Show that S_K bounds a region with finite area A, and calculate A.

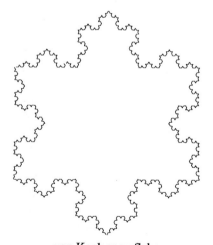

von Koch snowflake

Figure 4.9

4.   a.   Show that the Menger sponge M that appears in Figure 4.7(b) is self-similar.
     b.   Show that $\dim_c M = (\ln 20)/(\ln 3) \approx 2.73$.

5.   a.   Show that the Sierpinski carpet is self-similar.
     b.   Show that the capacity dimension of the Sierpinski carpet in Figure 4.1 is $(\ln 8)/(\ln 3) \approx 1.89$.

6.   Let C_5 be the Cantor set in $[0, 1]$ obtained by deleting the middle 5th segment in $[0, 1]$, then deleting the middle 5th segment of each remaining line segment, and continuing the process indefinitely. Find $\dim_c C_5$.

7.   Let $0 < r < 1$ and $0 < s < 1$. Denote by C_s the Cantor set in $[0, 1]$ obtained by deleting the middle segment of length s from $[0, 1]$, then deleting the middle sth portion of each remaining line segment, and continuing this process indefinitely. Determine s so that $\dim_c C_s = r$.

8.   Let $S = \{1, 1/2, 1/3, 1/4, \ldots\}$. Show that $\dim_c S = 1/2$.

9.   Let $S = \{1, 1/2, 1/2^2, 1/2^3, \ldots\}$. Find $\dim_c S$.

10. Let $S = \{1, 1/4, 1/9, 1/16, \ldots\}$. Find $\dim_c S$.

11. Assume that $\dim_c E$ and $\dim_c F$ exist, and that $E \supseteq F$. Show that $\dim_c E \geq \dim_c F$.

12. a.   Suppose that S is in R^2 and that $\dim_c S$ exists. Prove that

$$\dim_c S = \lim_{\varepsilon \to 0} \frac{N^*(\varepsilon)}{\ln(1/\varepsilon)}$$

where $N^*(\varepsilon)$ = the smallest number of disks of radius ε that are needed to cover S. (Hint: If B_ε = the square of side length ε, and if D_δ = the disk of radius δ, then find positive constants a and b such that $B_{a\varepsilon} \supseteq D_\varepsilon \supseteq B_{b\varepsilon}$.)
     b.   Prove a similar result if disks are replaced by congruent rectangles, the lengths of the sides of which have constant ratio as ε approaches 0.

13. Suppose that S is in R^3 and that $\dim_c S$ exists. Prove that

$$\dim_c S = \lim_{\varepsilon \to 0} \frac{N^*(\varepsilon)}{\ln(1/\varepsilon)}$$

where $N(\varepsilon)$ = the smallest number of balls of radius ε that are needed to cover S.

4.2 LYAPUNOV DIMENSION

The sets with which we illustrated the notion of capacity dimension in Section 4.1 have properties of self-similarity, so that we were able to ascertain the capacity dimension algebraically.

Suppose we try to approximate the capacity dimension of the Hénon attractor A_H, say, with the grid shown in Figure 4.10. Notice that the number of rectangles that intersect A_H is approximately 100. If we let ε equal the height of each rectangle, then $\varepsilon = 1/20$, so that

$$\frac{\ln N(\varepsilon)}{\ln (1/\varepsilon)} \approx \frac{\ln 100}{\ln 20} \approx 1.58$$

However, it is known that $\dim_L A_H \approx 1.26$, which can be confirmed by letting ε be microscopic and using self-similarity of the attractor near the saddle fixed point **p**. (Had we let ε equal the length of the longer side of the rectangles, we would have obtained a still rougher estimate of the Lyapunov dimension.)

A reason that one needs extremely fine grids in order to approximate the capacity dimension of A_H is that many of the squares in the grid meet portions of A_H that are clearly one-dimensional (such as those curves on the left portion of the attractor), and relatively few squares meet the the portion of the attractor close to the fixed point **p** which, however, is where the dimension rises markedly above 1.

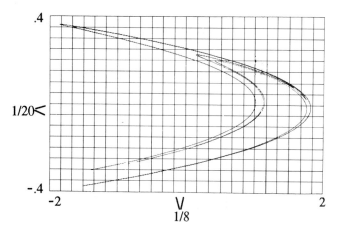

Figure 4.10

The capacity dimension of a given set is a "geometric dimension" because its value depends solely on the geometry of the set. However, if we wish to attach a dimension to an attractor of a map, we might wish the dimension to reflect the frequency with which points in the attractor are visited by orbits of various points. The Lyapunov dimension, which was defined by Kaplan and Yorke in 1978, has such a property.

Before continuing, we remark that except in special cases, we can only find a good approximation of the Lyapunov dimension by means of a computer. There are sophisticated programs such as *Dynamics*, by James Yorke (1990), that produce the Lyapunov dimensions of a variety of maps.

Since the Lyapunov dimension focuses on attractors of general maps, we first will give a formal definition of an attractor of a map.

DEFINITION 4.3. Let V be a subset of R^n and $F: V \to R^n$, where $n = 1, 2,$ or 3. Also let A be a subset of V. Then A is an **attractor** of F provided that the following conditions hold:

i. A is a closed, invariant subset of V.

ii.   There is a neighborhood U of A such that whenever $\mathbf{v}$ is in U, then $F^{[n]}(\mathbf{v}) \to A$ (in the sense that for each $\varepsilon > 0$, there is a positive integer N such that if $n \geq N$, there exists a $\mathbf{w}_n$ in A such that $\|F^{[n]}(\mathbf{v}) - \mathbf{w}_n\| < \varepsilon$).

The invariance means that the iterates of any point in A are also in A. By the definition of attractor, attracting cycles are attractors, as is A_H.

In preparation to define the Lyapunov dimension, let us recall from Section 2.1 that the Lyapunov exponent of a one-dimensional function f at x is given by

$$\lambda(x) = \lim_{n \to \infty} \frac{1}{n} \ln |(f^{[n]})'(x)| \tag{1}$$

provided that the limit exists. Moving on to two dimensions, we let V be a subset of R^2, and suppose that $F: V \to R^2$ has continuous partial derivatives. Assume also that $\mathbf{v}_0$ is in V, with orbit $\{\mathbf{v}_n\}_{n=0}^{\infty}$. For each $n = 1, 2, ...,$ we define $D_n F(\mathbf{v}_0)$ by the formula

$$D_n F(\mathbf{v}_0) = [DF(\mathbf{v}_{n-1})][DF(\mathbf{v}_{n-2})] \cdots [DF(\mathbf{v}_0)] \tag{2}$$

where $DF(\mathbf{v}_k)$ denotes the 2×2 matrix identified with the differential of F at $\mathbf{v}_k$. Then $D_n F(\mathbf{v}_0)$ is a 2×2 matrix (depending on n). If $D_n F(\mathbf{v}_0)$ has nonzero real eigenvalues, we denote their absolute values by $d_{n1}(\mathbf{v}_0)$ and $d_{n2}(\mathbf{v}_0)$. For convenience we will assume that $d_{n1}(\mathbf{v}_0) \geq d_{n2}(\mathbf{v}_0)$. Now we define the **Lyapunov numbers**

$\lambda_1(\mathbf{v}_0)$ and $\lambda_2(\mathbf{v}_0)$ of F at $\mathbf{v}_0$:

$$\lambda_1(\mathbf{v}_0) = \lim_{n\to\infty} [d_{n1}(\mathbf{v}_0)]^{1/n} \quad \text{and} \quad \lambda_2(\mathbf{v}_0) = \lim_{n\to\infty} [d_{n2}(\mathbf{v}_0)]^{1/n} \tag{3}$$

provided that the limits exist. The Lyapunov numbers were originally defined by the Russian mathematician V. I. Oseledec (1968). They measure the rate at which a circle of small radius, centered on the attractor is deformed through iteration. In particular, after n iterations a circle of radius ε centered at $\mathbf{v}_0$ is transformed into an ellipse whose major and minor radii are approximately $[\lambda_1(\mathbf{v}_0)]^n \varepsilon$ and $[\lambda_2(\mathbf{v}_0)]^n \varepsilon$, respectively (Figure 4.11).

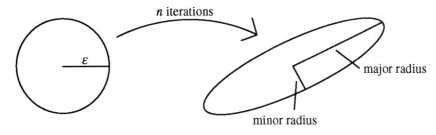

Figure 4.11

Since $d_{n1}(\mathbf{v}_0) \ge d_{n2}(\mathbf{v}_0)$ by prescription, it follows that $\lambda_1(\mathbf{v}_0) \ge \lambda_2(\mathbf{v}_0)$. Notice that if we take logarithms of both sides in (3), then (3) reduces in a natural way to the formula in (1) when the function is one-dimensional. Now we are ready to define the Lyapunov dimension.

DEFINITION 4.4. Let V be a subset of R^2, and let function $F:V \to R^2$ have coordinate functions with continuous partial derivatives. Also assume that F has an attractor A_F, and that $\mathbf{v}_0$ is in A_F. Finally, assume that $\lambda_1(\mathbf{v}_0) > 1 > \lambda_2(\mathbf{v}_0)$. Then the **Lyapunov dimension** of A_F at $\mathbf{v}_0$, denoted $\dim_L A_F(\mathbf{v}_0)$, is given by

$$\dim_L A_F(\mathbf{v}_0) = 1 - \frac{\ln \lambda_1(\mathbf{v}_0)}{\ln \lambda_2(\mathbf{v}_0)} \tag{4}$$

In the event that $\lambda_1(\mathbf{v}_0)$ and $\lambda_2(\mathbf{v}_0)$ are independent of $\mathbf{v}_0$ (except possibly for isolated points $\mathbf{v}_0$), we write λ_1 and λ_2 for $\lambda_1(\mathbf{v}_0)$ and $\lambda_2(\mathbf{v}_0)$, respectively. In that case we define the **Lyapunov dimension** of A_F by the formula

$$\dim_L A_F = 1 - \frac{\ln \lambda_1}{\ln \lambda_2} \tag{5}$$

By the definition of Lyapunov dimension and with the help of a computer, one can show that the Lyapunov dimension of the Hénon attractor satisfies $\dim_L A_H \approx$ 1.26. Thus $\dim_L A_H = \dim_c A_H$.

Figure 4.12 depicts the attractor for the map F given by

$$F\begin{pmatrix} x \\ y \end{pmatrix} = \begin{pmatrix} x^2 - y^2 + 0.9x - 0.6013y \\ 2xy + 2x + 0.5y \end{pmatrix}$$

which has been studied by James Yorke and named the **Tinkerbell attractor**. By computer we find that $\dim_L A_F \approx 1.40$. Notice that the Tinkerbell attractor appears to be more complex than the Hénon attractor. This is borne out in its larger Lyapunov dimension. The Tinkerbell attractor also appears in Color Plate 14. In the color plate, the black region is the basin of attraction for A_F. The outer colors represent points whose iterates are unbounded, color coded by the speed with which their iterates grow without bound. Notice that among the points whose iterates are unbounded are a few scattered points that reside inside the loops of the attractor.

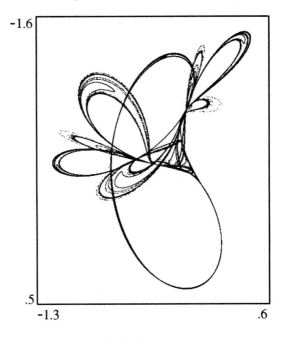

Tinkerbell attractor

Figure 4.12

Theoretical reasons why $\dim_L A_F$ should be approximately equal to $\dim_c A_F$ are discussed in Farmer, Ott and Yorke (1983). However, the Lyapunov dimension factors in the effect of iteration, as we suggested at the outset of the discussion.

Next we will find the Lyapunov dimension of a map without the aid of a computer.

EXAMPLE 1. Let M be the horseshoe map discussed in Section 3.5. Find the Lyapunov dimension of the attractor A_M.

Solution. If $\mathbf{v}$ and $\mathbf{w}$ are in the attractor A_M and are very near to each other, then M shrinks the distance between $\mathbf{v}$ and $\mathbf{w}$ vertically by a factor $a < 1/3$, and expands the distance horizontally by a factor $b = 3$. Therefore

$$DM(\mathbf{v}) = \begin{pmatrix} a & 0 \\ 0 & 3 \end{pmatrix}$$

For each $\mathbf{v}$ in A_M, the eigenvalues of $DM(\mathbf{v})$ are a and 3. Now fix $\mathbf{v}_0$ in A_M. Then the iterates of $\mathbf{v}_0$ are also in A_M, so by (3) and results of Section 3.1, the absolute values of the eigenvalues $d_{n1}(\mathbf{v}_0)$ and $d_{n2}(\mathbf{v}_0)$ of $D_n M(\mathbf{v}_0)$ are given by

$$d_{n1}(\mathbf{v}_0) = 3^n \quad \text{and} \quad d_{n2}(\mathbf{v}_0) = a^n$$

Notice that $d_{n1}(\mathbf{v}_0) > d_{n2}(\mathbf{v}_0)$ because $3 > 1 > a$ by hypothesis. Therefore the Lyapunov numbers $\lambda_1(\mathbf{v}_0)$ and $\lambda_2(\mathbf{v}_0)$ are given by

$$\lambda_1(\mathbf{v}_0) = [d_{n1}(\mathbf{v}_0)]^{1/n} = 3 \quad \text{and} \quad \lambda_2(\mathbf{v}_0) = [d_{n2}(\mathbf{v}_0)]^{1/n} = a$$

Since these numbers are independent of $\mathbf{v}_0$, we find that

$$\lambda_1 = 3 \quad \text{and} \quad \lambda_2 = a$$

Finally, by (5) the Lyapunov dimension of the attractor A_M is given by

$$\dim_L A_M = 1 - \frac{\ln 3}{\ln a} \qquad \square$$

We say that a map has a **strange attractor** if the attractor has a non-integer Lyapunov dimension. The Hénon map has a strange attractor, as does the horseshoe map whenever $(\ln 3)/(\ln a)$ is not an integer. In the same vein, a map has a **chaotic attractor** if the attractor has sensitive dependence on initial conditions or a Lyapunov number larger than 1. The Hénon map and the horseshoe map (with virtually any admissible value of a) have chaotic attractors. Although most attractors that are strange are also chaotic, Grebogi, Ott, Pelikan and Yorke (1984) exhibit attractors that are strange but not chaotic, and vice versa.

We end the section with two observations. The first is that the object of the Lyapunov dimension is to gather information as the iterates of a point run around the attractor. Thus there is no reason to expect that if $\mathbf{p}$ is a fixed point of the map F, then $\dim_L A_F(\mathbf{p})$ would equal $\dim_L A_F(\mathbf{v})$ for non-periodic points $\mathbf{v}$ in A_F. For example, consider the Hénon map H, with $a = 1.4$ and $b = 0.3$. Then $\dim_L A_H(\mathbf{v}) \approx 1.26$ for practically every $\mathbf{v}$ in the attractor A_H, yet if $\mathbf{v}$ is the fixed point $\mathbf{p}$ on the attractor, then $\dim_L A_H(\mathbf{p})$ is *not* approximately 1.26 (Exercise 2). Thus when determining the Lyapunov dimension of an attractor, one should use a non-periodic point whose orbit is expected to be dense in the attractor.

A more discomforting observation is that when one has an attractor that comes from observed data, then it may be very difficult to obtain a good measurement of the dimension for the attractor. This is especially true if the data is biological in nature.

Alan Garfinkel and several colleagues (1991) have studied electrocardiograms of cats under the influence of cocaine. Figure 4.13(a) displays a time series for the EKG before the cat ingested cocaine, and Figure 4.13(b) shows the time series after ingestion. In Figures 4.13(c)–(d) the EKG time series is transformed into a "phase-plane plot," with the derivative of the voltage (as a function of time) plotted against the voltage. The results are very interesting. The attractor on the left, which corresponds to the EKG before ingestion of cocaine, appears much less dispersed than the one on the right, which corresponds to the EKG after ingestion. The fractal dimension of the attractor on the left appears to be smaller than that for the attractor on the right. However, Garfinkel cautions that attempts to measure the Lyapunov dimension, among other fractal dimensions, have given very imprecise information.

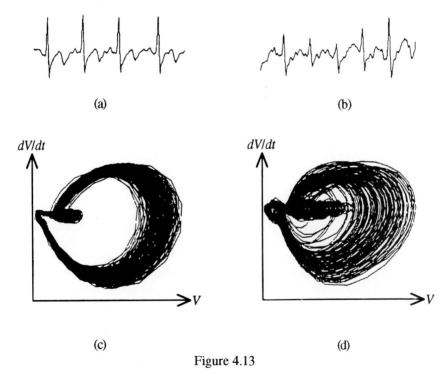

(a)   (b)

(c)   (d)

Figure 4.13

EXERCISES 4.2

1. Let ε = the length of the horizontal sides of the rectangles in Figure 4.10. Find an approximation to $\dim_c A_H$. How does your answer compare with the one obtained at the outset of the section, which uses the vertical sides?

2. Let **p** be the fixed point of the Hénon map H_{ab} given in (1) and (2) of Section 3.4. Assume that $a = 1.4$ and $b = 0.3$. Show that $\dim_L A_H$ (**p**) is *not* approximately 1.26.

3. Show that if we take logarithms of both sides of the equations in (3), they would reduce to the one in (1) if the function were one-dimensional.

4.3 JULIA SETS AND THE MANDELBROT SET

Many of the beautiful, exotic sets that have appeared recently on calendars and covers of magazines are sets that can be described in terms of iterates of complex functions. In this section we will define two well-known kinds of sets that are defined in terms of complex functions: Julia sets and the Mandelbrot set.

In order to set the stage, we need a brief introduction to complex numbers and functions. For a complete discussion of complex numbers and functions, you could consult the treatise by Burckel (1979).

If $z = x + yi$ and $w = u + vi$, then the sum and product of z and w are given by

$$z + w = (x + u) + (y + v)i \quad \text{and} \quad zw = (xu - yv) + (xv + yu)i$$

In particular $z^2 = (x^2 - y^2) + 2xyi$.

A complex number $x + yi$ can be considered as an ordered pair (x, y) of real numbers, as well as a point (x, y) in the plane. When the complex numbers are identified with points, we refer to them collectively as the **complex plane**. We will identify the real number x with the complex number $x + 0i$; in this way real numbers are considered as special complex numbers. As such, the real numbers lie on the horizontal axis in the complex plane, called the **real axis**; the vertical axis is the **imaginary axis**, consisting of pure imaginary numbers (Figure 4.14).

Let $z = x + iy$. Because of the identification of complex numbers with points in the plane, we denote the **absolute value** or **modulus** of z by $|z|$, where

$$|z| = \sqrt{x^2 + y^2}$$

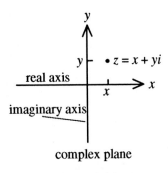

complex plane

Figure 4.14

Thus $|z|$ is the distance between the complex number z and the origin. Similarly, $|z - w|$ is the distance between z and w. The open disk about w with radius r consists of all z such that $|z - w| < r$. A set A in the complex plane is **bounded** if $A \subseteq U$ for some open disk about 0 with finite radius r. A sequence $\{z_n\}_{n=0}^{\infty}$ **converges** to w if and only if $|z_n - w| \to 0$ as n increases without bound. If A contains the limit point of each sequence in A, then A is **closed**.

The usual rules for absolute value are valid for complex numbers:

$$|z + w| \leq |z| + |w|, \quad |z - w| \geq \left| |z| - |w| \right|, \quad \text{and} \quad |zw| = |z||w|$$

The first of the inequalities is known as the **triangle property**, from the associated geometric interpretation (Figure 4.15(a)).

In order to find roots of complex numbers, we write z in the **polar form** as

$$re^{i\theta}, \quad \text{or equivalently,} \quad r(\cos \theta + i \sin \theta)$$

Here $r = |z|$; θ is the angle shown in Figure 4.15(b), and is called the (principal) **argument** of z (Figure 4.15(b)). For example, consider $z = 3i$. Since $r = 3$ and the argument is $\pi/2$, the polar form of $3i$ is

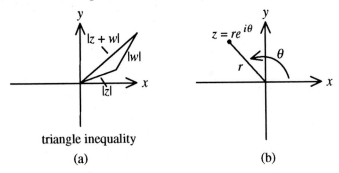

triangle inequality

(a)

(b)

Figure 4.15

$$3e^{i\pi/2}, \quad \text{or equivalently,} \quad 3\left(\cos\frac{\pi}{2} + i\sin\frac{\pi}{2}\right)$$

For later use we remark that the unit circle C consists of all complex numbers of the form $e^{i\theta}$, that is, those complex numbers with $r = 1$.

Products of complex numbers have a particularly simple form when we use polar notation. More specifically, let

$$z_1 = r_1(\cos\theta_1 + i\sin\theta_1) \quad \text{and} \quad z_2 = r_2(\cos\theta_2 + i\sin\theta_2)$$

By the sum formulas for $\sin(\theta_1 + \theta_2)$ and $\cos(\theta_1 + \theta_2)$, we find that

$$z_1 z_2 = [r_1(\cos\theta_1 + i\sin\theta_1)][r_2(\cos\theta_2 + i\sin\theta_2)]$$

$$= r_1 r_2 [(\cos\theta_1\cos\theta_2 - \sin\theta_1\sin\theta_2) + i(\sin\theta_1\cos\theta_2 + \cos\theta_1\sin\theta_2)]$$

$$= r_1 r_2 [\cos(\theta_1 + \theta_2) + i\sin(\theta_1 + \theta_2)]$$

Therefore

$$z_1 z_2 = r_1 r_2 [\cos(\theta_1 + \theta_2) + i\sin(\theta_1 + \theta_2)] \tag{1}$$

The formula in (1) says that the modulus of a product is the product of the moduli, and the argument of the product is the sum of the arguments. An application of (1) and the law of induction yields De Moivre's Theorem, named after one of the early fathers of probability theory, Abraham De Moivre (1667–1754).

THEOREM 4.5 (De Moivre's Theorem). Let $z = r(\cos\theta + i\sin\theta)$. Then

$$z^n = r^n(\cos n\theta + i\sin n\theta) \quad \text{for any integer } n \geq 2 \tag{2}$$

Proof. We proceed by induction. For $n = 2$ the formula is valid by (1) if we let $z_1 = z_2 = z$. Now assume that the formula in (2) is true for a given $n \geq 3$. Then by the induction hypothesis and by another application of (1), we obtain

$$z^{n+1} = z^n z = [r^n(\cos n\theta + i\sin n\theta)][r(\cos\theta + i\sin\theta)]$$

$$= r^{n+1}[\cos(n+1)\theta + i\sin(n+1)\theta]$$

By the Law of Induction, (2) is valid for all $n \geq 2$. ■

De Moivre's Theorem helps us to determine a square root of a nonzero complex number $z = re^{i\theta}$, that is, to find a number w such that $w^2 = z$. Indeed, by De Moivre's Theorem,

$$w_1 = \sqrt{r}\, e^{i\theta/2} \quad \text{and} \quad w_2 = \sqrt{r}\, e^{i(\theta/2 + \pi)}$$

are two square roots of z. We conclude that *every* nonzero complex number has two square roots, located on opposite sides of the origin (Figure 4.16(a)); the argument of one square root is half that of the given complex number. For example, to find a square root of $3i$, we first write it in polar form:

$$3i = 3\left(\cos\frac{\pi}{2} + i\sin\frac{\pi}{2}\right)$$

Then we take the square root of 3, and half the argument $\pi/2$, which yields

$$\sqrt{3}\,(\cos\frac{\pi}{4} + i\sin\frac{\pi}{4}) = \sqrt{3}\,(\frac{1}{2}\sqrt{2} + \frac{1}{2}\sqrt{2})i = \frac{\sqrt{6}}{2}(1 + i)$$

The other square root, symmetric with respect to the origin, is $-\sqrt{6}\,(1+i)/2$.

In general, the n nth roots of a nonzero complex number z lie on the circle of radius $|z|^{1/n}$ about the origin. If $z = |z|e^{i\theta}$, then the nth roots of z are equally spaced around the circle, with one located at an angle θ/n with respect to the positive real axis (Figure 4.16(b)).

A **complex function** is a function whose domain and range consist of complex numbers (which may be real numbers, of course). Polynomial functions are written in the usual way, but with z's instead of x's. Thus if

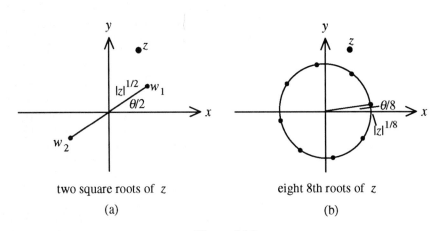

two square roots of z eight 8th roots of z

(a)   (b)

Figure 4.16

$$f(z) = a_n z^n + a_{n-1} z^{n-1} + \cdots + a_1 z + a_0 \tag{3}$$

then f is a (complex) polynomial. Limits and derivatives of complex functions are defined as they are for real-valued functions of a real variable. In particular,

$$f'(z_0) = \lim_{z \to z_0} \frac{f(z) - f(z_0)}{z - z_0}$$

provided that the limit exists. Thus if f is the polynomial function in (3), then

$$f'(z) = na_n z^{n-1} + (n-1)a_{n-1} z^{n-2} + \cdots + 2a_2 z + a_1$$

Thus if c is a (complex) constant and if $f(z) = z^2 + c$, then $f'(z) = 2z$.

Let f be a complex function. A complex number p is a **fixed point** of f if $f(p) = p$. The **Fundamental Theorem of Algebra** says that if g is a complex polynomial function, then there is a complex number z such that $g(z) = 0$. This is equivalent to saying that every complex polynomial function f has a fixed point. Indeed, let f be a polynomial function, and $g(z) = f(z) - z$. Then

$$f(z) = z \text{ if and only if } g(z) = f(z) - z = 0$$

Therefore zeros of g are fixed points of f. Finding the fixed point is easy if f is the function defined by $f(z) = z^2 + c$, because then z is a fixed point of f provided that $z = z^2 + c$, or equivalently, $z^2 - z + c = 0$. By the quadratic formula, the fixed points of f are given by

$$z = \frac{1}{2} \pm \frac{1}{2} \sqrt{1 - 4c} \tag{4}$$

The definitions of attracting and repelling fixed points for complex functions are adaptations of the definitions for real-valued functions in Chapter 1. A fixed point p of a complex function f is **attracting** provided that there is a disk U centered at p such that if z is in the domain of f and in U, then $|f^{[n]}(z) - p| \to 0$ as n increases without bound. In the same manner, a fixed point p is **repelling** if there is a disk U centered at p such that if z is in the domain of f and in U, then $|f(z) - p| > |z - p|$. With only trivial modifications, Theorem 1.6 and its proof carry over to complex functions.

THEOREM 4.6. Let f be a differentiable complex function with fixed point p.
a.   If $|f'(p)| < 1$, then p is attracting.
b.   If $|f'(p)| > 1$, then p is repelling.
c.   If $|f'(p)| = 1$, then p may be attracting, repelling, or neither.

As before, a point p is an attracting (or repelling) period-n point of the complex function f if p is an attracting (or repelling) fixed point of $f^{[n]}$.

The notion of attractor for a complex function corresponds to that for functions of two variables. The set A in the complex plane is an **attractor** of a function f if it is closed and invariant, and if it attracts the iterates (for f) of all points in some neighborhood of A. To illustrate an attractor, we consider the family $\{f_\mu\}$ of complex functions that models an optical switch for a laser system, given by

$$f_\mu(z) = a + bze^{ik - [i\mu/(1 + |z|^2)]}$$

where a, b, and k are constants and μ is a parameter denoting the amplitude of the laser pulse entering the optical switch. Specifically, let $a = 0.85$, $b = 0.9$, and $k = 0.4$. As μ increases past the bifurcation value $\mu_0 \approx 7.26994894$, it turns out that the attractor, which appears in Color Plate 13, explodes. The points colored yellow represent the attractor when μ is a trifle smaller than μ_0; those points colored red are points added to the attractor as μ becomes slightly larger than μ_0.

In the earlier chapters we saw that the attracting periodic points of a function hold decisive information about the dynamical behavior of the function. In the discussion that follows it will be the repelling periodic points that we will focus on.

Julia Sets

In Section 1.4 we analyzed the family $\{g_c\}$ of real-valued functions defined by

$$g_c(x) = x^2 + c$$

where c is a real parameter. For the remainder of the section we will consider the complex analog of the family: g_c as a complex function and c a complex parameter. Thus for each complex number c, g_c is defined by

$$g_c(z) = z^2 + c$$

We know from (4) that unless $c = 1/4$, g_c has two fixed points. Presently we will prove that at least one of the two fixed points must be repelling.

THEOREM 4.7. Let $c \neq 1/4$. Then at least one fixed point of g_c is repelling.

Proof. By (4) the fixed points of g_c are given by

$$z = \frac{1}{2} \pm \frac{1}{2}\sqrt{1 - 4c}$$

Since $g'(z) = 2z$, it follows that

$$g_c'(\frac{1}{2} \pm \frac{1}{2}\sqrt{1-4c}) = 1 \pm \sqrt{1-4c}$$

If $w = \sqrt{1-4c}$, then either w or $-w$ lies on or to the left of the imaginary axis, so that $|1 - w| > 1$ or $|1 + w| > 1$. It follows that either

$$\left| 1 + \sqrt{1-4c} \right| > 1 \quad \text{or} \quad \left| 1 - \sqrt{1-4c} \right| > 1$$

which means that

$$\left| g_c'(\frac{1}{2} + \frac{1}{2}\sqrt{1-4c}) \right| > 1 \quad \text{or} \quad \left| g_c'(\frac{1}{2} - \frac{1}{2}\sqrt{1-4c}) \right| > 1$$

By Theorem 4.6, g_c has a repelling fixed point. ∎

If $c = 1/4$, then the single fixed point of g_c is neither attracting nor repelling. However, g_c has a repelling 2-cycle (Exercise 4).

By Theorem 4.7, g_c has a repelling fixed point whenever $c \neq 1/4$. However, g_c may have multitudes of other repelling periodic points, as Example 1 shows.

EXAMPLE 1. Consider g_0, defined by $g_0(z) = z^2$, and let C denote the unit circle in the complex plane. Show that the set of repelling periodic points of g_c is dense in C.

Solution. We can consider the unit circle as the collection of all z of the form

$$z = e^{2\pi i \theta}$$

where $0 \leq \theta < 1$. Then $g_0(z) = e^{2\pi i(2\theta)}$, so we can interpret g_0 as the function that doubles every number in the half-open interval $[0, 1)$ (modulo 1). Thus formulated, g_0 is reminiscent of the baker's function, defined in Section 1.3 by

$$B(x) = \begin{cases} 2x & \text{for } 0 \leq x \leq 1/2 \\ 2x - 1 & \text{for } 1/2 < x \leq 1 \end{cases}$$

If we disregard the dyadic rationals in $[0, 1]$, which are exactly the numbers whose iterates are eventually $1/2$, the two functions g_0 and B are equivalent. In Section 1.3 we noted that if p is an odd, positive integer, then k/p is periodic for B, for $k = 1, 2, \ldots, p - 1$. Since the collection of such rational numbers is dense in the interval $[0, 1)$, it follows that the collection of periodic points of g_0 is dense in C. The fact that $|g_c'(z)| = |2z| = 2$ for each z in C implies that every periodic point in

C is repelling. Therefore the repelling periodic points are dense in C.   ❑

  The identification of g_0 with the baker's function yields additional information about g_0. The fact that B has sensitive dependence on initial conditions and has an element with dense orbit in its domain is inherited by g_0 on C. Therefore g_0 is strongly chaotic on the unit circle.

  Continuing the analysis of the dynamics of g_0, we notice that

  i. if $|z| < 1$, then $|g_0^{[n]}(z)| = |z|^{2^n} \to 0$ as n increases without bound

  ii. if $|z| > 1$, then $|g_0^{[n]}(z)| = |z|^{2^n} \to \infty$ as n increases without bound

By (i), the basin of attraction of the fixed point 0 consists of all z such that $|z| < 1$. If we consider ∞ as an attractor of g_0, then its basin of attraction consists of all z such that $|z| > 1$. Therefore the unit circle C has two important qualities: it is the boundary of the basins of attraction of 0 and ∞, and as such it is called a **basin boundary.** Also, it contains all repelling periodic points of g_0 (and indeed it contains all periodic points except 0). It is this latter quality that leads us to make the following definition.

DEFINITION 4.8. Let c be any complex number. The smallest closed set in the complex plane that contains all repelling periodic points of g_c is called the **Julia set** of g_c, and is denoted J_c.

  Julia sets are named for the French mathematician Gaston Julia, who studied iterates of complex rational functions in depth during the early part of this century. By definition, the Julia set J_0 is the unit circle C.

  Let c be an arbitrary complex number. By definition, J_c is a closed set. It is also nonempty by virtue of Theorem 4.7 and the comment following it. Next we will prove that J_c is a bounded set.

THEOREM 4.9. If $|z| > |c| + 1$, then the orbit of z for g_c is unbounded.

Proof. Assume that $|z| > |c| + 1$. Then

$$|g_c(z)| = |z^2 + c| = |z| \left| z + \frac{c}{z} \right| \geq |z| \left| |z| - \frac{|c|}{|z|} \right| \geq |z| \left| |c| + 1 - \frac{|c|}{|c| + 1} \right|$$

Let

$$r = |c| + 1 - \frac{|c|}{|c| + 1}$$

Since $|g_c(z)| > r|z|$ and $r > 1$, it follows that $|g_c^{[n]}(z)| > r^n|z| \to \infty$ as n increases without bound. Therefore if $|z| > |c| + 1$, then the orbit of z is unbounded. ∎

COROLLARY 4.10. If z is in J_c, then $|z| \le |c| + 1$, so that J_c is a bounded subset of the complex plane.

Proof. If z is a periodic point, then the iterates of z are bounded, so that by Theorem 4.9, $|z| \le |c| + 1$. Since J_c is the smallest closed closed set containing all repelling periodic points, any z in J_c also has the property that $|z| \le |c| + 1$. Consequently J_c is bounded in the complex plane. ∎

Obviously the periodic points in J_c have bounded orbits. Our next goal is to show that the orbit of *each* number z in J_c is bounded. First we have a preliminary result.

THEOREM 4.11. Suppose the orbit of z for g_c is unbounded. Then there is a disk U centered at z such that the orbit of each element in U is unbounded.

Proof. Let the positive integer k be so large that $\left| g_c^{[k]}(z) \right| > |c| + 1$. Since g_c and thus $g_c^{[k]}$ are continuous, we can find a disk U centered at z such that if w is in U, then $|g_c^{[k]}(w)| > |c| + 1$. By Theorem 4.9, it follows that the orbit of w is unbounded. ∎

It follows from Theorem 4.11 that limits of sequences all of whose members have bounded orbits also have bounded orbits. Thus we have the following corollary.

COROLLARY 4.12. For any complex number c, the collection of all complex numbers z whose orbit for g_c is bounded is a closed subset of the complex plane.

THEOREM 4.13. For any complex number z in J_c, the orbit of z is bounded.

Proof. Suppose that z is in J_c and has an unbounded orbit. By Theorem 4.11 this is true for each element in some disk U centered at z. By the definition of J_c, the repelling periodic points of g_c are dense in J_c, so that there is a repelling periodic point w in U. However, w cannot be periodic *and* have an unbounded orbit. This contradiction implies that if z is in J_c, then z has a bounded orbit. ∎

The theory of complex variables implies that for any complex number c, the Julia set J_c contains no open disks; usually it has fractal dimension. Moreover, since $g_c'(0) = 0$, the number 0 is a critical point of g_c. The orbit of 0 tells us much about the Julia set J_c. In particular, if the iterates of 0 approach

i.   an attracting fixed point, then J_c is a simple closed curve that is non-differentiable if $c \ne 0$.
ii. an attracting cycle, then J_c is a more complicated connected set.
iii. ∞, then J_c is a totally disconnected set called **Fatou dust**, after the French mathematician Pierre Fatou, who along with Julia wrote extensively on the geometric properties of complex rational functions.

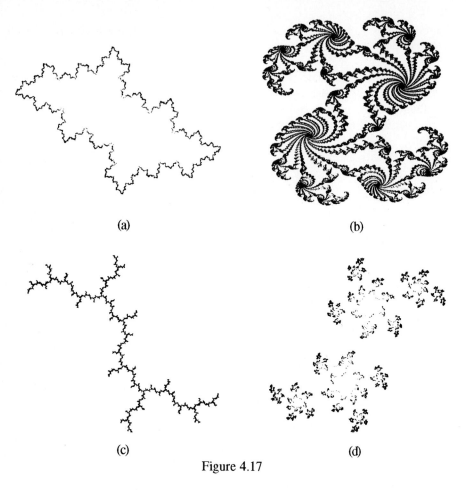

(a)   (b)

(c)   (d)

Figure 4.17

Figure 4.17 presents a sample of the variety possible in Julia sets:

In (a), J_c is a simple closed curve; g_c has an attracting fixed point.
In (b), J_c is a more complicated connected set; g_c has an attracting 11-cycle.
In (c), J_c does not enclose a figure with interior, and is called a **dendrite**.
In (d), J_c is totally disconnected, and is an example of Fatou dust.

Each Julia set is symmetric with respect to the origin, and if c is on the x axis, then J_c is symmetric with respect to both axes as well. In addition, in Julia sets there is generally some form of self-similarity, and correspondingly they normally have fractal dimension. Additional representative Julia sets occupy Color Plates 15–17.

How can one obtain Julia sets on the computer screen? One way that is frequently effective is to use the fact that the repelling periodic points of g_c are dense in J_c. If we select (almost) any point in the complex plane, then the backward iterates approach points of J_c, and frequently after the first few backwards iterates they begin tracing out a set that is indistinguishable from J_c. To start the procedure, choose a complex number z_0 that is not a periodic point of g_c. Since $g_c(z) = z^2 + c$, it follows that z_0 has two distinct pre-images,

$$\sqrt{z_0 - c} \quad \text{and} \quad -\sqrt{z_0 - c}$$

Either pre-image works equally well for the choice of initial backward iterate of z_0. Choose the pre-image *randomly*. Then the process is repeated indefinitely, with a randomly chosen pre-image at each stage in the process. The computer program JULIA at the end of the book is based on these ideas.

The Mandelbrot Set

Now we turn to the celebrated Mandelbrot set, which is related to Julia sets. To begin with, let

K_c = the collection of complex numbers z whose orbit under g_c is bounded

Then Corollary 4.12 implies that K_c is a closed set in the complex plane, and Theorem 4.13 indicates that $K_c \supseteq J_c$. What is actually true is that J_c is the boundary of K_c. The set K_c is therefore called the **filled-in Julia set** for g_c.

If the critical point 0 of g_c is in K_c, then a recent, deep theorem due to A. Douady and John Hubbard (1982) proves that K_c is a connected set. It can be shown, moreover, that the converse is true: if K_c is connected, then the orbit of 0 for g_c is bounded. Devaney (1989) has a more detailed discussion of these results. The special values of c that have these properties constitute a very special set.

DEFINITION 4.14. The collection of complex numbers c such that the orbit of 0 for g_c is bounded is called the **Mandelbrot set**, and is denoted by M.

The set M is named for the mathematician Benoit Mandelbrot, who in the late 1970's brought the set wide acclaim with the help of images on a high-speed computer.

THEOREM 4.15. The following hold for the Mandelbrot set M.
a. M contains all c such that $|c| \leq 1/4$.
b.   If c is in M, then $|c| \leq 2$.

Proof. To prove (a), let $|c| \leq 1/4$. Then

$$|g_c^{[2]}(0)| = |g_c(c)| = |c^2 + c| \leq |c|^2 + |c| \leq \frac{1}{4} + \frac{1}{4} = \frac{1}{2}$$

To apply induction, we assume that $|g_c^{[n]}(0)| \leq 1/2$. Then

$$|g_c^{[n+1]}(0)| = |g_c(g_c^{[n]}(0))| = |[g_c^{[n]}(0)]^2 + c| \leq |g_c^{[n]}(0)|^2 + |c| \leq \frac{1}{4} + \frac{1}{4} = \frac{1}{2}$$

Therefore the Law of Induction implies that $|g_c^{[n]}(0)| \leq 1/2$ for all n, so that c is in M. Thus part (a) is proved.

To prove (b), suppose that $|c| > 2$, and let $r = |c| - 1$. If $|z| \geq |c|$, then

$$|g_c(z)| = |z^2 + c| = |z|\,|z + \frac{c}{z}| \geq |z|\,(|z| - \frac{|c|}{|z|}) \geq |c|\,(|c| - 1) \geq |z|r \qquad (5)$$

Since $|g_c(0)| = |c|$, it follows that $|g_c^{[2]}(0)| \geq |c|r$. In order to apply induction, assume that $|g_c^{[n]}(0)| \geq |c|r^{n-1}$. Then by (5) and the induction hypothesis,

$$|g_c^{[n+1]}(0)| = |g_c(g_c^{[n]}(0))| \geq |g_c^{[n]}(0)|r \geq (|c|r^{n-1})\,r = |c|r^n$$

Therefore by induction, $|g_c^{[n]}(0)| \geq |c|r^n$ for all n. Since $r > 1$, the iterates of 0 are unbounded, so that c is not in M. This completes the proof of (b). ■

Next we will use a complex version of Singer's Theorem (see Section 1.8). It was proved by Julia seventy years ago, and appears in Devaney (1989, p. 281).

THEOREM 4.16. Assume that p is an attracting periodic point of a polynomial function f. Then there is a critical point that lies in the basin of attraction of p.

COROLLARY 4.17. If g_c has an attracting periodic point, then c is in M.

Proof. Since $g_c'(z) = 2z$, 0 is the lone critical point of g_c. If g_c has an attracting periodic point p, then by Theorem 4.16, the iterates of 0 for g_c converge to p and its iterates. Hence the iterates of 0 are bounded, so that c is in M. ■

Let us now recall that the fixed points of g_c are $(1 \pm \sqrt{1 - 4c})/2$, so that

$$g_c'\left((1 \pm \sqrt{1 - 4c})/2\right) = 1 \pm \sqrt{1 - 4c}$$

It follows that g_c has an attracting fixed point only if

$$\text{either } |1 + \sqrt{1 - 4c}| < 1 \quad \text{or} \quad |1 - \sqrt{1 - 4c}| < 1$$

We are now ready to show that M contains a robust cardioid.

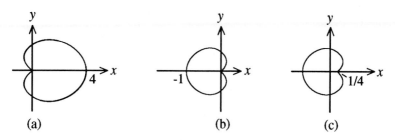

Figure 4.18

THEOREM 4.18. The set of all c such that g_c has an attracting fixed point is the region bounded by the cardioid appearing in Figure 4.18(c).

Proof. Let $w = \sqrt{1-4c}$. We will show that the set of complex numbers w for which either $|1 - w| = 1$ or $|1 + w| = 1$ is a cardioid. Notice that in polar coordinates, $|1 - w| = 1$ if and only if $r = 2 \cos \theta$, so that $w = (2 \cos \theta, \theta)$. Then $w^2 = (4 \cos^2\theta, 2\theta)$. Next, by the half-angle formula, $2 \cos^2\theta = 1 + \cos 2\theta$, which means that $4 \cos^2\theta = 2 + 2 \cos 2\theta$. Therefore in polar coordinates, w^2 satisfies $(2 + 2 \cos 2\theta, 2\theta)$, or equivalently, w^2 satisfies $(2 + 2 \cos \theta, \theta)$. In other words, w^2 has polar coordinates satisfying $r = 2 + 2 \cos \theta$. However, this is an equation of a cardioid. A similar conclusion is drawn if we consider w such that $|1 + w| = 1$. We will use these conclusions in a moment.

Now if $w = \sqrt{1-4c}$, then $c = (1 - w^2)/4$. By the preceding paragraph, w^2 is on the cardioid in Figure 4.18(a), so that $- w^2/4$ is on the cardioid in Figure 4.18(b), and hence $c = (1 - w^2)/4$ is on the cardioid in Figure 4.18(c). This is what we desired to prove. ■

Figure 4.19(a) depicts the entire Mandelbrot set, the large central cardioid consisting of all c 's for which g_c has an attracting fixed point. As you can see, large disks and smaller protrusions and tendrils emanate from the cardioid. The various numbered bulbs appearing in Figure 4.19(b) represent the values of c for which g_c has an attracting orbit of period n , for $n = 1, 2, 3, \ldots$. When one zooms in on the Mandelbrot set with a high enough magnification, baby Mandelbrot-like sets appear. Color Plates 18–24 and 25–30 show two sequences of zooms from different points on the boundary of the Mandelbrot set. Other shapes of an incredible variety also appear under magnification, a few of which appear in Color Plates 31–36. The Mandelbrot set has been called the most complicated set ever scrutinized by mathematicians.

There is no way to sketch the Mandelbrot set, or even describe it, without a computer. There is a simple program that fills in the Mandelbrot set on the computer screen, although the program takes times to run. The principle behind the program is that if $|c| > 2$, then c is not in the Mandelbrot set (by Theorem 4.15(b)). Thus to obtain a reasonable image of the Mandelbrot set M , we select a

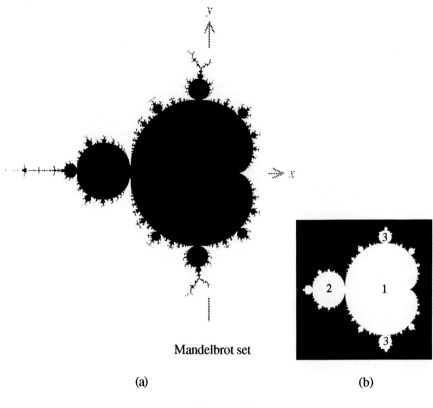

Mandelbrot set

(a)   (b)

Figure 4.19

relatively dense collection S of c's in the disk of radius 2 centered at the origin. Those numbers in S all of whose first, say, 50 iterates are bounded in absolute value by 2 give a good approximation to M. The program MANDELBROT is designed to do this.

Art Matrix (1990) has produced a 20-minute videotape, *Focus on Fractals*, that includes zooms toward four different points on the boundary of the Mandelbrot set. The videotape shows some of the fantastic patterns that arise as the magnification increases, and illustrates the emergence of baby Mandelbrot-like figures that come into view when one penetrates the boundary. The film also depicts the various Julia sets corresponding to points on the boundary of the Mandelbrot set cardioid.

The book *The Beauty of Fractals*, by H.-O. Peitgen and P. H. Richter (1986), contains not only much information about Julia sets and the Mandelbrot set, but also contains exquisite color plates of Julia sets, zooms of the Mandelbrot set and other fractals.

In addition to the videotape and book mentioned above, there are many others that you may wish to consult as you learn more about the fascinating Julia sets and Mandelbrot set.

EXERCISES 4.3

1. Let $z = re^{i\theta}$. Use De Moivre's Theorem to show that for any positive integer n, $r^{1/n} e^{i(\theta/n + 2k\pi/n)}$ is an nth root of z, for $k = 0, 1, ..., n-1$.

2. Let $z = e^{2\pi i\theta}$ with $\theta = k/m$, where m and k are positive integers such that m is a prime number greater than 2 and $1 \le k < m$. Show that z is periodic for g_0.

3. Show that g_0 is transitive and has sensitive dependence on the set J_0.

4. a. Show that $g_{1/4}$ has a single fixed point that is neither attracting nor repelling.
    b. Find the repelling 2-cycle of $g_{1/4}$.

5. Determine whether J_c is symmetric with respect to the origin for every complex number c.

6. Show that if θ is an irrational number in $(0, 1)$, then $e^{2\pi i\theta}$ is not eventually periodic for g_0.

7. Show that if $c = x + yi$ in M, then $\bar{c} = x - yi$ is also in M.

8. Use the program JULIA to determine Julia sets for the given values of c.
    a. $-.11 + .67i$ b. $-.194 + .6557i$ c. $-.74 + .113i$
    d. -1.2 e. -1.8 f. -2

9. Take several values of c near -1.25, and by using the program JULIA, determine how the corresponding Julia sets are related to one another.

10. Use the program MANDELBROT to obtain an image of the Mandelbrot set.

11. Alter the program MANDELBROT so as to focus in on various points close to the boundary of the Mandelbrot set.

12. Consider the function $f_c(z) = cz(1 - z)$. Let Q denote the collection of complex numbers c such that the iterates of the critical point $1/2$ are bounded. Alter the program MANDELBROT to obtain an image of Q.

4.4   ITERATED FUNCTION SYSTEMS

In the first four chapters we have analyzed iterates of a single point for a given function. In this final section of Chapter 4 we will discuss iterates of *sets* in R^2 for *several* functions used simultaneously. Each function under consideration will be a **contraction**, by which we mean a function $F : R^2 \to R^2$ such that

$$\|F(\mathbf{v}) - F(\mathbf{w})\| < \|\mathbf{v} - \mathbf{w}\|$$

for all $\mathbf{v}$ and $\mathbf{w}$ in R^2. For example, let F_1, F_2, and F_3 be the functions defined on R^2 by

$$F_1\begin{pmatrix} x \\ y \end{pmatrix} = \begin{pmatrix} 1/2 & 0 \\ 0 & 1/2 \end{pmatrix}\begin{pmatrix} x \\ y \end{pmatrix}, \quad F_2\begin{pmatrix} x \\ y \end{pmatrix} = \begin{pmatrix} 1/2 & 0 \\ 0 & 1/2 \end{pmatrix}\begin{pmatrix} x \\ y \end{pmatrix} + \begin{pmatrix} 1/2 \\ 0 \end{pmatrix},$$

$$F_3\begin{pmatrix} x \\ y \end{pmatrix} = \begin{pmatrix} 1/2 & 0 \\ 0 & 1/2 \end{pmatrix}\begin{pmatrix} x \\ y \end{pmatrix} + \begin{pmatrix} 1/4 \\ 1/2 \end{pmatrix}$$

Then for any $\mathbf{v}$ and $\mathbf{w}$ in R^2,

$$\|F_k(\mathbf{v}) - F_k(\mathbf{w})\| = \frac{1}{2}\|\mathbf{v} - \mathbf{w}\| \text{ for } k = 1, 2, \text{ and } 3$$

This means that F_1, F_2, and F_3 are contractions.

Next, let $\mathcal{K}$ denote the collection of all closed and bounded subsets of R^2, and let the function $\mathcal{F}$ be defined by

$$\mathcal{F}(A) = F_1(A) \cup F_2(A) \cup F_3(A), \text{ for each } A \text{ in } \mathcal{K}$$

Then $F_1(A)$, $F_2(A)$, and $F_3(A)$ are closed and bounded in R^2, and so is their union $\mathcal{F}(A)$ (Exercise 12). We will call $\mathcal{F}$ the **union** of the functions F_1, F_2, and F_3.

What can we learn about the iterates of $\mathcal{F}$?

EXAMPLE 1. Let A be the equilateral triangle in Figure 4.20(a), placed so that the bottom lies on the unit interval $[0, 1]$ on the x axis. Using F_1, F_2, F_3, and also using $\mathcal{F}$ as defined above, show that

$$\mathcal{F}^{[n]}(A) \supseteq \mathcal{F}^{[n+1]}(A) \quad \text{for all } n \tag{1}$$

and that $\bigcap_{n \geq 1} \mathcal{F}^{[n]}(A)$ is a Sierpinski gasket.

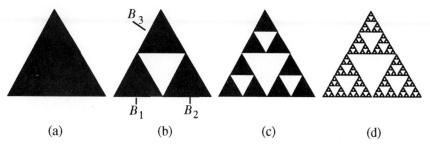

$$B_3$$

$$B_1 \qquad B_2$$

(a)   (b)   (c)   (d)

Figure 4.20

Solution. A moment's reflection shows that $F_i(A) = B_i$ in Figure 4.20(b), for $i =$ 1, 2, 3. Thus $\mathcal{F}(A)$ is the shaded region in Figure 4.20(b). Similarly $\mathcal{F}^{[2]}(A)$ is the region shaded in Figure 4.20(c). It becomes evident that (1) is true, and that in addition, $\underset{n \geq 1}{\cap}\ \mathcal{F}^{[n]}(A)$ is a Sierpinski gasket that appears in Figure 4.20(d). ❏

Suppose that the initial set A were a disk or other more complicated region in R^2. Is there any way of telling what limit, if any, $\mathcal{F}^{[n]}(A)$ would approach? An important result called the Contraction Mapping Theorem, which we will prove at the end of the section, tells us that if $F_1, F_2, \ldots, F_n$ are contractions defined on R^2 and if $\mathcal{F}$ is defined by the formula

$$\mathcal{F}(A) = F_1(A) \cup F_2(A) \cup \cdots \cup F_n(A)$$

for all closed, bounded subsets A of R^2, then there is a unique closed, bounded subset $A_{\mathcal{F}}$ in R^2 such that

$$A_{\mathcal{F}} = \mathcal{F}(A_{\mathcal{F}})$$

The set $A_{\mathcal{F}}$ is called an **attractor** for $\mathcal{F}$. The Contraction Mapping Theorem indicates, moreover, that in a sense to be described more precisely below, through $\mathcal{F}$ the set $A_{\mathcal{F}}$ attracts *every* closed, bounded subset of R^2. Thus for the group of three functions given in Example 1, the iterates of each closed, bounded subset of R^2 converge to the Sierpinski gasket. That is a potent result. It says that whether the initial set A is a single point or a large closed, bounded set in R^2, its iterates (for the trio of funtions) converge to the Sierpinski gasket!

Apparently the first person to systematically study groups of contractions was John Hutchinson (1981). More recently Michael Barnsley has introduced the terminology **iterated function system** for such groups, and has applied the theory to the field of data compression and transmission.

Iterated function systems can be effectively used in order to draw intriguing patterns on a computer screen. Since the attractor for a given iterated function system attracts iterates of *every* bounded and closed subset of R^2, it attracts the iterates of any single point that we might select. Thus the Contraction Mapping

Theorem implies that we can obtain a representation of $A_{\mathcal{F}}$ by evaluating the sequence $\{\mathcal{F}^{[n]}(\mathbf{v})\}_{n=1}^{\infty}$ for any given $\mathbf{v}$ in R^2. However, $\mathcal{F}^{[n]}(\mathbf{v})$ becomes an ever larger set of points as n increases, so this method of approximating $A_{\mathcal{F}}$ can be time-consuming for the computer. Barnsley has recently proposed an alternative method called the **random iteration algorithm** that can be more economical.

The algorithm of Barnsley utilizes the ability of the computer to produce "random" numbers between 0 and 1. Each time the statement RND is executed, a new such number is produced which, although not truly random, is for all practical purposes a random number. For the system $\mathcal{F}$ consisting of the contractions F_1, F_2, ..., F_n, the algorithm proceeds as follows:

STEP 1. Choose an arbitrary point $\mathbf{v}$ in R^2.
STEP 2. Generate a random number r in $(0, 1)$. Then rn is a "random" number lying in the interval $(0, n)$.
STEP 3. If $k < rn \leq k + 1$, then plot the point $F_k(\mathbf{v})$.
STEP 4. Let $\mathbf{v} = F_k(\mathbf{v})$.
STEP 5. With the new point $\mathbf{v}$, repeat Steps 2–4, and then repeat the process as often as needed in order to generate a reasonable representation of the attractor $A_{\mathcal{F}}$.

You can experiment with various iterated function systems and different initial points by using the computer program ITERATED FUNCTION SYSTEM.

Now we turn the process around, and ask how we can determine an iterated function system whose attractor is (approximately) a given shape in the plane. Assume that A is a given closed and bounded subset of R^2. If we can find contractions F_1, F_2, ..., F_n such that

$$A = F_1(A) \cup F_2(A) \cup \cdots \cup F_n(A)$$

then A is the attractor of the union $\mathcal{F}$ of the functions, that is, $A = A_{\mathcal{F}}$. Although in general it is difficult to determine such functions F_1, F_2, ..., F_n, in many cases we can. The functions we will use henceforth in the iterated function systems will have the form

$$F\begin{pmatrix} x \\ y \end{pmatrix} = \begin{pmatrix} a & b \\ c & d \end{pmatrix}\begin{pmatrix} x \\ y \end{pmatrix} + \begin{pmatrix} e \\ f \end{pmatrix} \tag{2}$$

Such a function is called an **affine function**, and is the composite of a linear function G and a translation H, where

$$G\begin{pmatrix} x \\ y \end{pmatrix} = \begin{pmatrix} a & b \\ c & d \end{pmatrix}\begin{pmatrix} x \\ y \end{pmatrix} \quad \text{and} \quad H\begin{pmatrix} x \\ y \end{pmatrix} = \begin{pmatrix} x \\ y \end{pmatrix} + \begin{pmatrix} e \\ f \end{pmatrix}$$

Thus $F = H \circ G$. Since we will be studying collections of affine functions, it is useful to streamline the way we identify them. The function F in (2) is characterized by the six constants a, b, c, d, e, and f, which constitute the **code** of F. By their definitions, the three functions F_1, F_2, and F_3 exhibited at the outset of the section have codes given by

$$
\begin{array}{ccccccc}
 & a & b & c & d & e & f \\
F_1: & 1/2 & 0 & 0 & 1/2 & 0 & 0 \\
F_2: & 1/2 & 0 & 0 & 1/2 & 1/2 & 0 \\
F_3: & 1/2 & 0 & 0 & 1/2 & 1/4 & 1/2
\end{array}
\tag{3}
$$

An affine function is a contraction if the eigenvalues λ and μ of $\begin{pmatrix} a & b \\ c & d \end{pmatrix}$ satisfy $|\lambda| < 1$ and $|\mu| < 1$. In particular, this is the case if the matrix can be written in the form

$$
\begin{pmatrix}
r \cos \theta & -r \sin \theta \\
r \sin \theta & r \cos \theta
\end{pmatrix}
$$

where $|r| < 1$. The reason is that if

$$
F\begin{pmatrix} x \\ y \end{pmatrix} = \begin{pmatrix}
r \cos \theta & -r \sin \theta \\
r \sin \theta & r \cos \theta
\end{pmatrix}
\tag{4}
$$

then F reduces distances by a factor of r and rotates any shape counterclockwise through an angle of θ about the positive x axis (Figure 4.21). In code, the function in (4) becomes

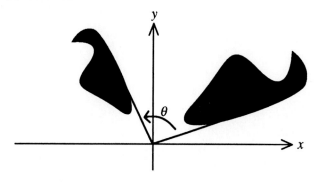

Figure 4.21

$$F: \quad r\cos\theta \quad -r\sin\theta \quad r\sin\theta \quad r\cos\theta \quad 0 \quad 0 \tag{5}$$

Suppose that F and its code are given by

$$F: \quad \frac{1}{3} \quad -\frac{1}{3}\sqrt{3} \quad \frac{1}{3}\sqrt{3} \quad \frac{1}{3} \quad 0 \quad 0$$

You can check that F contracts distances by a factor of $2/3$ and rotates by an angle of $\pi/3$.

In the next example, we will specify contractions that yield the von Koch curve.

EXAMPLE 2. Find an iterated function system $\mathcal{F}$ that has the von Koch curve K as its attractor (Figure 4.22(a)).

Solution. The von Koch curve K is the union of the four contracted copies K_1, K_2, K_3, and K_4 identified in Figure 4.22(b). Assume that the left-hand point of K is the origin, and that the curve "lies" on the unit interval $[0, 1]$ of the x axis. In particular, we obtain the four contracted copies from K as follows:

K_1 : contract by a factor of $\dfrac{1}{3}$

K_2 : contract by a factor of $\dfrac{1}{3}$, rotate by $\dfrac{\pi}{3}$, and shift by $\begin{pmatrix} 1/3 \\ 0 \end{pmatrix}$

K_3 : contract by a factor of $\dfrac{1}{3}$, rotate by $-\dfrac{\pi}{3}$, and shift by $\begin{pmatrix} 1/2 \\ \sqrt{3}/6 \end{pmatrix}$

K_4 : contract by a factor of $\dfrac{1}{3}$, and shift by $\begin{pmatrix} 2/3 \\ 0 \end{pmatrix}$

Therefore the codes of the four functions that yield K_1, K_2, K_3, and K_4 are given by

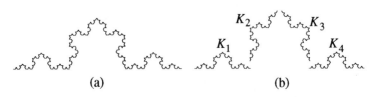

(a)   (b)

Figure 4.22

$$
\begin{array}{lllllll}
F_1: & 1/3 & 0 & 0 & 1/3 & 0 & 0 \\
F_2: & 1/6 & -\sqrt{3}/6 & \sqrt{3}/6 & 1/6 & 1/3 & 0 \\
F_3: & 1/6 & \sqrt{3}/6 & -\sqrt{3}/6 & 1/6 & 1/2 & \sqrt{3}/6 \\
F_4: & 1/3 & 0 & 0 & 1/3 & 2/3 & 0
\end{array}
$$

Thus we have found functions comprising an iterated function system that generates K. ❏

After one executes the program ITERATED FUNCTION SYSTEM, the picture on the screen is the attractor of the system. However, generally it takes many thousands of iterations to obtain a reasonable rendition of the attractor. For example, let us execute the program in order to obtain the Sierpinski carpet. Figure 4.23(a) shows the result after 5000 iterates. Figures 4.23(b)–(c) display the results after 20,000 and 50,000 iterates, respectively. Taking even more iterates would make the picture more precise.

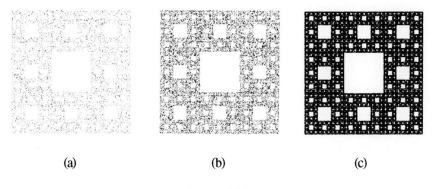

      (a)   (b)   (c)

Figure 4.23

Perhaps the most celebrated fractal obtained by iterated function systems is the fern that Barnsley has studied, and appears in Figure 4.24. (See Barnsley, 1988, for more details.) A fern similar to his can be considered to be the attractor for a system consisting of four affine functions coded as follows:

$$
\begin{array}{lllllll}
F_1: & 0 & 0 & 0 & .17 & 0 & 0 \\
F_2: & .85 & .04 & -.04 & .85 & 0 & 3 \\
F_3: & .2 & -.26 & .23 & .22 & 0 & 1.4 \\
F_4: & -.25 & .28 & .26 & .24 & 0 & .4
\end{array}
$$

Figure 4.24

To see how the union of the four functions yield the fern, we first notice that

$$F_1\begin{pmatrix} x \\ y \end{pmatrix} = \begin{pmatrix} 0 & 0 \\ 0 & .17 \end{pmatrix}\begin{pmatrix} x \\ y \end{pmatrix}$$

Therefore the image of F_1 is merely the bottom segment of the fern stem. By contrast, the image of F_2 is the whole fern except the bottom two leaves. The images of F_3 and F_4 are, respectively, the left bottom and right bottom leaves.

With practice and ingenuity, one can create all kinds of fascinating fractals by means of iterated function systems. However, the rest of the section is devoted to a theoretical analysis of iterated function systems.

Theoretical Considerations

The several ideas leading up to the Contraction Mapping Theorem and its application to iterated function systems are interesting and important. We mention, however, that the remainder of this section is independent of the rest of the book.

Let A be an arbitrary nonempty closed subset of R^2. If a sequence in A converges to some element of R^2, then by the definition of a closed set, the sequence converges to an element of A. In the event that A is also bounded, then every sequence in A has a subsequence that converges to an element of A. That is a consequence of the famous Bolzano-Weierstrass Theorem.

THEOREM 4.19 (Bolzano-Weierstrass Theorem). Let $\{w_n\}_{n=1}^{\infty}$ be a bounded sequence in R^2. Then there is a convergent subsequence.

We will apply the Bolzano-Weierstrass Theorem in the next two theorems. Before we state and prove the first of the two theorems, we need additional notation. If A is any bounded set in R^2 and $\varepsilon > 0$, we will denote by $A + \varepsilon$ the set of all $\mathbf{v}$ in R^2 such that $\|\mathbf{v} - \mathbf{a}\| \le \varepsilon$ for some $\mathbf{a}$ in A. In other words, $A + \varepsilon$ can be considered to be an ε-neighborhood of A.

THEOREM 4.20. Suppose that A is closed and bounded in R^2. Then $A + \varepsilon$ is also closed and bounded.

Proof. Choose M so that $\|\mathbf{a}\| \le M$ for all $\mathbf{a}$ in A, and let $\mathbf{v}$ be an arbitrary element of $A + \varepsilon$. Then there is an $\mathbf{a}$ in A such that $\|\mathbf{v} - \mathbf{a}\| \le \varepsilon$. It follows from the triangle inequality that

$$\|\mathbf{v}\| = \|\mathbf{v} - \mathbf{a} + \mathbf{a}\| \le \|\mathbf{v} - \mathbf{a}\| + \|\mathbf{a}\| \le \varepsilon + M$$

Therefore $A + \varepsilon$ is bounded. To show that $A + \varepsilon$ is closed, let us suppose that $\{\mathbf{v}_n\}_{n=1}^{\infty} \subseteq A + \varepsilon$ and that $\mathbf{v}_n$ converges to $\mathbf{v}$. We will show that $\mathbf{v}$ is also in $A + \varepsilon$. For each n, $\mathbf{v}_n$ is in $A + \varepsilon$, so that there is an $\mathbf{a}_n$ in A with $\|\mathbf{a}_n - \mathbf{v}_n\| \le \varepsilon$. Since A is closed and bounded, the Bolzano-Weierstrass Theorem implies that there is a subsequence $\{\mathbf{a}_{n_k}\}_{k=1}^{\infty}$ that converges to an element $\mathbf{a}$ in A. Notice that

$$\|\mathbf{a} - \mathbf{v}\| \le \|\mathbf{a} - \mathbf{a}_{n_k}\| + \|\mathbf{a}_{n_k} - \mathbf{v}_{n_k}\| + \|\mathbf{v}_{n_k} - \mathbf{v}\|$$

The first and third terms on the right side converge to 0 as k increases without bound, and the middle term is bounded by ε. Consequently

$$\|\mathbf{a} - \mathbf{v}\| \le \varepsilon$$

so that $\mathbf{v}$ is in $A + \varepsilon$. ∎

Next we will show that there is a minimal distance between a point and a closed, bounded subset of R^2.

THEOREM 4.21. Suppose that A is a closed, bounded subset of R^2, and let $\mathbf{v}$ be in R^2 but not in A. Then there is an element $\mathbf{a}^*$ in A closest to $\mathbf{v}$.

Proof. Let

$$\alpha = \text{the largest number less than or equal to } \|\mathbf{v} - \mathbf{a}\| \text{ for all } \mathbf{a} \text{ in } A$$

and let $\{\mathbf{a}_n\}_{n=1}^{\infty}$ be a sequence in A such that $\lim_{n \to \infty} \|\mathbf{v} - \mathbf{a}_n\| = \alpha$ (Figure 4.25). The sequence is bounded since it is contained in A. Therefore the Bolzano-Weierstrass Theorem says that there is a subsequence $\{\mathbf{a}_{n_k}\}_{k=1}^{\infty}$ that converges to, say, $\mathbf{a}^*$. Since A is closed, $\mathbf{a}^*$ is in A. We observe that

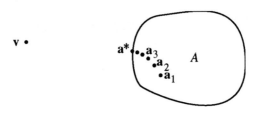

Figure 4.25

$$\|\mathbf{v} - \mathbf{a}^*\| \leq \|\mathbf{v} - \mathbf{a}_{n_k}\| + \|\mathbf{a}_{n_k} - \mathbf{a}^*\|$$

As k increases without bound, the first term on the right approaches α and the second converges to 0. Therefore $\|\mathbf{v} - \mathbf{a}^*\| \leq \alpha$. Since $\|\mathbf{v} - \mathbf{a}^*\| \geq \alpha$ by the definition of α, it follows that $\|\mathbf{v} - \mathbf{a}^*\| = \alpha$. Consequently $\mathbf{a}^*$ is an element of A that is closest to $\mathbf{v}$. ■

Early in the section we asserted that for an iterated function system, the iterates of closed and bounded sets in R^2 "converge" to the attractor. In order to give precise meaning to this convergence, we will define the notion of distance between any two closed, bounded subsets of R^2. The definition comes in two parts.

First we define the distance $d(\mathbf{v}, A)$ between any point $\mathbf{v}$ and a given closed bounded subset A of R^2:

$$d(\mathbf{v}, A) = \text{the minimum value of } \|\mathbf{v} - \mathbf{a}\| \text{ for } \mathbf{a} \text{ in } A$$

If $\mathbf{v}$ is in A, then $d(\mathbf{v}, A) = 0$. By contrast, if $\mathbf{v}$ is not in A, then $d(\mathbf{v}, A)$ is defined and nonzero, by virtue of Theorem 4.21.

Next we are prepared to define a distance d from an arbitrary, closed, and bounded subset A in R^2 to another such set B. Using the Bolzano-Weierstrass Theorem, one can show that there is an $\mathbf{a}^*$ in A such that $d(\mathbf{a}^*, B)$ is the maximum value of $d(\mathbf{a}, B)$ for all $\mathbf{a}$ in A (Exercise 13). Therefore we make the following definition:

$$d(A, B) = \text{the maximum value of } d(\mathbf{a}, B), \text{ for } \mathbf{a} \text{ in } A$$

We will employ this notion of distance in the following example.

EXAMPLE 3. Let A and B be the disks shown in Figure 4.26. Find $d(A, B)$ and $d(B, A)$.

Solution. If $\mathbf{a}$ is in A, then $d(\mathbf{a}, B) \geq 1$. Moreover, if $\mathbf{a}^* = \begin{pmatrix} -2 \\ 0 \end{pmatrix}$, then $\mathbf{a}^*$

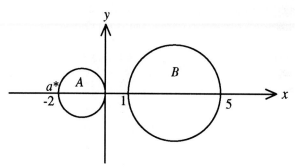

Figure 4.26

is in A, and $d(\mathbf{a}^*, B) = 3$. From Figure 4.26 it is evident that 3 is the maximum value of $d(\mathbf{a}, B)$ for $\mathbf{a}$ in A. We conclude that $d(A, B) = 3$. In a similar fashion we find that $d(B, A) = 5$.   ❏

In general, $d(A, B) \neq d(B, A)$, as occurs for the disks in Example 3. In order to have a notion of distance between two closed, bounded subsets A and B that is symmetric, we define the **Hausdorff distance**, or just plain **distance**, by the formula

$$D(A, B) = \text{the larger of the two numbers } d(A, B) \text{ and } d(B, A)$$

The association of the name Hausdorff refers to the German mathematician Felix Hausdorff, who was one of the pioneers in the area of geometry called topology during the early part of this century.

By the preceding definition of distance, if A and B are the two sets in Example 2, then $D(A, B) = 5$. Next, we will show that on the collection $\mathcal{K}$ of all closed, bounded subsets of R^2, the new concept of distance has the properties of a metric. In other words, we will show that if A and B are in $\mathcal{K}$, then

i. $D(A, B) = D(B, A)$
ii. $D(A, B) \geq 0$, and $D(A, B) = 0$ if and only if $A = B$
iii. $D(A, B) \leq D(A, C) + D(C, B)$ for every C in $\mathcal{K}$

THEOREM 4.22. Hausdorff distance is a metric on $\mathcal{K}$.

Proof. By the definition of D, $D(A, B) = D(B, A)$, $D(A, B) \geq 0$, and $D(A, A) = 0$. Thus (i) and part of (ii) are valid. Next, suppose that $A \neq B$; without loss of generality assume that there is an $\mathbf{a}$ in A that is not in B. Then

$$0 < d(\mathbf{a}, B) \leq d(A, B) \leq D(A, B)$$

Consequently the remainder of (ii) is proved. Now we will show that (iii) is valid.

Let **a** be an arbitrary element of A. By the triangle inequality and the definition of d, we find that

$$d(\mathbf{a}, B) \leq \|\mathbf{a} - \mathbf{b}\| \leq \|\mathbf{a} - \mathbf{c}\| + \|\mathbf{c} - \mathbf{b}\| \text{ for all } \mathbf{b} \text{ in } B \text{ and } \mathbf{c} \text{ in } C$$

Thus

$$d(\mathbf{a}, B) \leq \|\mathbf{a} - \mathbf{c}\| + d(\mathbf{c}, B) \leq \|\mathbf{a} - \mathbf{c}\| + D(C, B) \text{ for all } \mathbf{c} \text{ in } C$$

It follows that

$$d(\mathbf{a}, B) \leq d(\mathbf{a}, C) + D(C, B) \leq D(A, C) + D(C, B)$$

Since **a** is an arbitrary element of A, we deduce that

$$d(A, B) \leq D(A, C) + D(B, C) \tag{6}$$

Similarly,

$$d(B, A) \leq D(B, C) + D(C, A) \tag{7}$$

Combining (6) and (7) with the definition of D, we conclude that (iii) is valid. Therefore the proof of the theorem is complete. ∎

Now we know that $\mathcal{K}$ is a metric space. In terms of the metric, a sequence $\{A_n\}_{n=1}^{\infty}$ in $\mathcal{K}$ converges to B if $D(A_n, B) \to 0$ as n increases without bound. Geometrically this means that as n increases, the sets A_n become essentially coincident with B.

In order to satisfy the hypotheses of the upcoming Contraction Mapping Theorem, $\mathcal{K}$ must have an additional property called completeness – a property shared by many metric spaces. To describe completeness, we first need to define the notion of Cauchy sequence. Let X be a metric space with metric d. A sequence $\{x_n\}_{n=1}^{\infty}$ in X is a **Cauchy sequence** if for every $\varepsilon > 0$ there is a positive integer N such that whenever $m, n \geq N$, we have $d(x_m, x_n) < \varepsilon$. Cauchy sequences are named for the French mathematician Augustin-Louis Cauchy, who two hundred years ago helped to mold the calculus into the subject as we know it today. Geometrically a sequence is a Cauchy sequence if eventually members of the sequence are uniformly close together. Convergent sequences are Cauchy sequences, as we will see presently.

THEOREM 4.23. Let X be a metric space with metric d. Convergent sequences in X are Cauchy sequences.

Proof. Let $\{x_n\}_{n=1}^{\infty}$ be a sequence that converges to x, and let $\varepsilon > 0$. Then there is a positive integer N such that if $n \geq N$, then $d(x_n, x) < \varepsilon/2$. If in addition we have $m \geq N$, then

$$d(x_m, x_n) \leq d(x_m, x) + d(x, x_n) < \frac{\varepsilon}{2} + \frac{\varepsilon}{2} = \varepsilon$$

Therefore the sequence is a Cauchy sequence.   ■

It follows from Theorem 4.23 that in a metric space, a convergent sequence is a Cauchy sequence. If the converse is true, we say that the space is complete.

DEFINITION 4.24. Let X be a metric space. If every Cauchy sequence in X converges (to an element of X), then X is **complete**.

If one assumes the usual axioms for the real numbers, then the real numbers, as well as R^2, are complete metric spaces. (See p. 46 of Ross, 1980, for details.) Moreover, any closed subset of a complete metric space is complete (Exercise 14). A metric space that is *not* complete is the open interval $(0, 1)$. The reason is that a sequence such as $\{1/n\}_{n=1}^{\infty}$ is a Cauchy sequence. However, it converges to 0, which is not in $(0, 1)$.

The Contraction Mapping Theorem concerns metric spaces that are complete. In order to apply the Contraction Mapping Theorem to $\mathcal{K}$, we need to show that $\mathcal{K}$ is complete. Suppose that $\{A_n\}_{n=1}^{\infty}$ is a Cauchy sequence in $\mathcal{K}$. We must find a closed and bounded set A such that $D(A_n, A) \to 0$ as n increases without bound. To that end, we define A by

$$A = \{\mathbf{v} \text{ in } R^2 : \text{there is a Cauchy sequence } \{\mathbf{a}_n\}_{n=1}^{\infty} \text{ in } R^2$$
$$\text{converging to } \mathbf{v}, \text{ such that } \mathbf{a}_n \text{ is in } A_n \text{ for each } n\} \qquad (8)$$

The proof that A is in $\mathcal{K}$ and is the limit of the sequence $\{A_n\}_{n=1}^{\infty}$ is long. Therefore we have divided the proof into six lemmas. The first lemma involves enhancing a Cauchy sequence into another Cauchy sequence, and will be used in showing that A is nonempty.

LEMMA 1 (Enhancement Lemma). Suppose that $\{n_k\}_{k=1}^{\infty}$ is an increasing sequence of positive integers. If $\{\mathbf{a}_{n_k}\}_{k=1}^{\infty}$ is a Cauchy sequence in R^2 such that $\mathbf{a}_{n_k}$ is in A_{n_k} for each k, then $\{\mathbf{a}_{n_k}\}_{k=1}^{\infty}$ is a subsequence of a Cauchy sequence $\{\mathbf{a}_n\}_{n=1}^{\infty}$ in R^2 with the property that $\mathbf{a}_n$ is in A_n for all n.

Proof. Let k be a given positive integer, and define $n_0 = 1$. For each n such that $n_{k-1} < n < n_k$, we need to select $\mathbf{a}_n$ in A_n such that the sequence $\{\mathbf{a}_n\}_{n=1}^{\infty}$ that results is a Cauchy sequence. We accomplish this by letting

$$r_{k-1} = \text{the maximum value of } D(A_{n_k}, A_n) \text{ for } n_{k-1} < n < n_k$$

and by choosing $\mathbf{a}_n$ in A_n such that

$$\|\mathbf{a}_{n_{k-1}} - \mathbf{a}_n\| \leq r_{k-1}, \text{ for } n_{k-1} < n < n_k \tag{9}$$

To show that the sequence $\{\mathbf{a}_n\}_{n=1}^{\infty}$ is a Cauchy sequence, let $\varepsilon > 0$. Since by hypothesis $\{\mathbf{a}_{n_k}\}_{k=1}^{\infty}$ is a Cauchy sequence, there is a k such that if i, j, p and q $\geq n_k$, then

$$\|\mathbf{a}_{n_p} - \mathbf{a}_{n_q}\| < \frac{\varepsilon}{3} \text{ and } D(A_i, A_j) < \frac{e}{3} \tag{10}$$

Now let m, $n \geq n_k$. Suppose that n_p is the largest integer in $\{n_k\}_{k=1}^{\infty}$ smaller than m, and n_q is the largest integer in $\{n_k\}_{k=1}^{\infty}$ smaller than n. Then two applications of the triangle property, and a reference to (9) and (10), yields the following inequalities:

$$\|\mathbf{a}_m - \mathbf{a}_n\| \leq \|\mathbf{a}_m - \mathbf{a}_{n_p}\| + \|\mathbf{a}_{n_p} - \mathbf{a}_{n_q}\| + \|a_{n_q} - \mathbf{a}_n\|$$

$$< \frac{\varepsilon}{3} + \frac{\varepsilon}{3} + \frac{\varepsilon}{3} = \varepsilon \tag{11}$$

Therefore the sequence $\{\mathbf{a}_n\}_{n=1}^{\infty}$ is a Cauchy sequence.   ■

The inequality in (11) is a variety common in proofs that sequences are Cauchy sequences. In fact, that inequality is sometimes referred to as the "$\varepsilon/3$ argument."

Next, recall that A is defined in (8) by means of a Cauchy sequence $\{A_n\}_{n=1}^{\infty}$. We are ready to prove that A is nonempty, closed, and bounded.

LEMMA 2. A is nonempty.

Proof. To show that A is nonempty, we need to find a Cauchy sequence $\{\mathbf{a}_n\}_{n=1}^{\infty}$ in R^2 such that $\mathbf{a}_n$ is in A_n for all n. Since $\{A_n\}_{n=1}^{\infty}$ is a Cauchy sequence by prescription, there is an increasing sequence of positive integers $\{n_k\}_{k=1}^{\infty}$ such that

$$D(A_n, A_{n_k}) < \frac{1}{2^k} \text{ for all } n \geq n_k$$

Choose $\mathbf{a}_1$ in A_{n_1}. Because $D(A_n, A_{n_1}) < 1/2$ for all $n \geq n_1$, we can find $\mathbf{a}_2$ in A_{n_2} such that $\|\mathbf{a}_2 - \mathbf{a}_1\| < 1/2$. By induction, for every integer $k > 2$, we choose $\mathbf{a}_k$ in A_{n_k} such that

$$\|\mathbf{a}_k - \mathbf{a}_{k-1}\| < \frac{1}{2^{k-1}}$$

We will show that $\{\mathbf{a}_k\}_{k=1}^{\infty}$ is a Cauchy sequence in R^2. Let $\varepsilon > 0$, and choose k

so large that $1/2^k < \varepsilon$. If $n > m > k$, then by multiple applications of the triangle inequality, we find that

$$\|\mathbf{a}_m - \mathbf{a}_n\| \le \|\mathbf{a}_m - \mathbf{a}_{m+1}\| + \|\mathbf{a}_{m+1} - \mathbf{a}_{m+2}\| + \cdots + \|\mathbf{a}_{n-1} - \mathbf{a}_n\|$$

$$< \frac{1}{2^m} + \frac{1}{2^{m+1}} + \cdots + \frac{1}{2^{n-1}} < \frac{1}{2^{m-1}} \le \frac{1}{2^k} < \varepsilon$$

Therefore $\{\mathbf{a}_k\}_{k=1}^{\infty}$ is a Cauchy sequence in R^2, which by the Enhancement Lemma can be enlarged to be a Cauchy sequence associated with the full sequence $\{A_n\}_{n=1}^{\infty}$. Since R^2 is complete, the full sequence converges to, say, $\mathbf{v}$. It follows that $\mathbf{v}$ is in A. Consequently A is nonempty. ∎

LEMMA 3. A is closed.

Proof. Let $\{\mathbf{a}_n\}_{n=1}^{\infty}$ be a sequence in A that converges to $\mathbf{a}$ in R^2. We will prove that $\mathbf{a}$ is in A. We can assume without loss of generality that $\|\mathbf{a}_n - \mathbf{a}\| < 1/2^n$ for each n, that $\{\mathbf{a}_{n,k}\}_{k=1}^{\infty}$ is a Cauchy sequence in R^2 such that $\mathbf{a}_{n,k}$ is in A_k for each k, and $\lim_{k\to\infty} \mathbf{a}_{n,k} = \mathbf{a}_n$. For each n, select a positive integer $k(n)$ such that $\|\mathbf{a}_{n,k(n)} - \mathbf{a}_n\| < 1/2^n$. Then

$$\|\mathbf{a}_{n,k(n)} - \mathbf{a}\| \le \|\mathbf{a}_{n,k(n)} - \mathbf{a}_n\| + \|\mathbf{a}_n - \mathbf{a}\| < \frac{1}{2^n} + \frac{1}{2^n} = \frac{1}{2^{n-1}}$$

Consequently $\{\mathbf{a}_{n,k(n)}\}_{n=1}^{\infty}$ converges to $\mathbf{a}$, and has the property that $\mathbf{a}_{n,k(n)}$ is in $A_{k(n)}$. By the Enhancement Lemma, the sequence can be enlarged to be a Cauchy sequence whose nth member is in A_n for each n. Since the given sequence converges to $\mathbf{a}$, so does the enlarged sequence. We conclude that $\mathbf{a}$ is in A, so that A is closed. ∎

LEMMA 4. A is bounded.

Proof. Let $\varepsilon = 1$. Since $\{A_n\}_{n=1}^{\infty}$ is a Cauchy sequence there is a positive integer N such that if $m, n \ge N$, then $D(A_m, A_n) < 1$. Since A_N is bounded, there is an $M > 0$ such that A_N is contained in the disk of radius M centered at the origin. Thus A_n is contained in the disk of radius $M + 1$ centered at the origin, for all $n \ge N$. Consequently the same is true of A. Therefore A is bounded. ∎

Since A is closed and bounded, A is in $\mathcal{K}$. In order to confirm that $\mathcal{K}$ is complete, we need to show that the sequence $\{A_n\}_{n=1}^{\infty}$ from which A was defined in (8) actually converges to A. In proving this convergence, we will use the following lemma.

LEMMA 5. $D(A, B) \le \varepsilon$ if and only if $A \subseteq B + \varepsilon$ and $B \subseteq A + \varepsilon$.

Proof. Assume that $D(A, B) \le \varepsilon$, and let $\mathbf{a}$ be in A. Then $d(\mathbf{a}, B) \le \varepsilon$, so that there exists a $\mathbf{b}$ in B such that $\|\mathbf{a} - \mathbf{b}\| \le \varepsilon$. Thus $\mathbf{a}$ is in $B + \varepsilon$. Consequently $A \subseteq B + \varepsilon$. Since $D(A, B) = D(B, A)$, it follows that $B \subseteq A + \varepsilon$. Conversely, assume that $A \subseteq B + \varepsilon$ and $B \subseteq A + \varepsilon$. If $\mathbf{a}$ is in A, then $d(\mathbf{a}, B) \le \varepsilon$. Since this is true of every element of A, we conclude that $d(A, B) \le \varepsilon$. Similarly, the hypothesis that $B \subseteq A + \varepsilon$ implies that $d(B, A) \le \varepsilon$. Therefore $D(A, B) \le \varepsilon$. ∎

LEMMA 6. $\mathrm{Lim}_{m \to \infty} D(A_m, A) = 0$.

Proof. Let $\varepsilon > 0$. Since $\{A_n\}_{n=1}^{\infty}$ is a Cauchy sequence, there is an N such that if $m, n \ge N$, then $D(A_m, A_n) \le \varepsilon$. Let $m \ge N$. We will show that $D(A_m, A) \le \varepsilon$. By Lemma 5 we can show equivalently that $A \subseteq A_m + \varepsilon$ and $A_m \subseteq A + \varepsilon$. First we will prove that $A \subseteq A_m + \varepsilon$. Since $D(A_m, A_n) \le \varepsilon$ for all $n \ge m$, Lemma 5 implies that $A_n \subseteq A_m + \varepsilon$, for all $n \ge m$. Thus any sequence $\{\mathbf{a}_n\}_{n=m}^{\infty}$ such that $\mathbf{a}_n$ is in A_n for all $n \ge m$ also lies in $A_m + \varepsilon$. By Theorem 4.20, $A_m + \varepsilon$ is closed. Therefore the limit of such a sequence also lies in $A_m + \varepsilon$. Consequently by the definition of A, $A \subseteq A_m + \varepsilon$.

Next we will show that there is an $N^* \ge N$ such that if $m \ge N^*$, then $A_m \subseteq A + \varepsilon$. To that end, let N^* be so large that if $m, n \ge N^*$, then $D(A_m, A_n) < \varepsilon/4$. Let $m \ge N^*$, and assume that $\mathbf{a}$ is in A_m. We will prove that $\mathbf{a}$ is in $A + \varepsilon$. Since $\{A_n\}_{n=1}^{\infty}$ is a Cauchy sequence, for each integer $k > 1$ there is a positive integer $n_k > n$ such that $D(A_n, A_{n_{k-1}}) < \varepsilon/2^k$. We can assume that the sequence $\{n_k\}_{k=1}^{\infty}$ is strictly increasing. Next we choose $\mathbf{a}_1$ in A_{n_1} such that $\|\mathbf{a} - \mathbf{a}_1\| < \varepsilon/4$. Inductively we select $\mathbf{a}_k$ in A_{n_k} such that $\|\mathbf{a}_k - \mathbf{a}_{k-1}\| < \varepsilon/2^k$. Then $\{\mathbf{a}_k\}_{k=1}^{\infty}$ is a Cauchy sequence, and so converges to some $\mathbf{v}$ in R^2. The Enhancement Lemma implies that there is an enlargement of the sequence that is a Cauchy sequence and thus also converges to $\mathbf{v}$. Therefore by the definition of A, $\mathbf{v}$ is in A. Suppose that k is so large that $\|\mathbf{a}_k - \mathbf{v}\| < \varepsilon/4$. By multiple applications of the triangle inequality we find that

$$\|\mathbf{a} - \mathbf{v}\| \le \|\mathbf{a} - \mathbf{a}_1\| + \|\mathbf{a}_1 - \mathbf{a}_2\| + \cdots + \|\mathbf{a}_{k-1} - \mathbf{a}_k\| + \|\mathbf{a}_k - \mathbf{v}\|$$

$$< \frac{\varepsilon}{4} + \frac{\varepsilon}{2^2} + \frac{\varepsilon}{2^3} + \cdots + \frac{\varepsilon}{2^k} + \frac{\varepsilon}{4} < \varepsilon$$

Consequently $\mathbf{a}$ is in $A + \varepsilon$. ∎

We have completed the preparation for the following theorem.

THEOREM 4.25. $\mathcal{K}$ is a complete metric space.

Proof. By Theorem 4.22, $\mathcal{K}$ is a metric space, and by Lemma 6 it is complete. ∎

In a metric space X with metric d, a **contraction** f that has a scaling factor $s < 1$ has the property that

$$d(f(y), f(z)) \leq sd(y, z) \text{ for all } y \text{ and } z \text{ in } X$$

Such a function f is continuous on X, because if y and z are close together, then $f(y)$ and $f(z)$ are even closer together.

The stage is set for the statement and proof of the Contraction Mapping Theorem and the application to iterated function systems. The theorem will be presented in the context of general complete metric spaces, but will apply to iterated function systems on $\mathcal{K}$.

THEOREM 4.26 (Contraction Mapping Theorem). Let X be a complete metric space with metric d, and assume that $f : X \to X$ is a contraction map with scaling factor $s < 1$. Then f has exactly one fixed point p, whose basin of attraction is X.

Proof. Let x be any element of X. We will show that $\{f^{[n]}(x)\}_{n=1}^{\infty}$ is a Cauchy sequence in X. Let $m < n$. Then

$$d(x, f^{[n-m]}(x)) \leq d(x, f(x)) + d(f(x), f^{[2]}(x)) + \cdots + d(f^{[n-m-1]}(x), f^{[n-m]}(x))$$

$$\leq d(x, f(x)) + sd(x, f(x)) + s^2 d(x, f(x)) + \cdots + s^{n-m-1} d(x, f(x))$$

$$= \left(\sum_{k=0}^{n-m-1} s^k \right) d(x, f(x))$$

Since

$$\sum_{k=0}^{n-m-1} s^k = \frac{1 - s^{n-m}}{1 - s}$$

it follows that

$$d(x, f^{[n-m]}(x)) \leq \frac{1 - s^{n-m}}{1 - s} d(x, f(x)) \tag{12}$$

If $m \leq n$, then

$$d(f^{[m]}(x), f^{[n]}(x)) \leq sd(f^{[m-1]}(x), f^{[n-1]}(x)) \leq \cdots \leq s^m d(x, f^{[n-m]}(x)) \tag{13}$$

Together (12) and (13) imply that if $m < n,$ then

$$d(f^{[m]}(x), f^{[n]}(x)) \leq s^m \left(\frac{1 - s^{n-m}}{1 - s} \, d(x, f(x)) \right) = \frac{s^m - s^n}{1 - s} \, d(x, f(x)) \quad (14)$$

Since $0 < s < 1,$ the right-hand side of (14) approaches 0 as m and n increase without bound. Consequently $\{f^{[n]}(x)\}_{n=1}^{\infty}$ is a Cauchy sequence. The fact that X is a complete metric space implies that the sequence converges to an element of X which we will denote by $p.$ Since contractions are automatically continuous and $f^{[n]}(x) \to p$ as n increases without bound, we also find that $f(f^{[n]}(x)) \to f(p).$ However, $f(f^{[n]}(x)) = f^{[n+1]}(x) \to p.$ Consequently $f(p) = p,$ so that p is a fixed point of $f.$ Finally, let x be any element in $X.$ Since $p = f^{[n]}(p),$ we find by (13) with $m = n$ that as n increases without bound,

$$d(p, f^{[n]}(x)) = d(f^{[n]}(p), f^{[n]}(x)) \leq s^n d(p, x) \to 0$$

We conclude that the iterates of all elements of X converge to $p.$ Thus the basin of attraction of p is X and p is the only fixed point of $f.$ ■

Before we can apply the Contraction Mapping Theorem to iterated function systems on $\mathcal{K},$ we need to show that the union of contractions is a contraction on $\mathcal{K}.$ We will prove this result in Theorem 4.28. First we have preliminary results.

LEMMA 1. Let $A,$ $B,$ and C be in $\mathcal{K}.$ Then $D(A \cup B, C) \leq$ the maximum of $D(A, C)$ and $D(B, C).$

Proof. By definition,

$$D(A \cup B, C) = \text{the maximum value of } d(\mathbf{v}, C) \text{ for all } \mathbf{v} \text{ in } A \cup B$$
$$\text{and of } d(\mathbf{c}, A \cup B) \text{ for all } \mathbf{c} \text{ in } C$$

Without loss of generality let us assume that the maximum value of $d(\mathbf{v}, C)$ for $\mathbf{v}$ in $A \cup B$ occurs for $\mathbf{a}$ in $A.$ Then

$$d(\mathbf{a}, C) \leq d(A, C) \leq \text{the maximum of } D(A, C) \text{ and } D(B, C)$$

Since $d(\mathbf{c}, A \cup B) \leq d(\mathbf{c}, A) \leq d(C, A) \leq D(C, A)$ and in a similar way we have $d(\mathbf{c}, A \cup B) \leq d(C, B),$ the result follows. ■

LEMMA 2. Let $A,$ $B,$ $C,$ and E be in $\mathcal{K}.$ Then

$$D(A \cup B, C \cup E) \leq \text{the maximum of } D(A, C) \text{ and } D(B, E)$$

Proof. We leave this to Exercise 15. ■

THEOREM 4.27. Let A and B be in $\mathcal{K}$, and assume that F is a contraction on R^2 with contraction factor $s < 1$. Then

$$D(F(A), F(B)) \leq s\, D(A, B)$$

Proof. Let **a** be in A, and let **b** be in B such that $\|\mathbf{a} - \mathbf{b}\| = d(\mathbf{a}, B)$. Then

$$\|F(\mathbf{a}) - F(\mathbf{b})\| \leq s\, \|\mathbf{a} - \mathbf{b}\| = s\, d(\mathbf{a}, B) \leq s\, D(A, B)$$

Therefore

$$d(F(\mathbf{a}), F(B)) \leq s\, D(A, B), \text{ so that } d(F(A), F(B)) \leq sD(A, B)$$

Similarly, $d(F(B), F(A)) \leq sD(A, B)$. Combining the last two inequalities yields

$$D(F(A), F(B)) \leq s\, D(A, B) \quad \blacksquare$$

THEOREM 4.28. Assume that $F_1, F_2, ..., F_n$ are contractions on R^2 with contraction factors bounded above by $s < 1$. Let $\mathcal{F}$ be their union. Then $\mathcal{F}$ is a contraction with contraction factor less than or equal to s.

Proof. Let $n = 2$, and let A and B be in $\mathcal{K}$. Then by Lemma 2 and Theorem 4.27,

$$D(\mathcal{F}(A), \mathcal{F}(B)) = D(F_1(A) \cup F_2(A),\ F_1(B) \cup F_2(B))$$

$$\leq \text{ the maximum of } D(F_1(A), F_1(B)) \text{ and } D(F_2(A), F_2(B))$$

$$\leq s\, D(A, B)$$

The result for an arbitrary positive integer n can be proved by induction. Therefore $\mathcal{F}$ is a contraction with contraction factor less than or equal to s. $\blacksquare$

Theorem 4.28 is the missing ingredient that allows us to apply the Contraction Mapping Theorem to $\mathcal{F}$ on $\mathcal{K}$.

THEOREM 4.29. Let $f_1, f_2, ..., f_n$ be contractions defined on R^2 and $\mathcal{F}$ their union. Then there is a unique closed and bounded set A such that the iterates of every set in $\mathcal{K}$ converge (in the Hausdorff metric) to A.

Proof. By Theorem 4.28, $\mathcal{F}$ is a contraction. It is defined on the complete metric space $\mathcal{K}$, with range contained in $\mathcal{K}$. By the Contraction Mapping Theorem, $\mathcal{F}$ has a unique fixed point in $\mathcal{K}$, that is, a closed and bounded set A. Moreover, that theorem asserts that the iterates of every closed and bounded set converge in the Hausdorff metric to A. $\blacksquare$

Theorem 4.29 completes our mission of showing that an iterated function system consisting of contractions has a unique attractor. Much more is known about such systems, and about fractal geometry. The book by Barnsley (1988) is devoted to this topic.

EXERCISES 4.4

1. Find an affine function F such that the image of the circle $x^2 + y^2 = 1$ is the ellipse $2x^2 + 4y^2 = 1$.

2. Let S be the square with vertices $(0, 0)$, $(1, 0)$, $(1, 1)$, and $(0, 1)$. Find an affine function F such that $F(S)$ is
    a. the rectangle with vertices $(1, 0)$, $(3, 0)$, $(3, 1)$, and $(1, 1)$.
    b. the rectangle in part (a) rotated $\pi/4$ radians counterclockwise about the positive x axis.

3. Find an iterated function system for the Cantor ternary set.

4. Find an iterated function system for the Sierpinski carpet that is placed so that its bottom left corner is the origin and two of its sides lie on the axes.

5. a. Use the program ITERATED FUNCTION SYSTEM to determine how many iterations are needed in order to obtain a good visual image of the Sierpinski gasket.
    b. Try various initial points to show that, within reason, it does not make any difference to the number of iterates needed in order to obtain a decent rendition of the Sierpinski gasket.

6. Let $\mathcal{F}$ have the code

$$F_1: \quad .5 \quad 0 \quad 0 \quad .5 \quad 0 \quad 0$$
$$F_2: \quad .5 \quad 0 \quad 0 \quad .5 \quad .5 \quad 0$$
$$F_3: \quad .5 \quad 0 \quad 0 \quad .5 \quad 1 \quad .5$$

Use the program ITERATED FUNCTION SYSTEM to determine the shape of the attractor.

In Exercises 7–9, determine an iterated function system whose attractor is the given shape.

7.   8.   9.

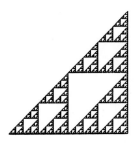   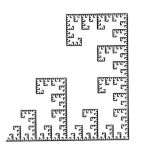   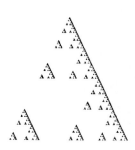

Figure 4.27   Figure 4.28   Figure 4.29

10. a.   Find the Hausdorff distance between the circle C of radius 1 centered at the origin and a square S of side length 1 centered at the origin.

b.   Let A consist of all points of the form $(x, 1/x)$ for $x \geq 1$, and let B be the x axis. Find $D(A, B)$.

11. Let T be the equilateral triangle whose vertices are $(0, 0)$, $(1, 0)$, and $(1/2, \sqrt{3}/2)$, and let S be the square whose vertices are $(1, 0)$, $(3, 0)$, $(3, 3)$ and $(1, 3)$. Find the Hausdorff distance between T and S.

12. a.   Suppose that F is continuous on R^2, and let A be a closed, bounded subset of R^2. Show that $F(A)$ is also closed and bounded.

b.   Show that the union of a finite collection of closed subsets of R^2 is closed.

c.   Show that the union of a finite collection of bounded subsets of R^2 is bounded.

13. Suppose that A and B are closed, bounded subsets of R^2. Show that there exists an element $\mathbf{a}^*$ of A such that $d(\mathbf{a}^*, B)$ is the maximum value of $d(\mathbf{a}, B)$ for all $\mathbf{a}$ in A.

14. Show that a closed subset of a complete metric space is also complete.

15. Let A, B, C, and E be in $\mathcal{K}$. Prove that

$$D(A \cup B, C \cup E) \leq \text{ the maximum of } D(A, C) \text{ and } D(B, E)$$

16. Prove Theorem 4.28 for arbitrary values of $n > 2$.

CHAPTER
5

SYSTEMS OF DIFFERENTIAL EQUATIONS

Objects in motion, like satellites in space, water molecules in a stream, and electrons in a diamond, have long intrigued scientists. During the past few centuries, scientists were under the impression that the motion of an object could, at least in theory, be completely understood if one could formulate accurately enough a set of equations that described certain attributes of the object such as its position, velocity and acceleration. That this premise was not universally true was observed around 1900 by Henri Poincaré in his studies of the three-body problem.

However, in the early 1960's the meteorologist Edward Lorenz discovered that weather patterns modeled by a simple system of differential equations could be profoundly unpredictable. It was this discovery, along with the power of modern-day computers, that launched the new field of chaotic dynamics. One goal of this final chapter is to describe and interpret the Lorenz system of differential equations.

Section 5.1 is a review of the important concepts relating to systems of differential equations, and includes an analysis of constant solutions of such systems. In Section 5.2 we apply the results of Section 5.1 to systems that are not linear but are what we will call almost linear systems. In the final sections we turn to two celebrated almost linear systems: the pendulum system that describes the motion of a pendulum, and the Lorenz system that has been applied to the study of weather.

5.1 REVIEW OF SYSTEMS OF DIFFERENTIAL EQUATIONS

Systems of differential equations are employed to describe a wide variety of physical phenomena, including the motion of a pendulum and particles in a weather system. Before we can analyze these applications, we need to understand the structure

240

of solutions of such systems. That is the goal of the present section. If you do not have prior knowledge of differential equations, then you should probably consult one of the standard texts on differential equations, such as Boyce and DiPrima (1986) or Sanchez, Allen and Kyner (1983) for further information.

We begin with terminology.

DEFINITION 5.1. Let x_1, x_2, ..., x_n be differentiable functions of time t on an interval J of real numbers, and let f_1, f_2, ..., and f_n be functions of $x_1, x_2, ...,$ and x_n as well as t. The n differential equations

$$\frac{dx_1}{dt} = f_1(x_1, x_2, \ldots, x_n, t)$$

$$\frac{dx_2}{dt} = f_2(x_1, x_2, \ldots, x_n, t)$$

$$\cdots \cdots \cdots \cdots \cdots \tag{1}$$

$$\frac{dx_n}{dt} = f_n(x_1, x_2, \ldots, x_n, t)$$

form a **system of differential equations**. If $X: J \to R^n$ is defined by

$$X(t) = \begin{pmatrix} x_1(t) \\ x_2(t) \\ . \\ . \\ x_n(t) \end{pmatrix} \tag{2}$$

and if X satisfies (1), then X is a **solution** of the system. If t_0 is in R and X is a solution for all $t \geq t_0$, then the element $X(t_0)$ of R^n is an **initial condition** of a solution X. Usually t_0 will be 0.

Since x_1, x_2, ..., x_n are differentiable functions of t, it follows that as t increases, $X(t)$ traces out a curve in R^n, called the **trajectory**, or **orbit**, of X.

If $n = 1$ in (1), the system is simply $dx/dt = f(x, t)$, and we will write a solution as x. One well-known example of such an equation appears in calculus:

$$\frac{dx}{dt} = kx \tag{3}$$

where k is nonzero constant. Solutions of the equation in (3) are functions of the form $x(t) = ce^{kt}$, which represents exponential growth if $k > 0$ and exponential decay if $k < 0$.

If $n = 2$ in (1), we write the system as

$$\frac{dx}{dt} = f(x, y, t)$$

$$\frac{dy}{dt} = g(x, y, t)$$

For such a system, $x(t)$ might denote the position and $y(t)$ the velocity of an object at time t. Analogously, if $n = 3$, then we write

$$\frac{dx}{dt} = f(x, y, z, t)$$

$$\frac{dy}{dt} = g(x, y, z, t)$$

$$\frac{dz}{dt} = h(x, y, z, t)$$

In this case, $x(t)$, $y(t)$, and $z(t)$ might represent the coordinates of an object at time t, or perhaps the position, velocity, and acceleration of the object at time t.

   In the same way that fixed points play a significant role in the analysis of the dynamics of a function, special solutions assist in the analysis of solutions of systems of differential equations.

DEFINITION 5.2. A **critical point** (or **equilibrium point**, or **stationary point**) of a system of differential equations is a constant solution, that is, a solution X such that $X(t) = X(t_0)$ for all t. If X is a critical point, then we identify the critical point with the vector $X(t_0)$.

   It can be shown that if X is a solution and $dX/dt = 0$ for some $t = t_0$, then X is constant for all t, and hence is a critical point. For example, if

$$\frac{dx}{dt} = (x - 2)(x + 3)$$

then the critical points of the equation are the constant solutions 2 and -3.

Linear Differential Equations

A differential equation of the form

$$\frac{dx}{dt} = a(t)x + g(t) \qquad (4)$$

where a and g are continuous functions of t on some given interval J, is called a **linear differential equation,** because the right-hand side of the equation is linear in x. Equation (4) can be solved completely, in the sense that we can find a formula that describes every solution of (4). Such a formula is called a **general solution.**

In the special case

$$\frac{dx}{dt} = ax + b \tag{5}$$

in which a and b are constant with $a \neq 0$, the general solution is particularly easy to identify:

$$x(t) = -\frac{b}{a} + ce^{at} , \text{ for all } t \text{ in } J \tag{6}$$

where c is any real constant. To check that x given in (6) is indeed a solution of (4), we just take the derivative of x in (6) and then see that (5) is satisfied:

$$\frac{dx}{dt} = ace^{at} = a\left(-\frac{b}{a} + ce^{at}\right) + b = ax + b$$

That every solution to (5) has the form of (6) follows from the theory of differential equations. We conclude that (6) gives the general solution of (5). The graph of an arbitrary solution is given in Figure 5.1; the graph curves up if $a > 0$, and curves down if $a < 0$. It is obvious from Figure 5.1 that the system in (5) has no critical point if $a \neq 0$.

The general solution of the general linear differential equation

$$\frac{dx}{dt} = a(t)x + g(t)$$

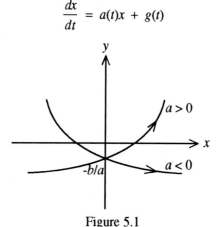

Figure 5.1

is given by

$$x(t) = e^{A(t)}\left(\int_{u_0}^{t} e^{-A(u)}g(u)du + c\right), \text{ where } A(t) = \int_{u_0}^{t} a(u)\, du \qquad (7)$$

where c is any real number. You can check that (7) defines a solution to the differential equation. Only in rare cases is it possible to find a reasonable formula for $x(t)$ if $a(t)$ is not constant.

Systems of Two Linear Differential Equations

We turn to those systems of two linear differential equations with constant coefficients a, b, c, and d that have the form

$$\frac{dx}{dt} = ax + by$$
$$\qquad\qquad\qquad\qquad (8)$$
$$\frac{dy}{dt} = cx + dy$$

If $X(t) = \begin{pmatrix} x(t) \\ y(t) \end{pmatrix}$ is a solution, then the system in (8) can be rewritten as

$$\frac{dX}{dt} = \begin{pmatrix} a & b \\ c & d \end{pmatrix}\begin{pmatrix} x \\ y \end{pmatrix} \qquad (9)$$

The matrix $\begin{pmatrix} a & b \\ c & d \end{pmatrix}$ is the **associated matrix** for the system.

Evidently if $X(t) = 0$ for all t, then X is a solution of the system in (8). The solution is called the **zero solution**, and is denoted by $\mathbf{0}$, as is the corresponding critical point.

By (6), the general solution of the linear differential equation $dx/dt = ax$ is given by $x(t) = ce^{at}$. Thus we might hope that a general solution of the system in (8) would have the form

$$X(t) = \begin{pmatrix} x(t) \\ y(t) \end{pmatrix} = \begin{pmatrix} re^{\lambda t} \\ se^{\lambda t} \end{pmatrix} \qquad (10)$$

for appropriate constants λ, r, and s. For this to be the case, X must satisfy

$$\begin{pmatrix} \lambda r e^{\lambda t} \\ \lambda s e^{\lambda t} \end{pmatrix} = \begin{pmatrix} dx/dt \\ dy/dt \end{pmatrix} \underset{(8)}{=} \begin{pmatrix} ax + by \\ cx + dy \end{pmatrix} = \begin{pmatrix} are^{\lambda t} + bse^{\lambda t} \\ cre^{\lambda t} + dse^{\lambda t} \end{pmatrix} = \begin{pmatrix} a & b \\ c & d \end{pmatrix} \begin{pmatrix} re^{\lambda t} \\ se^{\lambda t} \end{pmatrix}$$

so that

$$\lambda \begin{pmatrix} re^{\lambda t} \\ se^{\lambda t} \end{pmatrix} = \begin{pmatrix} a & b \\ c & d \end{pmatrix} \begin{pmatrix} re^{\lambda t} \\ se^{\lambda t} \end{pmatrix}, \text{ or equivalently, } \lambda \begin{pmatrix} r \\ s \end{pmatrix} = \begin{pmatrix} a & b \\ c & d \end{pmatrix} \begin{pmatrix} r \\ s \end{pmatrix} \qquad (11)$$

Consequently if $X(t) = \begin{pmatrix} re^{\lambda t} \\ se^{\lambda t} \end{pmatrix}$ and $X(t)$ is a solution of the system in (8), then λ

is an eigenvalue and $X(t)$ an eigenvector of the associated matrix. Since λ is an eigenvalue,

$$0 = \det \begin{pmatrix} a - \lambda & b \\ c & d - \lambda \end{pmatrix} = (a - \lambda)(d - \lambda) - bc$$

Therefore λ satisfies the **characteristic equation**

$$\lambda^2 - (a + d)\lambda + ad - bc = 0 \qquad (12)$$

If the characteristic equation has a real solution λ, then we see immediately that

$X(t) = \begin{pmatrix} re^{\lambda t} \\ se^{\lambda t} \end{pmatrix}$ furnishes a solution of the system in (8). However, as we will see

later in the section, the system in (8) has nonzero solutions even when the characteristic equation has no real solutions. In fact, the theory of differential equations tells us that there are always two solutions of the system in (8) that are not constant multiples of one another, that is, there are always two **independent solutions** of the system.

The system in (8) has a critical point, namely **0**. Are there any others? The answer is provided by Theorem 5.3.

THEOREM 5.3. **0** is the only critical point of the system in (8) if and only if 0 is *not* an eigenvalue of the associated matrix.

Proof. Let $\lambda = 0$ be an eigenvalue of the associated matrix. Then there is a nonzero

eigenvector $\begin{pmatrix} r \\ s \end{pmatrix}$. If X is defined by $X(t) = \begin{pmatrix} x(t) \\ y(t) \end{pmatrix} = \begin{pmatrix} r \\ s \end{pmatrix}$ for all t, then $dx/dt =$

$0 = dy/dt$, so that

$$\frac{dX}{dt} = \begin{pmatrix} dx/dt \\ dy/dt \end{pmatrix} = \begin{pmatrix} 0 \\ 0 \end{pmatrix} = \begin{pmatrix} a & b \\ c & d \end{pmatrix} \begin{pmatrix} r \\ s \end{pmatrix}$$

Thus X is a solution, so is a critical point of the system. Conversely, let $\begin{pmatrix} r \\ s \end{pmatrix}$ be

a nonzero critical point. If we let $X(t) = \begin{pmatrix} x(t) \\ y(t) \end{pmatrix} = \begin{pmatrix} r \\ s \end{pmatrix}$, then by (9),

$$\begin{pmatrix} 0 \\ 0 \end{pmatrix} = \begin{pmatrix} dx/dt \\ dy/dt \end{pmatrix} = \frac{dX}{dt} = \begin{pmatrix} a & b \\ c & d \end{pmatrix} \begin{pmatrix} r \\ s \end{pmatrix}$$

This implies that the associated matrix is not invertible. Hence 0 is an eigenvalue of the matrix. This completes the proof of the theorem. ∎

In our analysis of linear systems we will study those systems with only one critical point, namely 0. This means, by Theorem 5.3, that the associated matrices we will encounter will not have an eigenvalue equal to zero.

Fixed points of maps correspond to equilibrium solutions of systems of differential equations. The attributes of attracting and repelling fixed points of maps relate to the notion of stability for systems of differential equations.

DEFINITION 5.4. Consider the linear system in (8), with critical point 0.
a. 0 is **stable** if for each $\varepsilon > 0$ there is a $\delta > 0$ such that if X is any solution such that $\|X(0)\| < \delta$, then $\|X(t)\| < \varepsilon$ for all $t > 0$.
b. 0 is **asymptotically stable** if 0 is stable, and if there is a $\delta > 0$ such that for each solution X, if $\|X(0)\| < \delta$, then $\lim_{t \to \infty} \|X(t)\| = 0$.
c. 0 is **unstable** if it is not stable.

Intuitively, 0 is stable if any solution X that is close enough to 0 at time $t = 0$ remains close to 0 for all $t > 0$ (Figure 5.2(a)). In the same way, 0 is asymptotically stable if any solution X that is close enough to 0 at time $t = 0$ approaches 0 as t increases without bound (Figure 5.2(b)). By definition, if 0 is asymptotically stable, then it is stable. The converse, however, is not true, as we will note in Case 4 below.

We will classify the critical point 0 by the values of nonzero eigenvalues. If there are two eigenvalues, we denote them by λ and μ, and the corresponding eigenvectors by $\begin{pmatrix} re^{\lambda t} \\ se^{\lambda t} \end{pmatrix}$ and $\begin{pmatrix} ve^{\mu t} \\ we^{\mu t} \end{pmatrix}$, respectively. It follows from the theory of differential equations that the general solution of (8) is given by combinations of the two eigenvectors, that is,

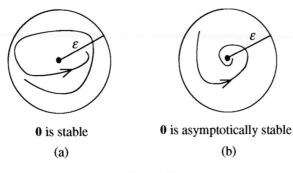

0 is stable   **0** is asymptotically stable

(a)   (b)

Figure 5.2

$$X(t) = p\begin{pmatrix} re^{\lambda t} \\ se^{\lambda t} \end{pmatrix} + q\begin{pmatrix} ve^{\mu t} \\ we^{\mu t} \end{pmatrix} = \begin{pmatrix} pre^{\lambda t} + qve^{\mu t} \\ pse^{\lambda t} + qwe^{\mu t} \end{pmatrix} \tag{13}$$

where p and q are arbitrary real numbers.

Case 1. λ and μ are real and distinct, with $\lambda\mu > 0$

The general solution is given by (13). Since $\lambda \neq \mu$, the two eigenvector solutions are not multiples of one another (Exercise 13).

If $0 > \lambda > \mu$, then $e^{\lambda t}$ and $e^{\mu t}$ converge to 0 as t increases without bound, so that $\lim_{t \to \infty} X(t) = \mathbf{0}$. Because all trajectories approach the origin as t increases, the critical point $\mathbf{0}$ is called an **asymptotically stable solution**, or an **asymptotically stable node**. An asymptotically stable node is analogous to an attracting fixed point of a map, since all solutions approach the node as t increases without bound. By hypothesis, $\lambda > \mu$, so that if t is large, then the terms containing $e^{\lambda t}$ dominate those containing $e^{\mu t}$. Thus, if X is any nonzero solution that is not a multiple of the eigenvector $\begin{pmatrix} ve^{\mu t} \\ we^{\mu t} \end{pmatrix}$, then for large t,

$$X(t) \approx p\begin{pmatrix} re^{\lambda t} \\ se^{\lambda t} \end{pmatrix}$$

for an appropriate nonzero constant p. But this vector has slope s/r. A diagram that shows the direction of solutions as t increases is often called the **portrait** of the solutions. If $\lambda > \mu > 0$, the portrait of the solutions is as in Figure 5.3(a). If $0 < \mu < \lambda$, then $\lim_{t \to \infty} \|X(t)\| = \infty$ for every solution. Therefore all trajectories point away from the origin (Figure 5.3(b)). In order to portray the behavior of the trajectories, we need only reverse the arrows. In this case, $\mathbf{0}$ is an **unstable node**.

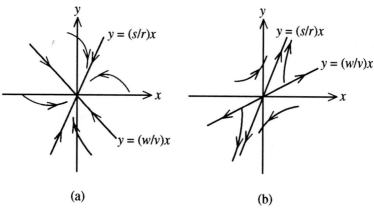

(a)   (b)

portrait when $0 > \lambda > \mu$ portrait when $0 < \mu < \lambda$

Figure 5.3

EXAMPLE 1. Consider the system

$$\frac{dx}{dt} = -4x + y$$

$$\frac{dy}{dt} = 3x - 2y$$

Show that **0** is asymptotically stable, and sketch a portrait of the solutions.

Solution. By (12), the characteristic equation is

$$\lambda^2 - (-4 - 2)\lambda + (-4)(-2) - (1)(3) = 0, \text{ that is, } \lambda^2 + 6\lambda + 5 = 0$$

Therefore the eigenvalues are -1 and -5, so that **0** is an asymptotically stable node. To find an eigenvector for the eigenvalue -1, we notice that

$$(-1)\begin{pmatrix} x \\ y \end{pmatrix} = \begin{pmatrix} -4 & 1 \\ 3 & -2 \end{pmatrix}\begin{pmatrix} x \\ y \end{pmatrix} = \begin{pmatrix} -4x + y \\ 3x - 2y \end{pmatrix}$$

Thus $-x = -4x + y$, so that $y = 3x$. Therefore $\begin{pmatrix} 1 \\ 3 \end{pmatrix}$ is an eigenvector for the eigenvalue -1. In a similar fashion, we find that $\begin{pmatrix} 1 \\ -1 \end{pmatrix}$ is an eigenvector for the eigenvalue − 5. Figure 5.3(a) gives an idea of the portrait of the solutions of the system, with the solutions approaching the origin asymptotically along the line $y = 3x$, with the exception of those solutions along the line $y = -x$. ❑

Case 2. $\mu < 0 < \lambda$

The fact that $\mu < 0$ implies that trajectories that begin near the line $y = wx/v$ tend to approach the origin before becoming slave to the term involving $e^{\lambda t}$ and moving away from the origin (Figure 5.4). Therefore if a solution is the form of (13) with $p \neq 0$, then for large values of t,

$$X(t) \approx p \begin{pmatrix} re^{\lambda t} \\ se^{\lambda t} \end{pmatrix}$$

The critical point **0** is called a **saddle point**, and is unstable. The line $y = wx/v$ is a stable manifold for the critical point, and the line $y = sx/r$ is an unstable manifold.

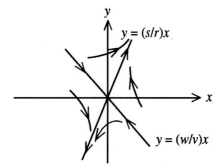

portrait when $\mu < 0 < \lambda$

Figure 5.4

EXAMPLE 2. Consider the system

$$\frac{dx}{dt} = -x + y$$

$$\frac{dy}{dt} = 3x + y$$

Show that **0** is a saddle point, and sketch a portrait of the solutions of the system.

Solution. By (12) the characteristic equation is

$$\lambda^2 - (-1 + 1)\lambda + (-1)(1) - (1)(3) = 0, \text{ that is, } \lambda^2 - 4 = 0$$

so that the eigenvalues are 2 and -2. By a straightforward calculation we find that an eigenvector for 2 is $\begin{pmatrix} 1 \\ 3 \end{pmatrix}$, and an eigenvector for -2 is $\begin{pmatrix} 1 \\ -1 \end{pmatrix}$. Figure 5.4 represents a portrait for the solutions of the system. $\square$

Case 3. $\lambda = \mu$

Here there is only one eigenvector, and as a result, the critical point is called a **degenerate node**. The discriminant of the characteristic equation $\lambda^2 - (a + d)\lambda + ad - bc = 0$ vanishes, which means that

$$(a + d)^2 - 4(ad - bc) = 0$$

and consequently $\lambda = (a + d)/2$. In the event that $b = 0$ and $c = 0$, the characteristic equation reduces to $\lambda^2 - (a + d)\lambda + ad = 0$. By assumption in Case 3, there is only one root for the characteristic equation, which implies that $a = d$. Consequently the system of differential equations reduces to

$$\frac{dx}{dt} = ax$$
$$\frac{dy}{dt} = ay$$

Evidently the two equations are independent of one another, and have solutions $x = pe^{at}$ and $y = qe^{at}$. It follows that the general solution to the system is given by

$$X(t) = \begin{pmatrix} pe^{at} \\ qe^{at} \end{pmatrix} = e^{at} \begin{pmatrix} p \\ q \end{pmatrix}$$

where p and q are arbitrary real constants. Thus all trajectories are linear, with slope q/p if $p \neq 0$. They are pointed toward the origin if $a < 0$ and away from the origin if $a > 0$ (Figures 5.5(a) and (b)). This quality of the solutions leads one to call **0** a **star solution**.

The other possibility is that $b \neq 0$ or $c \neq 0$. The general solution is more complicated to identify. One solution is given by

$$X_1(t) = \begin{pmatrix} re^{\lambda t} \\ se^{\lambda t} \end{pmatrix}$$

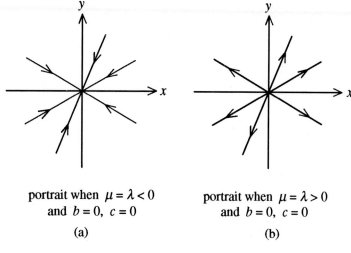

portrait when $\mu = \lambda < 0$
and $b = 0$, $c = 0$

(a)

portrait when $\mu = \lambda > 0$
and $b = 0$, $c = 0$

(b)

Figure 5.5

where $\begin{pmatrix} r \\ s \end{pmatrix}$ is an eigenvector corresponding to eigenvalue λ. To obtain a second solution the theory tells us that we must first find values for v and w such that

$$\begin{pmatrix} a - \lambda & b \\ c & d - \lambda \end{pmatrix} \begin{pmatrix} v \\ w \end{pmatrix} = \begin{pmatrix} r \\ s \end{pmatrix}$$

Then a second solution is given by

$$X_2(t) = \begin{pmatrix} (v + rt)e^{\lambda t} \\ (w + st)e^{\lambda t} \end{pmatrix}$$

Since X_1 and X_2 are not multiples of one another, the general solution is given by

$$X(t) = pX_1(t) + qX_2(t) \tag{14}$$

where p and q are arbitrary real constants.

One can show that if $\lambda < 0$, then a given nonzero trajectory approaches the origin along the line $y = wx/v$. Therefore $\mathbf{0}$ is an asymptotically stable degenerate node (Figure 5.6(a)). Analogously, if $\lambda > 0$, then the trajectory recedes from the origin along the same line, and $\mathbf{0}$ is an unstable degenerate node (Figure 5.6(b)).

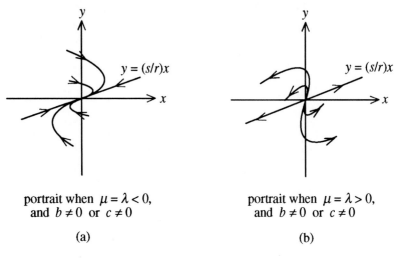

portrait when $\mu = \lambda < 0$,
and $b \neq 0$ or $c \neq 0$

(a)

portrait when $\mu = \lambda > 0$,
and $b \neq 0$ or $c \neq 0$

(b)

Figure 5.6

EXAMPLE 3. Consider the system

$$\frac{dx}{dt} = 3x - 4y$$

$$\frac{dy}{dt} = x - y$$

Find the general solution of the system, and sketch a portrait for the system.

Solution. Since

$$\det \begin{pmatrix} 3 - \lambda & -4 \\ 1 & -1 - \lambda \end{pmatrix} = \lambda^2 - 2\lambda + 1 = (\lambda - 1)^2$$

it follows that the eigenvalue is 1. To find an eigenvector, we calculate that

$$1 \begin{pmatrix} r \\ s \end{pmatrix} = \begin{pmatrix} 3 & -4 \\ 1 & -1 \end{pmatrix} \begin{pmatrix} r \\ s \end{pmatrix} = \begin{pmatrix} 3r - 4s \\ r - s \end{pmatrix}$$

which yields $r = 2s$. Thus $\begin{pmatrix} 2 \\ 1 \end{pmatrix}$ is an eigenvector for 1. Next we need to find v and w so that

$$\begin{pmatrix} 3-\lambda & -4 \\ 1 & -1-\lambda \end{pmatrix} \begin{pmatrix} v \\ w \end{pmatrix} = \begin{pmatrix} r \\ s \end{pmatrix}$$

Since $\lambda = 1$, $r = 2$, and $s = 1$, this reduces to

$$\begin{pmatrix} 2 & -4 \\ 1 & -2 \end{pmatrix} \begin{pmatrix} v \\ w \end{pmatrix} = \begin{pmatrix} 2 \\ 1 \end{pmatrix}$$

This means that $2v - 4w = 2$, so that $v = 1 + 2w$. As a result, we can choose $\begin{pmatrix} v \\ w \end{pmatrix} = \begin{pmatrix} 3 \\ 1 \end{pmatrix}$. By (14) the general solution of the given system is given by

$$X(t) = p \begin{pmatrix} 2e^t \\ e^t \end{pmatrix} + q \begin{pmatrix} (3+2t)e^t \\ (1+t)e^t \end{pmatrix}$$

for any real constants p and q.   Figure 5.6(b) is a portrait of the solutions.   ❏

We have completed the cases for which the eigenvalues are real and nonzero. The remaining general case involves complex eigenvalues.

Case 4. The eigenvalues are not real.

If the solutions of the characteristic equation $\lambda^2 - (a+d)\lambda + ad - bc = 0$ are not real, then they have the form $\alpha + i\beta$ and $\alpha - i\beta$, where α and β are real. In this case the general solution takes a little more work to produce. The theory tells us to find an eigenvector for $\alpha + i\beta$ of the form $\begin{pmatrix} r+iv \\ s+iw \end{pmatrix} = \begin{pmatrix} r \\ s \end{pmatrix} + i \begin{pmatrix} v \\ w \end{pmatrix}$, where r, s, v, and w are real numbers. If we let

$$X_1(t) = \left[\begin{pmatrix} r \\ s \end{pmatrix} \cos \beta t - \begin{pmatrix} v \\ w \end{pmatrix} \sin \beta t \right] e^{\alpha t} \text{ and } X_2(t) = \left[\begin{pmatrix} r \\ s \end{pmatrix} \sin \beta t + \begin{pmatrix} v \\ w \end{pmatrix} \cos \beta t \right] e^{\alpha t}$$

then the general solution is given by

$$X(t) = pX_1(t) + qX_2(t) \tag{15}$$

where p and q are arbitrary real constants. All trajectories spiral toward the origin

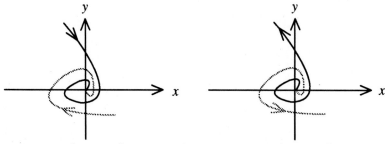

portrait when $\lambda = \alpha + i\beta$, and $\alpha < 0$ portrait when $\lambda = \alpha + i\beta$, and $\alpha > 0$

(a)   (b)

Figure 5.7

if $\alpha < 0$, and spiral away from the origin if $\alpha > 0$ (Figures 5.7(a)–(b)). For this reason the critical point **0** is called a **spiral point,** or a **focus,** when $\alpha \neq 0$.

EXAMPLE 4. Find the general solution of the system

$$\frac{dx}{dt} = x + 5y$$

$$\frac{dy}{dt} = -x - 3y$$

Solution. First we find that

$$\det \begin{pmatrix} 1-\lambda & 5 \\ -1 & -3-\lambda \end{pmatrix} = \lambda^2 + 2\lambda + 2 = 0 \text{ if and only if } \lambda = -1 \pm i$$

Therefore the eigenvalues are $-1 + i$ and $-1 - i$. To find an eigenvector, we notice that if

$$(-1 + i)\begin{pmatrix} y \\ z \end{pmatrix} = \begin{pmatrix} 1 & 5 \\ -1 & -3 \end{pmatrix}\begin{pmatrix} y \\ z \end{pmatrix} = \begin{pmatrix} y + 5z \\ -y - 3z \end{pmatrix}$$

then $-y + iy = y + 5z$, so $z = (-2 + i)y/5$. If $y = 5$, then $z = -2 + i$. Thus

$$\begin{pmatrix} 5 \\ -2 + i \end{pmatrix} = \begin{pmatrix} 5 \\ -2 \end{pmatrix} + i\begin{pmatrix} 0 \\ 1 \end{pmatrix}$$

is an eigenvector for the eigenvalue $-1 + i$. By (15), the general solution is

$$X(t) = p\left(\begin{pmatrix} 5 \\ -2 \end{pmatrix}\cos t - \begin{pmatrix} 0 \\ 1 \end{pmatrix}\sin t\right)e^{-t} + q\left(\begin{pmatrix} 5 \\ -2 \end{pmatrix}\sin t + \begin{pmatrix} 0 \\ 1 \end{pmatrix}\cos t\right)e^{-t}$$

where as usual, p and q are real constants.   ❑

If the eigenvalues of the associated matrix are $\lambda = i\beta$ and $\lambda = -i\beta$ (that is, if $\alpha = 0$), then the solution in (15) is periodic, with period $2\pi/\beta$. Moreover, the trajectories can be shown to be elliptical. In view of these observations, one calls the solution **0** a **center** (Figure 5.8). Notice that a center critical point is an example of a stable but not asymptotically stable critical point.

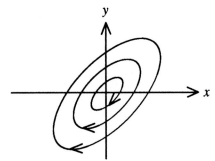

portrait when $\lambda = \alpha + i\beta$, and $\alpha = 0$

Figure 5.8

We have now completed our analysis of systems of two linear differential equations with constant coefficients. In the following table we recapitulate our results concerning critical points.

EIGENVALUES	TYPE OF CRITICAL POINT
$0 < \mu < \lambda$	unstable node (trajectories recede from **0**)
$\mu < \lambda < 0$	asymptotically stable node (trajectories converge to **0**)
$\mu < 0 < \lambda$	saddle point (some trajectories approach **0** before receding)
$0 < \lambda = \mu$	unstable degenerate node (trajectories recede from **0**)
$\lambda = \mu < 0$	asymptotically stable degenerate node (trajectories converge to **0**)
$\lambda = \alpha + i\beta, \ \alpha > 0$	unstable spiral point (trajectories spiral away from **0**)
$\lambda = \alpha + i\beta, \ \alpha < 0$	asymptotically stable spiral point (trajectories spiral toward **0**)
$\lambda = \alpha + i\beta, \ \alpha = 0$	stable center (trajectories are elliptical, centered at **0**)

In the next section we will study systems of differential equations that are not linear, but which are nearly linear.

EXERCISES 5.1

In Exercises 1–12, find the eigenvalues of the matrix associated with the given system and then classify the critical point **0**. Then sketch a portrait of the solutions of the system.

1. $\dfrac{dx}{dt} = -2x + 9y$

   $\dfrac{dy}{dt} = x - 2y$

2. $\dfrac{dx}{dt} = 2x - y$

   $\dfrac{dy}{dt} = 2x + 3y$

3. $\dfrac{dx}{dt} = -2x - y$

   $\dfrac{dy}{dt} = -2y$

4. $\dfrac{dx}{dt} = x + 5y$

   $\dfrac{dy}{dt} = -x - y$

5. $\dfrac{dx}{dt} = -3x + 2y$

   $\dfrac{dy}{dt} = x - 4y$

6. $\dfrac{dx}{dt} = x + y$

   $\dfrac{dy}{dt} = -9x + 3y$

7. $\dfrac{dx}{dt} = -2x - y$

   $\dfrac{dy}{dt} = 4x - 2y$

8. $\dfrac{dx}{dt} = -3x$

   $\dfrac{dy}{dt} = -3y$

9. $\dfrac{dx}{dt} = -2x - 3y$

   $\dfrac{dy}{dt} = 3x + 4y$

10. $\dfrac{dx}{dt} = x + y$

    $\dfrac{dy}{dt} = x + 2y$

11. $\dfrac{dx}{dt} = -x + 2y$

    $\dfrac{dy}{dt} = x + 3y$

12. $\dfrac{dx}{dt} = -2x - y$

    $\dfrac{dy}{dt} = x - 4y$

13. Let $X_1(t) = \begin{pmatrix} re^{\lambda t} \\ se^{\lambda t} \end{pmatrix}$ and $X_2(t) = \begin{pmatrix} ve^{\mu t} \\ we^{\mu t} \end{pmatrix}$. Show that X_2 is a constant multiple of X_1 (as a function) only if $\lambda = \mu$.

14. Consider the system

$$\frac{dx}{dt} = ax + 2y$$

$$\frac{dy}{dt} = x + 3y$$

   a. Determine a value of a so that **0** is a center.
   b. Is the value of a that you found in part (a) unique? Explain your answer.
   c. Is there a value of a for which **0** is a spiral? Explain your answer.

15.   Consider the so-called LRC circuit shown in Figure 5.9. By Kirchhoff's Law, the current I in the circuit and the charge Q on the capacitor are related by the equation

$$L\frac{dI}{dt} + RI + \frac{Q}{C} = 0 \tag{16}$$

where $dQ/dt = I$, and where C, L, and R are positive constants representing the capacitance, inductance, and resistance, respectively.

a.   Rewrite (16) as a system involving dI/dt and dQ/dt.

b.   Suppose that R is practically 0. Then the solutions are essentially periodic. Determine the (approximate) period in terms of C and L.

c.   Classify the critical point $\mathbf{0}$ if it is not a center.

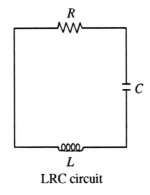

R

C

L

LRC circuit

Figure 5.9

5.2 ALMOST LINEARITY

In Section 5.1 we analyzed critical points for a general class of systems of linear differential equations. Now we relax the linearity condition and study a broader collection of systems called "almost linear" systems. As we will see in Section 5.3, the motion of such objects as a swinging pendulum can be described by an almost linear system of equations.

DEFINITION 5.5. Let V be a subset of R^2 that contains the origin in its interior. Assume that F and G are real-valued functions on V that vanish at the origin and whose partial derivatives are continuous and also vanish at the origin. Then the system

$$\frac{dx}{dt} = ax + by + F(x, y)$$

$$\frac{dy}{dt} = cx + dy + G(x, y)$$

(1)

is **almost linear** at the origin.

The systems we will discuss in the present section have the form of (1), in which F and G are continuous functions of x and y, but are independent of t. Such systems are **autonomous**.

Suppose that the system in (1) is almost linear at the origin, and let

$$H\left(\begin{matrix} x \\ y \end{matrix}\right) = \left(\begin{matrix} ax + by + F(x, y) \\ cx + dy + G(x, y) \end{matrix}\right)$$

The fact that the partials of F and G vanish at the origin implies that

$$DH(0) = \left(\begin{matrix} a & b \\ c & d \end{matrix}\right)$$

This is the **associated matrix** of the system in (1). We observe that the associated matrix is the same as the associated matrix for the corresponding linear system

$$\frac{dx}{dt} = ax + by$$

$$\frac{dy}{dt} = cx + dy$$

(2)

Acknowledging the relationship between the systems in (1) and (2), we call the linear system in (2) the **auxiliary system** for the system in (1).

The notions of asymptotical stability, stability and nonstability that we defined for linear systems in Section 5.1 carry forward without alteration for almost linear systems. Thus **0** is **stable** if for every $\varepsilon \to 0$ there is a $\delta > 0$ such that if X is any solution such that $\|X(0)\| < \delta$, then $\|X(t)\| < \varepsilon$ for all $t > 0$. Likewise, **0** is **asymptotically stable** if it is stable and any solution X such that $X(0)$ is suitably close to **0** has the property that $\lim_{t \to \infty} X(t) = 0$. Finally, **0** is **unstable** if it is not stable.

The relevance of these definitions is revealed in the following theorem, due to Lyapunov and proved in Boyce and Di Prima (1986).

THEOREM 5.6. Suppose that the system in (1) is almost linear at the origin. If

0 is asymptotically stable (respectively, unstable) for the auxiliary system, then **0** is asymptotically stable (respectively, unstable) for the almost linear system.

It can be shown in addition that if the critical point of the auxiliary system is a node, saddle point or spiral point, then the critical point of the almost linear system is of the same type. However, degenerate nodes and centers are not inherited by almost linear systems. More particularly, if a critical point of an auxiliary system is a center, then the corresponding critical point of the almost linear system may be asymptotically stable or unstable (Exercises 7–8). The reason that this can happen is that the eigenvalues of degenerate nodes and centers are very special numbers. Slight changes in the eigenvalues can alter the nature of the critical point.

EXAMPLE 1. Consider the system

$$\frac{dx}{dt} = -2x + y + 2xy$$

$$\frac{dy}{dt} = x + y$$

Show that the system is almost linear at the origin, and that the critical point **0** is a saddle point.

Solution. On the one hand, the second equation of the system is linear. On the other hand, the first equation has the form

$$\frac{dx}{dt} = -2x + y + F(x, y), \quad \text{where } F(x, y) = 2xy$$

Since F has continuous partial derivatives and $F_x(0, 0) = 0 = F_y(0, 0)$, the system is almost linear at the origin. The associated linear system is

$$\frac{dx}{dt} = -2x + y$$

$$\frac{dy}{dt} = x + y$$

and the associated matrix is

$$\begin{pmatrix} -2 & 1 \\ 1 & 1 \end{pmatrix}$$

Since the eigenvalues are $(-1 \pm \sqrt{13})/2$, it follows that **0** is an unstable saddle point for the linear system. We conclude from Theorem 5.6 and the comments following it that $\mathbf{0}$ is a saddle point for the given almost linear system. ❏

By Theorem 5.3, linear systems of differential equations can have critical points different from $\mathbf{0}$ only if 0 is an eigenvalue of the associated matrix. However, this is not true for other systems. For example, consider the system

$$\frac{dx}{dt} = 2x - y + 2xy$$

$$\frac{dy}{dt} = x + y + 1$$

(3)

Since $dx/dt = 0 = dy/dt$ if $x = 1$ and $y = -2$, it follows that $\begin{pmatrix} 1 \\ -2 \end{pmatrix}$ is a critical point of the system. You can check that the associated matrix has complex eigenvalues.

The simplest way to analyze the nature of critical points different from $\mathbf{0}$ is to perform a change of variables so that the translated critical point is the origin. We do this in the following way.

Suppose that $\mathbf{v}_0 = \begin{pmatrix} x_0 \\ y_0 \end{pmatrix}$ is a critical point of the system

$$\frac{dx}{dt} = f(x, y)$$

$$\frac{dy}{dt} = g(x, y)$$

We say that the system is **almost linear** at $\mathbf{v}_0$ if by making the change of variables $x = x_0 + u$, $y = y_0 + v$, the resulting system

$$\frac{du}{dt} = f(u, v)$$

$$\frac{dv}{dt} = g(u, v)$$

is almost linear at the origin.

EXAMPLE 2. Show that the system in (3) is almost linear at $\mathbf{v}_0 = \begin{pmatrix} 1 \\ -2 \end{pmatrix}$.

Solution. Since $x_0 = 1$ and $y_0 = -2$, we make the change of variables

$$x = 1 + u \quad \text{and} \quad y = -2 + v$$

Then the system in (3) becomes

$$\frac{du}{dt} = 2(1 + u) - (-2 + v) + 2(1 + u)(-2 + v) = -2u + v + 2uv$$

$$\frac{dv}{dt} = (1 + u) + (-2 + v) + 1 = u + v$$

Since this system is almost linear at **0** by Example 1, the system in (3) is almost linear at $\mathbf{v}_0$. ❏

Limit Cycles

Until this point, the solutions of linear and almost linear systems we have encountered have been fixed points or ellipses, or have orbits that converge to fixed points or are unbounded as t increases without bound. However, for systems of two differential equations there is yet another possibility. Consider the system

$$\frac{dx}{dt} = x - y - x^3 - xy^2$$

(4)

$$\frac{dy}{dt} = x + y - x^2y - y^3$$

You can check that the system is almost linear at the critical point **0**, which is unstable. What is the behavior of other solutions as t increases without bound? We will answer this question after converting the system in (4) into polar coordinates r and θ. In making such a conversion, we will need to find formulas for dr/dt and $d\theta/dt$ in terms of x and y and their derivatives.

THEOREM 5.7. Let x and y be differentiable functions, and let

$$x = r \cos \theta \quad \text{and} \quad y = r \sin \theta$$

Then

$$\frac{dr}{dt} = \frac{x}{r} \frac{dx}{dt} + \frac{y}{r} \frac{dy}{dt}$$

(5)

and

$$\frac{d\theta}{dt} = \frac{x}{r^2}\frac{dy}{dt} - \frac{y}{r^2}\frac{dx}{dt} \tag{6}$$

Proof. With the help of the Chain Rule, we find that

$$\frac{dx}{dt} = \frac{dx}{dr}\frac{dr}{dt} + \frac{dx}{d\theta}\frac{d\theta}{dt} = (\cos\theta)\frac{dr}{dt} - (r\sin\theta)\frac{d\theta}{dt} = \frac{x}{r}\frac{dr}{dt} - y\frac{d\theta}{dt}$$

$$\frac{dy}{dt} = \frac{dy}{dr}\frac{dr}{dt} + \frac{dy}{d\theta}\frac{d\theta}{dt} = (\sin\theta)\frac{dr}{dt} + (r\cos\theta)\frac{d\theta}{dt} = \frac{y}{r}\frac{dr}{dt} + x\frac{d\theta}{dt} \tag{7}$$

To solve for dr/dt we multiply the first equation in (7) by x and the second by y, and add. This yields

$$x\frac{dx}{dt} + y\frac{dy}{dt} = \frac{x^2}{r}\frac{dr}{dt} + \frac{y^2}{r}\frac{dr}{dt} = r\frac{dr}{dt}$$

from whence (5) arises. To obtain (6) we multiply the first equation in (7) by y and the second by x and subtract the first from the second. We obtain

$$x\frac{dy}{dt} - y\frac{dx}{dt} = x^2\frac{d\theta}{dt} + y^2\frac{d\theta}{dt}$$

Dividing both sides by r^2 yields (6). Thus the proof is complete. ∎

Now we are ready to convert the system in (4) into polar coordinates.

THEOREM 5.8. In polar coordinates the system in (4) is

$$\frac{dr}{dt} = r(1 - r^2)$$

$$\frac{d\theta}{dt} = 1 \tag{8}$$

Proof. To find dr/dt, we use (5), substituting from the equations in (4) for dx/dt and dy/dt, and recalling that $x^2 + y^2 = r^2$. We obtain

$$\frac{dr}{dt} = \frac{x}{r}(x - y - x^3 - xy^2) + \frac{y}{r}(x + y - x^2y - y^3)$$

$$= \frac{1}{r}[(x^2 + y^2) - (x^4 + 2x^2y^2 + y^4)] = \frac{1}{r}[r^2 - (x^2 + y^2)^2]$$

$$= \frac{1}{r} (r^2 - r^4) = r(1 - r^2)$$

Similarly, to find $d\theta/dt$ we use (6) and substitute from (4):

$$\frac{d\theta}{dt} = \frac{x}{r^2} (x + y - x^2 y - y^3) - \frac{y}{r^2} (x - y - x^3 - xy^2)$$

$$= \frac{1}{r^2} (x^2 + y^2) = 1 \quad \blacksquare$$

What can we say about the behavior of solutions of the system in (8)? The answer to this question relies on two results about orbits of almost linear systems, which you can find in the text by Boyce and DiPrima (1986).

i.   At most one orbit passes through any given point in the plane.
ii.   The orbit of a solution that is not a critical point either approaches a critical point or a closed orbit as t increases without bound, or it recedes arbitrarily far from the origin as t increases.

The first result effectively is the Uniqueness Theorem of differential equations. The second result will give us significant information about the system in (8).

Now we are prepared to survey the solutions of (8) whose orbits are neither the origin nor the unit circle.

Case 1. $\|X(0)\| = 1$

By hypothesis, $r = 1$, so that from (8), $dr/dt = r(1 - r^2) = 0$. Then for all $t > 0$ we have $r = 1$, so that the orbit of X lies on the unit circle centered at the origin. Since $d\theta/dt = 1$, the orbit encircles the unit circle with constant velocity, and travels in a counterclockwise fashion.

Case 2. $0 < \|X(0)\| < 1$

Since $0 < \|X(0)\| < 1$, the distance between $X(0)$ and the origin, which is r, satisfies $0 < r < 1$ for $t = 0$. Therefore $dr/dt = r(1 - r^2) > 0$ for all $t > 0$. Thus $\|X\|$ is an increasing function of t. By (ii), we deduce that $\lim_{t \to \infty} \|X(t)\| = 1$.

Case 3. $\|X(0)\| > 1$

In this case, $dr/dt = r(1 - r^2) < 0$, so that $\|X\|$ is decreasing, and consequently by (ii) we find that $\lim_{t \to \infty} \|X(t)\| = 1$.

Cases 2–3 tell us that the unit circle is a limit cycle, in the sense that if $0 <$ $\|X(0)\|$, then $X(t)$ approaches the unit circle as t increases without bound. More generally, a **limit cycle** in R^2 is a closed curve C that is periodic and attracts the orbit of any solution X such that $X(0)$ is near C. In Figure 5.10 the dashed ellipse-like curve portrays a limit cycle.

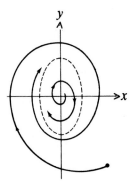

Figure 5.10

An **attractor** of a system of differential equations is a closed bounded set toward which the orbit of each solution approaches as t increases without bound. The results of Section 5.1 show that the origin is the attractor for any linear system in which **0** is an asymptotically stable critical point. The other linear systems described in Section 5.1 have *no* attractor in the plane, since their orbits are either unbounded or concentric ellipses.

If an almost linear system has an attractor, it can be an asymptotically stable critical point, a limit cycle, or the union of critical points and limit cycles. The discussion of the almost linear system in (4) and its polar representation in (8) indicates that it has an attractor consisting of the unit circle.

Is it possible for an almost linear system to have a another type of attractor? The celebrated Poincaré-Bendixson Theorem provides the answer, for any autonomous system whose solutions lie in the xy plane.

THEOREM 5.9 (Poincaré-Bendixson Theorem).   Consider the system

$$\frac{dx}{dt} = f(x, y)$$

$$\frac{dy}{dt} = g(x, y)$$

where f and g are continuous. If a trajectory remains bounded as t increases without bound, then the trajectory is a critical point or a limit cycle, or approaches a critical point or limit cycle.

The Poincaré-Bendixson Theorem implies that if an autonomous system in x and y has an attractor A, then A consists of a union of critical points and limit cycles. Consequently A cannot be a strange attractor. We will see in Section 5.4 that this result is no longer true if the solutions lie in three dimensions.

EXERCISES 5.2

In Exercises 1–2, determine whether the system is almost linear at $\mathbf{0}$. If it is, then determine its type (node, saddle point, or spiral) and stability, where possible.

1. $\dfrac{dx}{dt} = -x + \ln(1 + y^2)$

   $\dfrac{dy}{dt} = 2x - 3y - x^2 y$

2. $\dfrac{dx}{dt} = -x + 2y + 1 - e^{xy}$

   $\dfrac{dy}{dt} = x - y + x \sin y$

In Exercises 3–6, determine whether the system is almost linear at each of its critical points. If it is, then determine its type and stability, where possible.

3. $\dfrac{dx}{dt} = y$

   $\dfrac{dy}{dt} = x - x^3$

4. $\dfrac{dx}{dt} = y$

   $\dfrac{dy}{dt} = -x + x^3$

5. $\dfrac{dx}{dt} = 2x - x^2 - xy$

   $\dfrac{dy}{dt} = -y + xy$

6. $\dfrac{dx}{dt} = y - x^3$

   $\dfrac{dy}{dt} = 1 - xy$

7. Consider the system

$$\frac{dx}{dt} = y - x(x^2 + y^2)$$

$$\frac{dy}{dt} = -x - y(x^2 + y^2)$$

   a. Show that the system is almost linear at the critical point $\mathbf{0}$.
   b. Show that $\mathbf{0}$ is a center point of the auxiliary system.
   c. Show that $\mathbf{0}$ is asymptotically stable. (*Hint*: Use polar coordinates and show that if $r(0) < 1$, then $\lim_{t \to \infty} r(t) = 0$.)

8.   Consider the system

$$\frac{dx}{dt} = y + x(x^2 + y^2)$$

$$\frac{dy}{dt} = -x + y(x^2 + y^2)$$

   a.   Show that the system is almost linear at the critical point **0**.
   b.   Show that **0** is a center point of the auxiliary system.
   c.   Show that **0** is unstable.
   d.   Show that the given almost linear system has no attractor.

9.   A nonlinear electric oscillator can be modeled by the van der Pol equation

$$\frac{d^2x}{dt^2} + \varepsilon(x^2 - 1)\frac{dx}{dt} + x = 0$$

   where x is related to the voltage in the circuit, and ε is a positive parameter related to the given circuit.
   a.   Prove that the associated system of differential equations is almost linear at **0**. (*Hint*: To render the equation as a system, let one equation be $dx/dt = y$; the other equation becomes

$$\frac{dy}{dt} + \varepsilon(x^2 - 1) + x = 0$$

   b.   Show that **0** is unstable, whatever the value of ε is, and determine the values of ε for which **0** is a spiral point.

For the second-order differential equations appearing in Exercises 10–12, determine whether the associated system of differential equations is almost linear at the critical points. For any that are, classify the critical points where possible. Assume that all constants are positive.

10. $\dfrac{d^2x}{dt^2} + a\dfrac{dx}{dt} - \dfrac{1}{2}x(1 - x^2) = 0$, which arises in the study of double-well potentials

11. $\dfrac{d^2x}{dt^2} + a\dfrac{dx}{dt} - b\sin x = 0$, which arises in the study of electric force fields

12. $\dfrac{d^2x}{dt^2} + a\dfrac{dx}{dt} - x^3 = 0$, which arises in nonlinear electric circuits

5.3   THE PENDULUM

A grandfather clock runs by the motion of a pendulum. In this section we will use the theory of almost linear systems to study the dynamics of a pendulum, not only when there is no external force applied on it, but also when there is a sinusoidal external force.

Suppose that a pendulum of length L has mass m, and its bob swings back and forth. Let the angle of the pendulum with the vertical at time t be denoted by $\theta(t)$ (Figure 5.11). We will assume that g represents acceleration due to gravity, and for the present we will assume that there is no damping and no external force on the pendulum. This would be the case if the pendulum is located in a vacuum, under no influence save acceleration due to gravity. Under these conditions, the motion of the pendulum is governed by the second-order differential equation

$$mL^2 \frac{d^2\theta}{dt^2} + mgL \sin \theta = 0 \tag{1}$$

By letting $x = \theta$ and $y = d\theta/dt$, we transform (1) into the system

$$\frac{dx}{dt} = y$$

$$\frac{dy}{dt} = -\frac{g}{L} \sin x \tag{2}$$

We will call the system in (2) the **pendulum system.**   Notice that the pendulum system is not linear because $\sin x$ does not have the form cx. Finding exact solu-

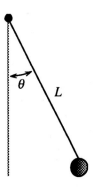

pendulum

Figure 5.11

tions for the pendulum system in (2) is no easier than finding exact solutions for the differential equation in (1). Indeed, one cannot solve either exactly!

Although the pendulum system is not linear, it is almost linear, as the following theorem shows.

THEOREM 5.10. The pendulum system in (2) is almost linear at the origin.

Proof. We observe that the first equation in (2) is linear. Next, we rewrite the second equation as

$$\frac{dy}{dt} = -\frac{g}{L} x + G(x, y)$$

where

$$G(x, y) = \frac{g}{L} x - \frac{g}{L} \sin x$$

It is apparent that G has continuous partial derivatives, and that $G(0, 0) = 0$. Since

$$G_x(x, y) = \frac{g}{L} - \frac{g}{L} \cos x \quad \text{and} \quad G_y(x, y) = 0$$

it follows that $G_x(0, 0) = 0 = G_y(0, 0)$. Consequently the pendulum system is almost linear at the origin.   ■

Because the pendulum system is almost linear at the origin, Theorem 5.6 tells us that we can use the auxiliary system

$$\frac{dx}{dt} = y$$

$$\frac{dy}{dt} = -\frac{g}{L} x \tag{3}$$

to analyze the nature of the critical point **0** of (2). First we observe that the associated matrix of (3) is

$$\begin{pmatrix} 0 & 1 \\ -g/L & 0 \end{pmatrix}$$

The eigenvalues of this matrix are $\lambda = \pm \sqrt{g/L}\, i$. It follows that the eigenvalues are pure imaginary, and thus **0** is a center for the auxiliary system in (3). We conclude that **0** is a stable critical point for the auxiliary system.

From the results in Case 4 of Section 5.1, we know that the orbits of solutions for the auxiliary system are ellipses, as shown in Figure 5.12(a). This figure, in which the vertical axis measures dx/dt, is called a **phase plane portrait**. (The expression "phase plane" comes from physics.) It is a reasonable representation of the motion of the undamped pendulum when the angular displacement and the angular velocity are small.

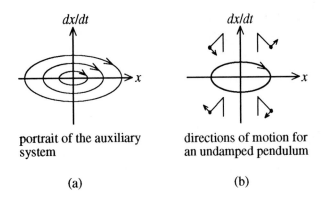

portrait of the auxiliary   directions of motion for
system   an undamped pendulum

(a)   (b)

Figure 5.12

The angular displacement x from the downward vertical (denoted by θ in (1)) is assumed to be positive when the pendulum lies to the right of the vertical and negative when to the left. It follows that the angular velocity dx/dt is positive when the pendulum bob moves to the right, and is negative when it moves to the left. Thus there are four possibilities, depending on the positivity and negativity of x and of dx/dt. They are detailed in Figure 5.12(b). One can show that with our conventions, the elliptical orbits appearing in Figure 5.12(a) are traversed in the clockwise direction (Exercise 1). The critical point **0** corresponds to a pendulum with the bob hanging motionless in the vertical position.

Next we will assume that the motion of the pendulum is damped. Damping can be caused by friction due to air or some other medium. Let us assume that the damping is constant through time and is represented by c. The resulting second-order differential equation for the motion of the pendulum is given by

$$ mL^2 \, \frac{d^2\theta}{dt^2} + cL \, \frac{d\theta}{dt} + mgL \sin \theta = 0 $$

Again letting $x = \theta$ and $y = d\theta/dt$, we transform the differential equation into the **pendulum system**

$$\frac{dx}{dt} = y$$

$$\frac{dy}{dt} = -\frac{g}{L}\sin x - \frac{c}{mL}y$$

(4)

The pendulum system is not linear, but is almost linear at the origin (Exercise 2). The auxiliary system is

$$\frac{dx}{dt} = y$$

$$\frac{dy}{dt} = -\frac{g}{L}x - \frac{c}{mL}y$$

(5)

with associated matrix

$$\begin{pmatrix} 0 & 1 \\ -g/L & -c/(mL) \end{pmatrix}$$

The eigenvalues are

$$\frac{-c/mL \pm \sqrt{c^2/(mL)^2 - 4g/L}}{2} = \frac{-c \pm \sqrt{c^2 - 4gm^2L}}{2mL}$$

Let the eigenvalues be denoted by λ and μ, where

$$\lambda = \frac{-c - \sqrt{c^2 - 4gm^2L}}{2mL} \quad \text{and} \quad \mu = \frac{-c + \sqrt{c^2 - 4gm^2L}}{2mL}$$

The nature of the critical point $\mathbf{0}$ depends on the values of λ and μ, which in turn depend on c, m, and L. For the analysis of the critical point when the damping parameter is allowed to vary, we will assume that m and L are fixed but positive numbers.

We already discussed the case $c = 0$, in which there is no damping, so we will assume that $c > 0$. Since the real parts of both λ and μ are negative numbers (Exercise 3), all solutions of the auxiliary system in (5) are asymptotically stable. By Theorem 5.6 the same is true for solutions of the pendulum system. In other words, all solutions tend to $\mathbf{0}$ with time (Figure 5.13). This is realistic, because if there is resistance due to air, then the pendulum winds down as time progresses. Anyone who regularly winds up a grandfather or cuckoo clock is aware of this phenomenon.

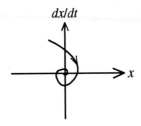

portrait of a damped pendulum

Figure 5.13

We remark that the tendency of the orbit of a solution to spiral while it approaches the zero solution diminishes as c increases from 0 toward $c^2/(mL)^2 - 4g/L$. As c increases further, the trajectory of a solution does not spiral noticeably as it converges to the zero solution.

The pendulum system in (4) has critical points different from the zero solution. In fact, let n be any integer, and $\mathbf{v}_n = \begin{pmatrix} n\pi \\ 0 \end{pmatrix}$. Then

$$\frac{dx}{dt} = 0 \quad \text{and} \quad \frac{dy}{dt} = -\frac{g}{L} \sin n\pi - \frac{c}{mL}(0) = 0$$

This shows that $\mathbf{v}_n$ is a critical point. In Theorem 5.11 we confirm the almost linearity of the system at $\mathbf{v}_n$.

THEOREM 5.11. The pendulum system

$$\frac{dx}{dt} = y$$

$$\frac{dy}{dt} = -\frac{g}{L} \sin x - \frac{c}{mL} y$$

(6)

is almost linear at $\mathbf{v}_n$, for each integer n.

Proof. We make the change of variables prescribed in Section 5.2:

$$x = n\pi + u \quad \text{and} \quad y = v$$

This transforms the system in (6) into

$$\frac{du}{dt} = v$$

$$\frac{dv}{dt} = -\frac{g}{L}\sin(n\pi + u) - \frac{c}{mL}v \tag{7}$$

The first equation in (7) is linear. The second equation becomes

$$\frac{dv}{dt} = -\frac{g}{L}\sin u - \frac{c}{mL}v \quad \text{if } n \text{ is even}$$

and

$$\frac{dv}{dt} = \frac{g}{L}\sin u - \frac{c}{mL}v \quad \text{if } n \text{ is odd}$$

Therefore if n is even, then the system in (7) is almost linear at the origin because it is (4) with x and y replaced by u and v. If n is odd, then again the system in (7) is almost linear, by an analogous proof (Exercise 4). Consequently the pendulum system is almost linear at $\mathbf{v}_n$ for each integer n. ∎

From Theorem 5.11 and the analysis of the critical point $\mathbf{0}$, we know that the critical point $\mathbf{v}_n$ for the pendulum system in (6) is asymptotically stable whenever n is an even integer. This is reasonable from a physical standpoint, since if n is even and if $x(0) = n\pi$ and $y(0) = 0$, then the pendulum is in the vertical position, pointed downward with the bob at rest.

The nature of the critical point $\mathbf{v}_n$ is far different if n is an odd integer. To be specific, let $n = 1$, so that the critical point is $\mathbf{v}_1 = \begin{pmatrix} \pi \\ 0 \end{pmatrix}$. Using the solution of Theorem 5.11, we find that the auxiliary system is

$$\frac{du}{dt} = v$$

$$\frac{dv}{dt} = \frac{g}{L}u - \frac{c}{mL}v \tag{8}$$

The associated matrix is

$$\begin{pmatrix} 0 & 1 \\ g/L & -c/(mL) \end{pmatrix}$$

Consequently the eigenvalues, which we denote by λ and μ, are given by

$$\lambda = \frac{-c - \sqrt{c^2 + 4gm^2L}}{2mL} \quad \text{and} \quad \mu = \frac{-c + \sqrt{c^2 + 4gm^2L}}{2mL}$$

Because the expression inside the square root is positive, λ and μ are real numbers. A moment's reflection shows that $\lambda < 0 < \mu$. Consequently the critical point **0** of (8) is an unstable saddle point, so that v_1 is also an unstable saddle point. The critical point v_1, in which the angle $x = \pi$, corresponds to the pendulum standing upright, with the bob above the anchor. Obviously this is an unstable position.

The results so far are predicated on the absence of an external applied force. Now suppose that there is an applied force F, such as a periodic push or impulse, thrust on the pendulum. Assume further that F is a function of t. Then the second-order equation becomes

$$mL^2 \frac{d^2\theta}{dt^2} + cL \frac{d\theta}{dt} + mgL \sin\theta = F(t) \tag{9}$$

The auxiliary system is

$$\frac{dx}{dt} = y$$

$$\frac{dy}{dt} = -\frac{g}{L} \sin x - \frac{c}{mL} y + \frac{1}{mL^2} F(t)$$

Assume that c, m and L are held constant. If F is small, then the critical points may well have the same form as when there is no external force. However, if the force F is substantial, then solutions may exhibit very complicated behavior.

For example, consider the pendulum equation

$$\frac{d^2\theta}{dt^2} + \alpha \frac{d\theta}{dt} + \sin\theta = \mu \cos t \tag{10}$$

which can be obtained from (9) by taking suitable values for c, m, and L, and by letting F denote a sinusoidal force. If $\alpha = 1/5$ and $\mu = 2$, then (10) becomes

$$\frac{d^2\theta}{dt^2} + \frac{1}{5} \frac{d\theta}{dt} + \sin\theta = 2 \cos t$$

It turns out that the corresponding system has two critical points. In Color Plate 11 all initial points whose orbits approach one of the critical points as time passes are colored red; those approaching the other critical point are colored blue. The faster the convergence is, the deeper the color is. If the amplitude of the external force is increased a little, so that μ increases from 2 to 2.06, then as you can see in Color Plate 12, red dots infiltrate into the region that was deep blue in Color Plate 11. Bifurcation has occurred! We mention that in Color Plate 12 the points colored red (and hence those colored blue) form a set with fractal dimension.

The damping is diminished but the external force is augmented if we let $\alpha =$ 1/10 and $\mu = 7/4$ in (10). In this case there are four critical points for the corresponding system. In Color Plate 3 the basin of attraction of each of the four critical points is assigned one of the colors red, yellow, green or blue. Color Plates 4–10 display increasing magnifications of the central region in Color Plate 3, each succeeding plate focusing on the central region of its predecessor. In the final Color Plate 10, the magnification is approximately 100,000. The plates show the self-similarity and fractal nature of the basins of attraction. These eight color plates are taken from a videotape "Chaos and Fractals in Simple Physical Systems,"produced by the University of Maryland Chaos Group, (1991). For more details about basins of attraction for the pendulum system, see the article of Grebogi, Ott, and Yorke (1986).

EXERCISES 5.3

1.   Show that under our conventions, orbits of an undamped pendulum are traversed clockwise.

2.   Show that the system in (4) is almost linear at the origin.

3.   Show that the eigenvalues of the auxiliary system associated with (4) have negative real parts.

4.   Show that the system in (7) is almost linear when n is an odd integer.

In Exercises 5–8 we will study further the motion of the undamped pendulum, given by

$$mL^2 \frac{d^2\theta}{dt^2} + mgL \sin \theta = 0 \tag{11}$$

5.   a.   Multiply both sides of (11) by $2\,\dfrac{d\theta}{dt}$ and integrate to obtain

$$\left(\frac{d\theta}{dt}\right)^2 = c + \frac{2g}{L}\cos\theta$$

b.   Suppose that the angle $\theta = 0$ and the angular velocity $d\theta/dt = \omega$ when $t = 0$. Show that

$$\frac{d\theta}{dt} = \omega \sqrt{1 - \frac{4g}{\omega^2 L} \sin^2 \frac{\theta}{2}}$$

6.   Assume that $\omega > 2\sqrt{g/L}$. Show that the pendulum rotates completely around its point of suspension, and has minimum angular velocity when $\theta = \pi$, that is, when the pendulum is in the upward vertical position.

7.   Assume that $\omega < 2\sqrt{g/L}$. Find the largest angle, θ_0, that the pandulum attains.

8.   Assume that $\omega = 2\sqrt{g/L}$. Show that the pendulum rises toward the upward vertical position, which it approaches as t increases without bound.

5.4   THE LORENZ SYSTEM

When a pot of water is heated on the stove, those particles near the bottom warm faster than the ones near the top. If the temperature difference is minimal, then the fluid near the bottom becomes lighter and rises in an orderly fashion (Figure 5.14(a)). However, if the temperature difference is larger, then the rising warmer water from the bottom and the falling cooler water from the top initiate what is called **convection**, which is a circulating flow (Figure 5.14(b)). This is the same phenomenon that occurs in our atmosphere. The air closer to the ground is heated by

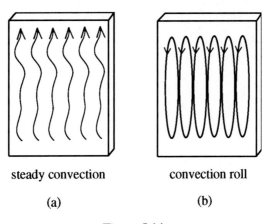

steady convection   convection roll

(a)   (b)

Figure 5.14

the sun's rays, and when the temperature differential is great enough, those air particles rise by convection. In this way warmer, polluted air escapes from the lower atmosphere to the upper atmosphere, where it is dispersed.

The almost linear system that we will study in this section describes convection. For a specific fluid enclosed in a fixed region with constant height, the rate of convection depends on several constants relative to the region and the fluid:

coefficient of thermal expansion, denoted by α
kinematic viscosity, denoted by ν
thermal conductivity, denoted by κ

These constants are automatically positive. In addition to the three listed above there is another constant: acceleration due to gravity, denoted by g. Finally, the rate of convection depends on the temperature difference between top and bottom of the region. We let

$$\Delta T = (\text{temperature at the bottom}) - (\text{temperature at the top})$$

In order to have convection, $\Delta T > 0$, which means that the bottom of the region must be warmer than the top.

Early in the twentieth century Lord Rayleigh (1916) systematically studied convective currents in a region with constant depth H. He discovered that convective motion develops if a quantity R_a, now called the **Rayleigh number** and given by

$$R_a = g\alpha H^3(\Delta T)\nu^{-1}\kappa^{-1}$$

exceeds a critical number $R_c = \pi^4 a^{-2}(1 + a^2)^3$. Here a is a number that is related to the region under consideration. The minimum possible value of R_c is $27\pi^4/4$, and occurs for $a = 1/\sqrt{2}$ (Exercise 1).

Over forty years later Barry Saltzman (1962) derived a system of differential equations that described convection. From that system a year later Edward Lorenz created the following famous stripped-down version:

$$\frac{dx}{dt} = \sigma y - \sigma x$$

$$\frac{dy}{dt} = rx - y - xz \tag{1}$$

$$\frac{dz}{dt} = xy - bz$$

We will refer to this system as the **Lorenz system**. The variable t refers to time. However, the variables x, y, and z do *not* refer to coordinates in space. In fact,

x is proportional to the intensity of convective motion

y is proportional to the temperature difference between ascending and descending currents

z is proportional to the distortion (from linearity) of the vertical temperature profile

The constants in (1) are given by

$$\sigma = \nu/\kappa, \text{ called the \textbf{Prandtl number}}$$

$$r = R_a/R_c$$

$$b = \frac{4}{1 + a^2}, \text{ a constant related to the given space}$$

For a given region and liquid, b and σ are positive constants. We will assume that $\sigma > 1$. The constant r depends, among other constants, on the temperature difference ΔT. We will assume that $r > 0$, which by the definition of r means that the bottom of the region is warmer than the top. By the comments accompanying the definitions of R_a and R_c, convection will occur if $R_a/R_c > 1$, that is, if $r > 1$.

Lorenz (1963) made a detailed study of the system in (1), with the help of a digital computer. As we study the Lorenz system, we will use the fact that the concepts pertaining to systems of two differential equations apply to systems of three differential equations, with only trivial modifications. With this in mind, we observe that **0** is a critical point of the Lorenz system. The proof that the system is almost linear at **0** is straightforward, as we will learn presently.

THEOREM 5.12. The Lorenz system is almost linear at **0**.

Proof. We write the system in (1) as follows:

$$\frac{dx}{dt} = -\sigma x + \sigma y + F(x, y, z)$$

$$\frac{dy}{dt} = rx - y + G(x, y, z)$$

$$\frac{dz}{dt} = -bz + H(x, y, z)$$

where $F(x, y, z) = 0$, $G(x, y, z) = -xz$, and $H(x, y, z) = xy$. Then $F(0, 0, 0) = G(0, 0, 0) = H(0, 0, 0) = 0$, and all first partials of F, G, and H vanish at the origin. Thus the Lorenz system is almost linear at **0**. ∎

The auxiliary system for the Lorenz system is

$$\frac{dx}{dt} = -\sigma x + \sigma y$$

$$\frac{dy}{dt} = rx - y \qquad\qquad (2)$$

$$\frac{dz}{dt} = -bz$$

and the associated matrix A_0 is given by

$$A_0 = \begin{pmatrix} -\sigma & \sigma & 0 \\ r & -1 & 0 \\ 0 & 0 & -b \end{pmatrix}$$

To evaluate the eigenvalues of A, we need to know how to find the determinant of a 3×3 matrix. If

$$B = \begin{pmatrix} a & b & c \\ d & e & f \\ g & h & j \end{pmatrix}$$

then by definition,

$$\det B = aej + bfg + cdh - ceg - bdj - afh$$

Applying this formula to $A_0 - \lambda I$ (where I is the 3×3 identity matrix), we find that

$$\det(A_0 - \lambda I) = \det \begin{pmatrix} -\sigma-\lambda & \sigma & 0 \\ r & -1-\lambda & 0 \\ 0 & 0 & -b-\lambda \end{pmatrix}$$

$$= (-\sigma - \lambda)(-1 - \lambda)(-b - \lambda) + r\sigma(b + \lambda)$$

$$= -(b + \lambda)[\lambda^2 + (\sigma + 1)\lambda + \sigma(1 - r)]$$

The eigenvalues of A_0 are the solutions of the equation $\det(A_0 - \lambda I) = 0$, of which there are (in theory) three:

$$\lambda_1 = \frac{-(\sigma + 1) + \sqrt{(\sigma - 1)^2 + 4r\sigma}}{2} \tag{3}$$

$$\lambda_2 = \frac{-(\sigma + 1) - \sqrt{(\sigma - 1)^2 + 4r\sigma}}{2} \tag{4}$$

$$\lambda_3 = -b \tag{5}$$

The eigenvalues of a 3×3 matrix indicate stability or instability of the critical point **0** in the same way they do for 2×2 matrices. In particular, if all the eigenvalues have negative real parts, then **0** is asymptotically stable. By contrast, if an eigenvalue has a positive real part, then **0** is unstable. We apply this information to the critical point **0** of the matrix associated with Lorenz system.

Case 1. $0 < r < 1$

Notice that if $0 < r < 1$, then $(\sigma - 1)^2 + 4r\sigma > (\sigma - 1)^2 + 4\sigma = (\sigma + 1)^2 \geq 0$, so that λ_1 and λ_2 are real numbers. The hypothesis that $r - 1 < 0$ implies that

$$\sqrt{(\sigma - 1)^2 + 4r\sigma} = \sqrt{(\sigma + 1)^2 + 4(r - 1)\sigma} < \sqrt{(\sigma + 1)^2} = \sigma + 1$$

so that all three eigenvalues are negative numbers. Thus if $0 < r < 1$, then **0** is an asymptotically stable critical point. Physically this means that as time passes, convection dies down and the system approaches a steady state of no convection. This conforms to the result of Rayleigh which states that convection begins as r rises above 1.

Case 2. $r = 1$

When $r = 1$, the eigenvalues simplify to $\lambda_1 = 0$, $\lambda_2 = -(\sigma + 1)$, and $\lambda_3 = -b$. Therefore two eigenvalues are negative, and the other is 0, which means that **0** is a **neutrally stable solution**. (This case is somewhat analogous to the case in which $\mu = 3$ for the quadratic family $\{Q_\mu\}$; $\mu = 3$ is a bifurcation point, at which the attracting fixed point becomes unstable.)

Case 3. $r > 1$

If $r > 1$, then by calculations similar to those in Case 1, we find that

$$\sqrt{(\sigma - 1)^2 + 4r\sigma} > \sigma + 1$$

It follows that $\lambda_1 > 0$, $\lambda_2 < 0$, and $\lambda_3 < 0$. Therefore **0** is an unstable solution.

In the terminology of hydrodynamics, convection occurs.

When r increases and passes through 1, the asymptotic stability of **0** gives way to instability. Thus 1 is a point of bifurcation for the family of Lorenz systems (with parameter r). Might the incipient instability of **0** foreshadow the emergence of other, new critical points when $r > 1$? Theorem 5.13 gives the answer.

THEOREM 5.13. If $r > 1$, then in addition to **0** there exist two critical points **p** and **q** for the system in (1), given by

$$\mathbf{p} = \begin{pmatrix} \sqrt{b(r-1)} \\ \sqrt{b(r-1)} \\ r-1 \end{pmatrix} \quad \text{and} \quad \mathbf{q} = \begin{pmatrix} -\sqrt{b(r-1)} \\ -\sqrt{b(r-1)} \\ r-1 \end{pmatrix}$$

Proof. Suppose that $\mathbf{v} = \begin{pmatrix} x \\ y \\ z \end{pmatrix}$ is a critical point. If $x = 0$, then by the first equation of (1) and by the hypothesis that $\sigma \neq 0$, we know that $y = 0$. Since $b \neq 0$, it follows from the third equation in (1) that $z = 0$. Therefore if $x = 0$, then $\mathbf{v} = \mathbf{0}$. Now suppose that $\mathbf{v}$ is a critical point and $\mathbf{v} \neq \mathbf{0}$. By our preceding comments, $x \neq 0$. Therefore from the first equation in (1),

$$0 = \frac{dx}{dt} = \sigma y - \sigma x = \sigma(y - x)$$

so that $y = x$. Consequently we deduce from the second equation in (2) that

$$0 = \frac{dy}{dt} = rx - y - xz = rx - x - xz = x(r - 1 - z)$$

Since $x \neq 0$ by assumption, it follows that $z = r - 1$. Therefore the third equation in (2) yields

$$0 = \frac{dz}{dt} = xy - bz = x^2 - bz = x^2 - b(r - 1)$$

Consequently $x^2 = b(r - 1)$. As a result,

$$x = y = \pm\sqrt{b(r - 1)} \quad \text{and} \quad z = r - 1$$

so that, by definition, **p** and **q** are critical points of the Lorenz system.   ■

One can show that the Lorenz system is almost linear at **p**, with associated matrix

$$A_{\mathbf{p}} = \begin{pmatrix} -\sigma & \sigma & 0 \\ 1 & -1 & -\sqrt{b(r-1)} \\ \sqrt{b(r-1)} & \sqrt{b(r-1)} & -b \end{pmatrix} \tag{6}$$

and eigenvalues given by

$$\lambda^3 + (\sigma + b + 1)\lambda^2 + b(\sigma + r)\lambda + 2b\sigma(r-1) = 0 \tag{7}$$

(Exercise 7).

Notice that the Lorenz system is symmetric with respect to the z axis. This means that if

$$X(t) = \begin{pmatrix} x(t) \\ y(t) \\ z(t) \end{pmatrix} \text{ and } Y(t) = \begin{pmatrix} -x(t) \\ -y(t) \\ z(t) \end{pmatrix}$$

and if X is a solution, then Y is a solution. Consequently the associated matrices $A_{\mathbf{p}}$ and $A_{\mathbf{q}}$ have identical characteristic equations, and hence the same eigenvalues for any given constants σ, b, and r (Exercise 5). Therefore both **p** and **q** are asymptotically stable (or unstable), or neither is. As a result, we will only analyze the dynamics of **p**.

Henceforth we will assume that $\sigma > b + 1$, so that $\sigma - b - 1 > 0$, and let

$$r^* = \frac{\sigma(\sigma + b + 3)}{\sigma - b - 1}$$

Notice that $r^* > 1$ since $\sigma > 1$ and $\sigma > b + 1$. It is known that if $1 < r < r^*$, then all three roots of (7) have negative real parts, so that **p** is asymptotically stable. By contrast, if $r > r^*$, then two roots of (7) have positive real parts, so that **p** is unstable.

It follows from the preceding comment that if $r > r^*$, then the three critical points **0**, **p**, and **q** are unstable critical points. A natural question to ask is what the orbits of solutions look like when $r > r^*$. Following Lorenz, we will let $\sigma = 10$ and $b = 8/3$ (which means that $\sigma > b + 1$), so that $r^* = 470/19 \approx 24.74$. The family of Lorenz systems in which $\sigma = 10$, $b = 8/3$, and $r > 1$ will be called the **Lorenz family**. Since **p** and **q** become unstable when r passes through r^*, it follows that the Lorenz family has a bifurcation at r^*.

To be able to visualize the behavior of solutions of the Lorenz system, let us follow the lead of Lorenz and fix $r = 28$, so that $r > r^*$. Since all three critical points are unstable for this value of r, it is difficult to conjecture what the orbit of an arbitrary solution might look like.

Assume that the initial point of a solution is near the unstable critical point **0**. The orbit is illustrated in Figure 5.15. To describe the dynamics of the orbit, we

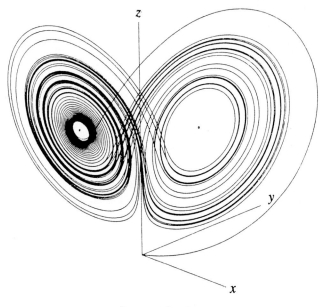

Lorenz attractor

Figure 5.15

observe that at first the three coordinates x, y, and z grow rapidly. Similar signs of x and y indicate that warmer fluid is rising and colder fluid is sinking. When convection becomes strong enough, there is a reversal, in which warmer fluid rises above the colder fluid at the top; the signs of y and x change. After the orbit comes close enough to **p** (the dot with positive x and y coordinates, around which the orbit circles), it spirals outward until it is deflected toward **q** (the other dot), and once again spirals outward until recaptured near **p**. This process continues indefinitely. Color Plate 2 gives a color rendition of an orbit in the Lorenz attractor.

The surface on which the orbit resides is (essentially) the attractor A_L of the Lorenz system, called the **Lorenz attractor**. It may appear that A_L lies in a single plane, but that is not quite true. In fact, each time the orbit spirals around the critical point **p** (or **q**), it travels on a different leaf, never retracing its steps. It has been shown that $\dim_c A_L \approx 2.07$. This demonstrates that the attractor does not lie in a single plane, and that it is a strange attractor. The fact that tiny differences in initial conditions can lead to far different values after a period of time means that the attractor is chaotic. This sensitive dependence led Lorenz to give the title "Predictability: Does the Flap of a Butterfly's Wings in Brazil Set Off a Tornado in Texas" for an address to the American Association for the Advancement of Sciences in 1979. The title, along with the shape of the Lorenz attractor, has led to the nickname **butterfly effect** for the Lorenz system and its attractor.

The Lorenz system has been widely studied during the past two decades, and various bifurcations of the system have been located for values of r much larger than 28. We will say that an orbit is of type ab^2 if it spirals around one critical point once (thus the single a) and around another critical point twice (thus the appearance of b^2). Francheschini (1980) has identified a stable ab^2-orbit when $r \approx$ 100.75 (Figure 5.16(a)), which represents a stable period-3 orbit because of the three loops. If r is lowered to 99.65, there is a stable a^2ba^2b-orbit (Figure 5.16(b)), which represents a stable period-6 orbit. Thus period-doubling has occurred. In fact, as r decreases further, there is a period-doubling cascade that terminates when r is approximately 99.24. Franceschini found that the ratios of successive bifurcation points approach the Feigenbaum constant, which is approximately 4.67.

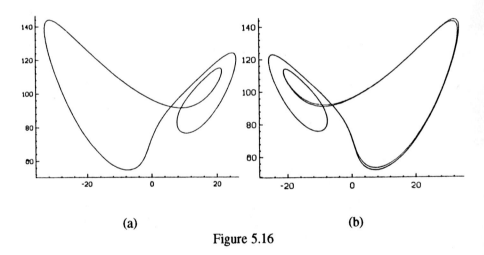

(a)   (b)

Figure 5.16

Conclusion

The system that Lorenz introduced to the world thirty years ago has promoted a vast amount of research. In 1982 Colin Sparrow published the book *The Lorenz Equations: Bifurcations, Chaos, and Strange Attractors*, which contains virtually everything announced in this section, plus a great deal more. However, there are still open questions concerning the Lorenz system, as there are in chaotic dynamics.

One might say that the appearance of the Lorenz system thirty years ago was the catalyst for modern research in chaotic dynamics, and for the emergence of a vast number of applications of chaotic dynamics in divergent fields. It will be interesting to see how the area of chaotic dynamics grows and is applied in the next thirty years.

EXERCISES 5.4

1. Let $f(a) = \pi^4 a^{-2}(1 + a^2)^3$. Show that $f(1/\sqrt{2}) = 27\pi^4/4$ is the minimum value of f.

2. Suppose that X is a solution of the Lorenz system and that $X(0)$ lies on the z axis. Show that the entire orbit lies on the z axis and that $\lim_{t \to \infty} X(t) = 0$.

3. Show that when viewed from above, the orbits revolving around the z axis do so in a clockwise direction.

4. Let $r > 1$. Show that the eigenvalues of A_0 satisfy $-\lambda_2 > \lambda_1 > -\lambda_3$ if and only if $r > 1 + b(\sigma + 1 + b)/\sigma$.

5. Let σ, b, and r be fixed. Show that A_p and A_q have the same characteristic equation, and hence the same eigenvalues.

6. Let $r > 1$. Show that the Lorenz system is almost linear at $\mathbf{p}$, with associated matrix given by (6).

7. Show that the eigenvalues of A_p satisfy (7).

8. Show that if (7) has complex roots, then the real root must be negative. (*Hint*: If $\alpha + i\beta$ is one complex root, then $\alpha - i\beta$ is a second complex root.)

9. a. Show that the characteristic equation of A_p has two real roots if and only if
$$r < \frac{\sigma^2 - b\sigma + b^2 + 2\sigma + 2b + 1}{3b}$$

    b. Let $\sigma = 10$ and $b = 8/3$. Find a number r_0 such that if $r > r_0$, then two of the eigenvalues of A_p are complex numbers.

10. Let $a > 0$ and $c > 0$. Show that the equation $\lambda^3 + a\lambda^2 + c\lambda + ac = 0$ has two pure imaginary roots.

11. Find a value of r_0 such that A_p has a pure imaginary eigenvalue for $r = r_0$.

12. Use the computer to determine a period-doubling window in the interval $145 < r < 166$. Can you say anything interesting about the periodic orbits?

APPENDIX

COMPUTER PROGRAMS

The following ten computer programs are written in TRUE BASIC (Version 2.02), and have been tested on the Macintosh IIsi. For other computers there may need to be minor modifications.

Program 1: ITERATE

```
10  Input  M, x, n
20  For  i = 1  to  n
30      Let  x = M*x*(1 – x)
40      Print  i; x
50  Next  i
60  End
```

ITERATE computes and prints the initial n iterates of Q_μ. We need to select the parameter μ (designated by M), initial point x, and number n of iterates to be evaluated.

Program 2: NUMBER OF ITERATES

10 Input M, x, p, d, n

20 For $i = 1$ to n

30   Let $x = M*x*(1 - x)$

40   Print x

50   If abs $(x - p) < d$ then go to 80

60 Next i

70 Print i

80 End

NUMBER OF ITERATES computes the iterates of Q_μ until an iterate is within d of p. We need to select the parameter μ (designated by M), initial point x, fixed point or periodic point p, the maximum allowed distance d between the desired iterate and p, and maximum number n of iterates to be evaluated.

Program 3: PLOT

```
10   Set window  − .2, 2, − .1, 1.5
20   Dim points1 (60,2)
30   Dim points2 (60,2)
40   Input M
50   Let n = 60
60   Let incx = 1/n
70   Plot lines:  − .1, 0;  1.1, 0
80   Plot lines:  0, − 1;  0, 1.1
90   Plot lines:  0, 1;  .1, 1
100  Plot lines:  1, 0;  1, .1
110  For i = 1 to n
120      Let x = (i − 1)*incx
130      Let points1 (i,1) = x
140      Let points1 (i,2) = M*x*(1 − x)
150      Let points2 (i,1) = x
160      Let points2 (i,2) = x
170  Next i
180  Mat plot lines: points1
190  Mat plot lines: points2
200  End
```

PLOT plots 60 points on the graph of the function Q_μ and the line $y = x$, and connects the plotted points. We need to select the parameter μ (designated M).

Program 4: BIFURCATION

10   Input j, n, p, q

20   Set window $p, q, - 0.1, 1.1$

30   Plot lines: $p, 0; q, 0$

40   For $M = p$ to q step 0.001

50   Let $x = .5$

60   For $i = 1$ to $j + n$

70   Let $x = M*x*(1 - x)$

80   If $i > j$ then plot M, x

90   Next i

100 Next M

110 End

BIFURCATION plots the bifurcation diagram of the quadratic family $\{Q_\mu\}$ for μ in $[p, q]$ by increments of 0.001, using iterates from the $(j+1)$st iterate to the nth iterate of .5. We need to select the values of $j, n, p,$ and q.

Program 5: HENON

10   Set window $-2, 2, -1.5, 1.5$

20   Plot lines: $-1.5, 0;$ $1.5, 0$

30   Plot lines: $0, -1.5; 0, 1.5$

40   Plot lines: $1, 0;$ $1, .05$

50   Plot lines: $0, 1;$ $.05, 1$

60   Input a, x, y

70   For $i = 1$ to 5000

80   Let $v = 1 - a*x*x + y$

90   Let $w = (.3)*x$

100   Let $x = v$

110   Let $y = w$

120   If $i > 50$ then plot x, y

130 Next i

140 End

HENON draws the Hénon attractor of $H_{a(.3)}$. It plots from the 51st iterate of the initial point (x, y) to the 5000th iterate. We need to select the parameter a and the initial point (x, y).

Program 6: JULIA

```
10    Set window  − 2.8, 2.8, − 2, 2
20    Input  a, b, x, y, n
30    For  i = 1  to  n
40        Let  u = x − a
50        Let  v = y − b
60        If  u = 0  then let  t = 3.14159 / 2
70        If  u > 0  then let  t = atn (v/u)
80        If  u < 0  then let  t = 3.14159 + atn (v/u)
90        Let  r = sqr(u*u + v*v)
100       Let  t = t/2 + int (2*rnd)*3.14159
110       Let  r = sqr(r)
120       Let  x = r*cos(t)
130       Let  y = r*sin(t)
140       If  i < 100  then go to 170
150       Plot  x, y
160       Plot  − x, − y
170   Next  i
180   End
```

JULIA plots the Julia set for $c = a + ib$. Let $g_c(z) = z^2 + c = w$. Then a backward iterate of w is z such that $z = \sqrt{w - c}$ or $-\sqrt{w - c}$. Let $w = x + iy$. If $u = x - a$ and $v = y - b$, then $z = \pm\sqrt{u + iv}$. In lines 60–90 we write $u + iv$ in its polar form re^{it}. In lines 100–110 we randomly choose one of the two square roots of $u + iv$. Disregarding the first 99 backward iterates (line 140), the program plots the 100th iterate through the nth iterate. Symmetry is used in line 160. We need to select a and b (for c), the initial point (x, y), and the number n of iterates to be plotted.

Program 7: MANDELBROT

```
10   Set window  − 2, 1, − 1.1, 1.1
20   Let incx = 1/300
30   Let incy = 1/400
40   For i = 0 to 400
50       For j = 0 to 900
60           Let s = − 2 + j*incx
70           Let t = 0 + i*incy
80           Let x = s
90           Let y = t
100          For k = 1 to 80
110              Let z = x*x* − y*y + s
120              Let w = 2*x*y + t
130              Let v = z*z + w*w
140              If v > 4 then go to 200
150              Let x = z
160              Let y = w
170          Next k
180          Plot points: s, t
190          Plot points: s, − t
200      Next j
210  Next i
220  End
```

MANDELBROT draws the Mandelbrot set. It uses the symmetry of the set about the x axis to draw the points below the x axis (line 190). The program analyzes iterates of $z = 0$ for $g_c(z) = z^2 + c$, where c is in $[− 2, 1] \times [0, 1]$. The number c begins at the point $(− 2, 0)$ and is altered, with the first coordinate increasing in increments of 1/300 and the second coordinate in increments of 1/400. In the program, if the first 80 iterates are no more than 2 in absolute value, then c is plotted. Otherwise c is not plotted.

Program 8: ITERATED FUNCTION SYSTEM

```
10    Set window  − .5, 1.6, − .1, 1.4
20    Input  x, y, n
30    For  i = 1  to  n
40        Let  p = 3*rnd
50        If  p < 1  then go to 100
60        If  p < 2  then go to 130
70        Let  x = .5*x
80        Let  y = .5*y
90        Go to 160
100       Let  x = .5*x + .5
110       Let  y = .5*y
120       Go to 160
130       Let  x = .5*x + .25
140       Let  y = .5*y + .5
150       Go to 160
160   If  i > 100  then plot  x, y
170   Next  i
180   End
```

ITERATED FUNCTION SYSTEM plots the iterates in the Iterated Function System given by three functions that are defined in lines 70–80, 100–110, and 130–140. For each k, one of the three given functions is chosen randomly by lines 40–60, and the kth iterate of the initial point (x, y) is plotted. The resulting figure is the Sierpinski gasket. We need to select the initial point (x, y), as well as the total number n of iterates to be evaluated.

Program 9: FERN LEAF

```
10   Set window:  − 14, 14, − 1, 22
20   Input  x, y, n
30   For  i = 1  to  n
40       Let  p = 100*rnd
50       If  p < 85  then go to 110
60       If  p < 92  then go to 140
70       If  p < 99  then go to 170
80       Let  x = 0
90       Let  y = .17*y
100      Go to 190
110      Let  x = .85*x + .04*y
120      Let  y = − .04*x + .85*y + 3
130      Go to 190
140      Let  x = .2*x − .26*y
150      Let  y = .23*x + .22*y + 1.4
160      Go to 190
170      Let  x = − .25*x + .28*y
180      Let  y = .26*x + .24*y + .4
190      If  i > 100  then plot  x, y
200  Next  i
210  End
```

FERN LEAF plots the iterates given by the four functions given in lines 80–90, 110–120, 140–150, and 170–180. The probability of selecting each of the four functions is given in lines 50–70. Notice that approximately 85% of the time the function in lines 110–120 will be chosen, and only about 1% of the time the function in lines 80–90 will be chosen. We need to select the initial point (x, y) and the total number of iterates n to be evaluated.

Program 10: CHAOS GAME

```
10    Set window  −.4, 1.4, −.1, .8
20    Input A1, A2, B1, B2, C1, C2, x, y, n
30    For  i = 1 to  n
40        Let  p = 3*rnd
50        If  p < 1  then go to 100
60        If  p < 2  then go to 130
70        Let  x = (x + A1)/2
80        Let  y = (y + A2)/2
90        Go to 150
100       Let  x = (x + B1)/2
110       Let  y = (y + B2)/2
120       Go to 150
130       Let  x = (x + C1)/2
140       Let  y = (y + C2)/2
150       If  i > 50  then plot  x, y
160   Next  i
170   End
```

CHAOS GAME selects the midpoint of the line between a given point and one of three given points $A = (A1, A2)$, $B = (B1, B2)$, and $C = (C1, C2)$. Which of A, B, or C is chosen is random (lines 40–60). The process continues, and eventually yields a (deformed) Sierpinski gasket. The first 50 iterates in the process are ignored. We need to select the three points A, B, and C, the initial point (x, y), and the total number n of iterates to be evaluated.

ANSWERS TO SELECTED EXERCISES

CHAPTER 1

Section 1.1

1. a. 0.7390851332   b. 0.7390851332
    c. 0.7390851332
3. They do; 0.7390851332   5. They oscillate between 1/3 and 2/3.
7. The limit exists precisely when $|a| < 1$; then its value is $b(1 - a)$.
9. 2.0946
11. a. Approximate zero of f: -0.2554228711
    b. Error message because $f'(1) = 0$
    c. Overflow message
    d. Approximate zero of f: -0.2554228711

Section 1.2

1. 0, 3; both are repelling   3. 0 is repelling; 1 is attracting
5. 0 is repelling   9. b. No; $f'(x) \approx 1$ if $x \approx 0$
11. No, because $f'(0)$ does not exist
13. a. $p = b/(1 - a)$; it is attracting if $|a| < 1$ and is repelling if $|a| > 1$
23. The interval $(-1, 1)$ 25. The interval $(-0.5, 1.5)$
29. a. $\dfrac{1}{2} - \dfrac{1}{4}\sqrt{2}$ is attracting, and $\dfrac{1}{2} + \dfrac{1}{4}\sqrt{2}$ is repelling

    b. $(\dfrac{1}{2} - \dfrac{1}{4}\sqrt{2}, \dfrac{1}{2} - \dfrac{1}{4}\sqrt{2})$

31. a. Attracting   b. $f(x) = \sin^2 x + x$

Section 1.3

5. \{0.5130445095, 0.7994554905\}
7. a. 1/7   b. 1/17   c. 1/11
11. Fixed points: 0 and 1; period-2 points: 1/3 and 2/3
13. 2^n

19. $b = a + 1$ with a in either $(-\dfrac{1}{4} - \dfrac{1}{4}\sqrt{17}, -\dfrac{1}{2})$ or in $(0, -\dfrac{1}{4}\sqrt{17})$

23. c. $f(x) = -x$

Section 1.4

1. a. Eventually periodic   b. Periodic
    c. Eventually periodic   d. Periodic
    e. Eventually fixed
3. a. 30   b. 2182
7. a. $2^{n-1}\mu^n$ b. μ
11. a. 1
    b. If $x \leq 1$, then $\lim_{n \to \infty} E_\mu^{[n]}(x) = 1$; if $x > 1$, then $\lim_{n \to \infty} E_\mu^{[n]}(x) = \infty$.
13. q_μ is attracting; p_μ is repelling

Section 1.5

5. $\mu^2(4 - \mu)/16$ 11. a. $\dfrac{3}{2} + \dfrac{1}{2}\sqrt{17}$

Section 1.6

1. Tangent bifurcation   3. 1; neither
5. The interval is approximately (3.626, 3.635).

Section 1.7

1. a. Approximately 3.89995
5. a. Let f have domain $\{0, 1, 2, 3, 4\}$, and let $f(0) = 0$, $f(1) = 2$, $f(2) = 1$,
        $f(3) = 4$, $f(4) = 5$, and $f(5) = 3$.
    b. No, by the Li-Yorke Theorem

Section 1.8

1. $-1/2$ 7. $a > 1$
9. $(-1/\sqrt{2}, 1/\sqrt{2})$
11. c. 1   d. No: $Sf(x) > 0$ for an x in $(0, 1)$
    f. 2–cycle
13. 1   15. 1

CHAPTER 2

Section 2.1

3. a. $\lambda(x) = \ln 2$ if x is not a dyadic rational
    b. $2 \ln 2$
7. b. $\ln |\mu - 2|$

Section 2.2

1. $B_n = [n, \infty)$ for $n = 1, 2, \ldots$
7.   a. x such that $h(x) = 0\ 100\ 1000\ 10000 \cdots$
13. a. $2/p,\ 4/p,\ 6/p,\ \ldots,\ (p-1)/p$

Section 2.3

1. $[1/\mu - 1,\ 1/\mu]$
5.   a.   0.165448   b.   0.292293
9.   No. If $h \circ Q_\mu = T_\lambda \circ h$ with h linear, then $h(x) = x$ or $h(x) = 1 - x$, neither of which is possible.
13. $g_\mu(x) = f_\mu(x + x_\mu) - x_\mu$

Section 2.4

5.   a.   Neither   b.   Totally disconnected
7. $D_n(C) = (\dfrac{2}{3})^n$; $\lim_{n \to \infty} D_n(C) = 0$

CHAPTER 3

Section 3.1

1.   Normal form: $\begin{pmatrix} 1 & 0 \\ 0 & -1 \end{pmatrix}$; eigenvalue 1 has eigenvector $\begin{pmatrix} 1 \\ 1 \end{pmatrix}$; eigenvalue
     -1 has eigenvector $\begin{pmatrix} 1 \\ -1 \end{pmatrix}$

3. Normal form: $\begin{pmatrix} 1 & 1 \\ -1 & 1 \end{pmatrix}$; eigenvalues are $1+i$ and $1-i$

5. $a^2 + b \geq 0$ 7. $a = d$ and $b = 0 = c$

Section 3.2

3. Eigenvalue $\dfrac{1}{3}$ has eigenvector $\begin{pmatrix} 1 \\ 0 \end{pmatrix}$; eigenvalue 3 has eigenvector $\begin{pmatrix} 0 \\ 1 \end{pmatrix}$

5. Eigenvalue 2 has eigenvector $\begin{pmatrix} 1 \\ 1 \end{pmatrix}$; eigenvalue 3 has eigenvector $\begin{pmatrix} 1 \\ 2 \end{pmatrix}$

15. b. It must be.

Section 3.3

1. fixed points: $\begin{pmatrix} 1/2 \\ 1 \end{pmatrix}$ and $\begin{pmatrix} 1/2 \\ -2 \end{pmatrix}$; area-expanding at both

3. fixed points: $\begin{pmatrix} 0 \\ 0 \end{pmatrix}$ and $\begin{pmatrix} 6 \\ 3 \end{pmatrix}$; area-contracting at $\begin{pmatrix} 0 \\ 0 \end{pmatrix}$, area-expanding at $\begin{pmatrix} 6 \\ 3 \end{pmatrix}$

5. Both are repelling.   7. $\begin{pmatrix} 0 \\ 0 \end{pmatrix}$ is attracting; $\begin{pmatrix} 6 \\ 3 \end{pmatrix}$ is repelling

9. No, since $L^{[2]}(\mathbf{v}) = 0$ for all $\mathbf{v}$

13. $\begin{pmatrix} 3/4 \\ 1 \end{pmatrix}$ 17. $\begin{pmatrix} 0 \\ 0 \end{pmatrix}$ and $\begin{pmatrix} 1/[2(1-c)] \\ 1 \end{pmatrix}$

Section 3.4

3. a. $\begin{pmatrix} 0.6314 \\ 0.1894 \end{pmatrix}$ b. $\lambda \approx 1.9239$ and $\mu \approx 0.1559$

    c.   Eigenvector for λ is $\begin{pmatrix} 1 \\ -\mu \end{pmatrix}$; eigenvector for μ is $\begin{pmatrix} 1 \\ -\lambda \end{pmatrix}$

7. a. $\begin{pmatrix} 1/(1-b) \\ b/(1-b) \end{pmatrix}$ b.   Attracting

c. For eigenvalue $\sqrt{b}$, an eigenvector is $\begin{pmatrix} 1 \\ \sqrt{b} \end{pmatrix}$; for eigenvalue $-\sqrt{b}$, an eigenvector is $\begin{pmatrix} -1 \\ \sqrt{b} \end{pmatrix}$

Section 3.5

3. b. Let $x_k = z_k$ for $|k| \leq n$; let $x_k = 0$ for $|k| > n$ and $z_k = 1$ for $|k| > n$.
    c. $x = \cdots \overline{0.0} \cdots$ and $z = \cdots \overline{1.1} \cdots$

CHAPTER 4

Section 4.1

3. b. $\dfrac{2}{5}L^2 \sqrt{3}$

7. $1 - e^{(r-1)(\ln 2)/r}$

Section 4.2

1. $\dfrac{\ln 100}{\ln 8} \approx 2.2146$; it is larger

Section 4.3

3. It is.

Section 4.4

1. $F\begin{pmatrix} x \\ y \end{pmatrix} = \begin{pmatrix} 1/\sqrt{2} & 0 \\ 0 & 1/2 \end{pmatrix}\begin{pmatrix} x \\ y \end{pmatrix}$

3. $F_1(x) = \dfrac{1}{3}\, x$, $F_2(x) = \dfrac{2}{3} + \dfrac{1}{3}\, x$

7. F_1:   1/2   0   0   1/2   0   0
    F_2:   1/2   0   0   1/2   0   1/2
    F_3:   1/2   0   0   1/2   1/2 1/2

9. F_1:   1/4   0   0   1/4   0   0
    F_2:   1/2   0   0   1/2   1/2 0
    F_3:   1/2   0   0   1/2   1/4 1/2

CHAPTER 5

Section 5.1

1. $-5, 1$; unstable saddle point
3. -2; asymptotically stable degenerate node
5. $-5, -2$; asymptotically stable node
7. $-2 + 2i, -2 - 2i$; asymptotically stable spiral point
9. 1; unstable degenerate node
11. $1 + \sqrt{6}, 1 - \sqrt{6}$; unstable saddle point
15. a. $\dfrac{dQ}{dt} = I$

        $\dfrac{dI}{dt} = -\dfrac{1}{LC} Q - \dfrac{R}{L} I$

    b. $\pi LC/2$
    c.   Asymptotically stable for all constants; spiral point if $R^2 < 4L/C$, and node if $R^2 > 4L/C$

Section 5.2

1. Almost linear; asymptotically stable node
3. Almost linear at **0**, which is an unstable critical point

    Almost linear at $\begin{pmatrix} 1 \\ 0 \end{pmatrix}$; cannot tell about stability

    Almost linear at $\begin{pmatrix} -1 \\ 0 \end{pmatrix}$; cannot tell about stability

5. Almost linear at **0**, which is an unstable saddle point

    Almost linear at $\begin{pmatrix} 1 \\ 1 \end{pmatrix}$, which is an asymptotically stable spiral point

Almost linear at $\begin{pmatrix} 2 \\ 0 \end{pmatrix}$, which is an unstable saddle point

9. b. $0 < \varepsilon < 2$

11. Critical points: $\begin{pmatrix} n\pi \\ 0 \end{pmatrix}$; if n is even, then the critical point is a saddle point;

if n is odd, then the critical point is asymptotically stable. It is a node if a^2 > 4b, a spiral point if $a^2 < 4b$, and is a degenerate node if $a^2 = 4b$.

Section 5.3

7. $\theta_0 = \arcsin\left(\dfrac{\omega}{2}\sqrt{L/g}\right)$

Section 5.4

9. b. $r_0 = \dfrac{961}{72}$

11. $r_0 = \dfrac{\sigma(\sigma + b + 3)}{\sigma - b - 1}$

REFERENCES

Art Matrix, "Focus on Fractals," videotape, Box 880, Ithaca, New York 14850.

Barnsley, Michael, *Fractals Everywhere*, Academic Press, San Diego, 1988.

Boyce, William, and Richard Di Prima, *Elementary Differential Equations and Boundary Value Problems*, Fourth Edition, Wiley, New York, 1986.

Buck, R. Creighton, *Advanced Calculus*, Third Edition, McGraw-Hill, New York, 1978.

Burckel, Robert B., *An Introduction to Classical Complex Analysis, Volume 1*, Birkhäuser, Basel, Switzerland, 1979.

Collet, Pierre, and Jean-Pierre Eckmann, *Iterated Maps on the Interval as Dynamical Systems*, Birkhäuser, Boston, 1980.

Derrida, B., A. Gervois, and Y. Pomeau, "Universal Metric Properties of Bifurcations of Endomorphisms," J. Physics A: Math. Gen., **12** No. 3 (1979), pp. 269–296.

Devaney, Robert L., *An Introduction to Chaotic Dynamical Systems*, Second Edition, Addison-Wesley, Menlo Park, California, 1989.

Devaney, Robert L., *Chaos, Fractals, and Dynamics*, Addison-Wesley, Menlo Park, California, 1990.

Douady, A., and J. H. Hubbard, "Itération des polynômes quadratiques complexes," C.R. Acad. Sci. Paris I, **294** (1982), pp. 123–126.

Du, Bau-Sen, "The Minimal Number of Periodic Orbits of Periods Guaranteed in Sharkovskii's Theorem," Bull. Austral. Math. Soc., **31** (1985), pp. 89–103.

Farmer, J. Doyne, Edward Ott, and James A. Yorke, "The Dimension of Chaotic Attractors", Physica **7D** (1983), pp. 153–180.

Fatou, Pierre, "Sur les equations fonctionnelles," Bull. Soc. Math. France **47** (1919), pp. 161–271.

Feigenbaum, Mitchell, "Quantitative Universality for a Class of Nonlinear Transformations," J. of Stat. Physics, **19** No. 1 (1978), pp. 25–52.

Feigenbaum, Mitchell, "Universal Behavior in Nonlinear Systems," Physica **5D**, (1983), pp. 16–39.

Franceschini, V., "A Feigenbaum Sequence of Bifurcations in the Lorenz Model," J. Stat. Physics, **22** No. 3 (1980), pp. 397–406.

Garfinkel, A. J., D. O. Walter, R. B. Trelease, R. K. Harper, and R. M. Harper, "Non-linear Dynamics of Electrocardiographic Waveforms Following Cocaine Administration," Life Sciences, **48** No. 22 (1991), pp. 2189–2193.

Gleick, James, *Chaos: Making a New Science*, Viking, New York, 1987.

Grebogi, Celso, Edward Ott, Steven Pelikan, and James A. Yorke, "Strange Attractors that are not Chaotic," Physica **13D** (1984), pp. 261–268.

Grebogi, Celso, Edward Ott, and James A. Yorke, "Fractal Basin Boundaries," Lecture Notes in Physics, Vol. 278, Springer-Verlag (1986), pp. 28–32.

Guckenheimer, John, and Philip Holmes, *Nonlinear Oscillations, Dynamical*

Systems, and Bifurcations of Vector Fields, Springer-Verlag, New York, 1986.

Hénon, Michel, "A Two-Dimensional Mapping with a Strange Attractor," Commun. Math. Phys. **5D** (1976), pp. 69–77.

Ho, Chung-Wu, and Charles Morris, "A Graph Theoretic Proof of Sharkovsky's Theorem on the Periodic Points of Continuous Functions," Pacific J. of Math. **96** No. 2 (1981), pp. 361–370.

Hutchinson, John, "Fractals and Self Similarity", Indiana Univ. J. Math. **30** (1981), pp. 713–747.

Julia, Gaston, "Memoire sur l'iteration des fonctions rationnelles," J. Math. **8** (1918), pp. 47–245.

Kaplan, James L., and James A. Yorke, "Chaotic Behavior of Multidimensional Difference Equations," Springer Lecture Notes, Vol. 730, H. O. Peitgen and H. O. Walther, Editors, Springer-Verlag, New York, 1978.

Li, Tien-Yien, and James A. Yorke, "Period Three Implies Chaos," Amer. Math. Monthly **82** (1975), pp. 985–992.

Lorenz, Edward N., "Deterministic Nonperiodic Flow," J. Atmospheric Sciences **20** (1963), pp. 130–141.

May, Robert M., "Simple Mathematical Models with Very Complicated Dynamics," Nature **261** (1976), pp. 459–467.

Moon, Francis C., *Chaotic Vibrations*, Wiley, New York, 1987.

Murray, Carl, "Is the Solar System Stable?" *New Scientist*, 25 November 1989, pp. 60–63.

Palmer, Tim, "A Weather Eye on Unpredictability," *New Scientist*, 11 November 1989, pp. 56–59.

Peitgen, H.-O., and P. H. Richter, *The Beauty of Fractals*, Springer-Verlag, Berlin, 1986.

Preston, Chris, *Iterates of Maps on an Interval*, Lecture Notes in Mathematics, Vol. 999, Springer-Verlag, New York, 1983.

Rasband, S. Neil, *Chaotic Dynamics of Nonlinear Systems*, Wiley, New York, 1990.

Rayleigh, Lord, "On Convective Currents in a Horizontal Layer of Fluid when the Higher Temperature is on the Under Side," Phil. Mag. **32** (1916), pp. 529–546.

Ross, Kenneth A., *Elementary Analysis: The Theory of Calculus*, Springer-Verlag, New York, 1980.

Saltzman, Barry, "Finite Amplitude Free Convection as an Initial Value Problem," J. Atmospheric Sci. **19** (1962), pp. 329–341.

Sanchez, David A., Richard C. Allen, and Walter T. Kyner, *Differential Equations*, Addison-Wesley, Reading, Massachusetts, 1983.

Schroeder, Manfred, *Fractals, Chaos, Power Laws*, W. H. Freeman, New York, 1991.

Scott, Stephen, "Clocks and Chaos in Chemistry," New Scientist, 2 Dec. 1989, pp. 50–59.

Shanks, Daniel, *Solved and Unsolved Problems in Number Theory*, Second Edition, Chelsea, New York, 1978.

Sharkovsky, A. N. "Co-existence of Cycles of a Continuous Mapping of a Line into Itself," Ukranian Math. Z., **16** (1964), pp. 61–71.

Singer, David, "Stable Orbits and Bifurcation of Maps of the Interval," SIAM J. Appl. Math., **35** No. 2 (1978), pp. 260–267.

Smale, Stephen, "Differentiable Dynamical Systems," Bull. Amer. Math. Soc., **73** (1967), pp. 747–817.

Sparrow, Colin, *The Lorenz Equations: Bifurcations, Chaos, and Strange Attractors*, Springer-Verlag, New York, 1982.

Straffin, P. D., Jr., "Periodic Points of Continuous Functions," Math. Mag., **51** (1978), pp. 99–105.

University of Maryland Chaos Group, "Chaos Fractals in Simple Physical Systems," videotape, University of Maryland, College Park, 1991.

Van Buskirk, Robert, and Carson Jeffries, "Observation of Chaotic Dynamics of Coupled Nonlinear Oscillators," Physical Review A **31** No. 5 (1985), pp. 3332–3357.

Wiggins, Stephen, *Global Bifurcations and Chaos*, Springer-Verlag, New York, 1988.

Wisdom, Jack, "Urey Prize Lecture: Chaotic Dynamics in the Solar System," *Icarus*, **72** (1987), pp. 241–275.

Yorke, James A., *Dynamics: An Interactive Program for IBM PC's and Compatibles*, Univ. of Maryland, College Park, 1991.

INDEX